Accreting Binaries

Nature, formation, and evolution

AAS Editor in Chief

Ethan Vishniac, Johns Hopkins University, Maryland, USA

About the program:

AAS-IOP Astronomy ebooks is the official book program of the American Astronomical Society (AAS) and aims to share in depth the most fascinating areas of astronomy, astrophysics, solar physics, and planetary science. The program includes publications in the following topics:

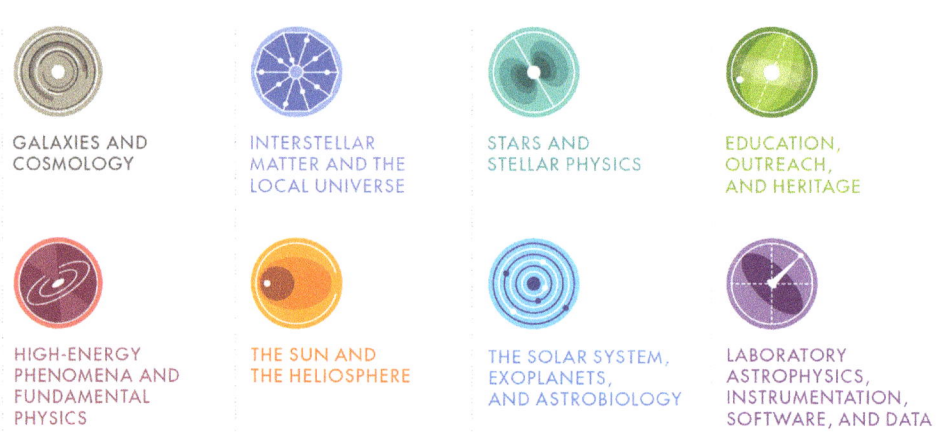

GALAXIES AND COSMOLOGY

INTERSTELLAR MATTER AND THE LOCAL UNIVERSE

STARS AND STELLAR PHYSICS

EDUCATION, OUTREACH, AND HERITAGE

HIGH-ENERGY PHENOMENA AND FUNDAMENTAL PHYSICS

THE SUN AND THE HELIOSPHERE

THE SOLAR SYSTEM, EXOPLANETS, AND ASTROBIOLOGY

LABORATORY ASTROPHYSICS, INSTRUMENTATION, SOFTWARE, AND DATA

Books in the program range in level from short introductory texts on fast-moving areas, graduate and upper-level undergraduate textbooks, research monographs, and practical handbooks.

For a complete list of published and forthcoming titles, please visit iopscience.org/books/aas.

About the American Astronomical Society

The American Astronomical Society (aas.org), established 1899, is the major organization of professional astronomers in North America. The membership (~7,000) also includes physicists, mathematicians, geologists, engineers, and others whose research interests lie within the broad spectrum of subjects now comprising the contemporary astronomical sciences. The mission of the Society is to enhance and share humanity's scientific understanding of the universe.

Accreting Binaries

Nature, formation, and evolution

Sylvain Chaty

Université Paris Cité, CNRS, Astroparticule et Cosmologie, F-75013 Paris, France

IOP Publishing, Bristol, UK

ISBN   978-0-7503-3887-5 (ebook)
ISBN   978-0-7503-3885-1 (print)
ISBN   978-0-7503-3888-2 (myPrint)
ISBN   978-0-7503-3886-8 (mobi)

DOI   10.1088/2514-3433/ac595f

Version: 20220901

AAS–IOP Astronomy
ISSN 2514-3433 (online)
ISSN 2515-141X (print)

British Library Cataloguing-in-Publication Data: A catalogue record for this book is available from the British Library.

Published by IOP Publishing, wholly owned by The Institute of Physics, London

IOP Publishing, Temple Circus, Temple Way, Bristol, BS1 6HG, UK

US Office: IOP Publishing, Inc., 190 North Independence Mall West, Suite 601, Philadelphia, PA 19106, USA

Cover image: Illustration of an X-ray binary. Image credit: Mark Garlick / Science Photo Library.

I dedicate this book to all the stars in the universe,

whether or not they belong to accreting binaries...

Contents

Preface

As soon as I discovered their existence in science books, I realized that accreting binaries were among the most intriguing objects in our universe. Consisting of a star in orbit around a compact object, such as a white dwarf, a neutron star or even a black hole, they contain all the ingredients to be the most mysterious celestial objects: located far away in our Galaxy, they are sources of energy, made of matter under extreme conditions, and above all, they simultaneously attract and eject matter... Even their names are intriguing: cataclysmic variable, kilonova, accretion disc, microquasar, superluminal ejections...

During my master's internship in 1994, I seized the opportunity to study more closely the nature of accreting binaries. I quickly understood that by trying to understand their formation, their behavior and their evolution, I would learn a lot about general topics of astrophysics: stellar formation, stellar evolution, stellar endpoints, but also related subjects, such as the properties of stellar winds, chemistry of the interstellar medium, ecology of galaxies, and even relativistic effects of matter orbiting compact objects... In fact, the most fundamental fact that I learned, is that the field of accreting binaries is a multidisciplinary research domain, within astrophysics.

At the end of the previous century, astronomers were still often characterized by their field of study, indicated by wavelength, they were radio-, infrared-, or high-energy astronomers. The field of high-energy astrophysics was still in its infancy. Now, the way of doing research in astrophysics has definitely changed, at the dawn of multimessenger astronomy, bringing together observations made by detection of photons, cosmic rays, neutrinos, and now even gravitational waves. Accreting binaries benefit greatly from being examined through all of these messengers, as most high-energy phenomena occurring within these celestial objects emit various messengers. Studying accreting binaries now involves keeping the astronomer's eyes wide open, to all possible messengers sent across the universe.

When I started my PhD, I entered ESO La Silla Observatory in Chile, to observe with the brand new NTT (New Technology Telescope), a 4m telescope as a prototype for the 8m telescopes that would later sit at the summit of Cerro Paranal near Antofagasta, northern Chile. At that time, there was no ESO/VLT observatory, only a few neutrinos had been detected, emanating from the 1987 supernova, no gravitational waves had yet been detected, and the first exoplanet around a star of solar type had not yet been discovered.

At the time of writing this book, the James Webb Space Telescope is releasing its first images, observing the sky with unprecedented sensitivity and resolution, while a new 38m diameter E-ELT telescope is under construction on Earth, the SKA radio telescope is being built, while at the other end of the electromagnetic spectrum, the CTA high-energy observatory is under development... Cosmic rays are routinely detected with the Pierre Auger observatory in Argentina, while neutrinos are detected with IceCube in Antarctica, and soon KM3NeT in Europe. The LIGO–Virgo–KAGRA interferometers are upgraded to detect gravitational waves even further, thanks to improved sensitivity. In the future, gravitational waves will be detected directly from space, with observatories such as the LISA interferometer... In parallel, numerical simulations have also improved a lot, with new computers, more efficient techniques, and more powerful clusters. In less than three decades, the sky has not changed much, but observations and simulations, put together, have allowed astronomers to radically change their vision of the sky.

This book is intended to give the reader an insight by observing through a new window open to our universe, focusing on one type of celestial objects—the accreting binaries—, hosting various physical phenomena and related to many fields in astrophysics. I sincerely hope that this book will be useful to newcomers to this rapidly evolving, but ever more interesting and intriguing field of accreting binaries. My attraction for their mysteries, which began 29 years ago, is today even stronger, now that I know them better...

Acknowledgement

I would like to thank all the people who accompanied me in the realm of accreting binaries: first, my PhD thesis director, Felix Mirabel, who discovered microquasars, with Luis Felipe Rodriguez, just two years before I started my PhD thesis. Then, my former PhD students: Farid Rahoui, Alexis Coleiro and Francis Fortin; my former post-docs: Peter Curran, Juan Antonio Zurita Heras, Alicia Lopez-Oramas and Federico Garcia; my colleagues in France: Eric Chassande-Mottin, Stéphane Corbel, Guillaume Dubus, Pierre-Alain Duc, Thierry Foglizzo, Isabelle Grenier, Jérôme Guilet, Antoine Kouchner, Pierre-Olivier Lagage and Jérôme Rodriguez; my colleagues in Spain: Josep Marti, Ignacio Negueruela, Josep Maria Paredes and Marc Ribo; my colleagues in Argentina: Jorge Combi, Federico Fogantini, Federico Garcia, Leonardo Pellizza, Gustavo Romero, Enzo Saavedra, Adolfo Simaz Bunzel and Florencia Vieyro; and my colleagues around the world: Qingzhong Liu and Jingzhi Yan in China, Thomas Tauris in Denmark, Carole Haswell in the United Kingdom, Rob Hynes and John Tomsick in the United States... Unfortunately, I cannot name everyone here, but I would like to thank all the astrophysicists met in many places on Earth, either during workshops or during long observing nights in observatories, from Paris to Kathmandu, from Oulan Bator to New York, and from Kiev to Santiago...

Author biography

Sylvain Chaty

Photography Sylvain Chaty, Credit Laurence Honnorat

Sylvain Chaty, Professor at Université Paris Cité in Paris, and Ecole polytechnique in Palaiseau, France, is an astrophysicist in the APC Laboratory (Astroparticle and Cosmology). His area of research is the multi-wavelength observational study of accreting binaries, in which two stars exchange matter and angular momentum. He is also focusing on stellar evolution in general, up to the merger phase of black holes and neutron stars, detected by gravitational waves. After a doctoral thesis in the Astrophysics Department of the CEA Saclay, obtained in 1998, he spent a year in the Astrophysics research laboratory of the University of Toulouse in France, before being recruited as a Leverhulme research fellow in the Department of Physics and Astrophysics from the Open University (United Kingdom), where he stayed for three years. In 2002, he became associate professor, and in 2011 full professor, at the University of Paris 7 - Diderot. Co-author of more than 360 research publications, he is a member of the LIGO-Virgo and LISA (gravitational wave detectors), Fermi and CTA (gamma-ray observatories) and Athena (X-ray satellite) collaborations. Vice-President of Université Paris Cité for Scientific Culture and its Outreach, Sylvain Chaty is also passionate about the dissemination of science, as evidenced by several popular works he has written, including *The Sky and the Stars* (Gallimard), *The Colonization of Space* (CNRS editions), originally in French and translated into other languages, and will soon publish *Great Challenges in Astrophysics* (Hachette).

Accreting Binaries
Nature, formation, and evolution
Sylvain Chaty

Chapter 1

Introduction

Stellar couples are common celestial sources in our universe: more than 70% of massive stars experience a common life with a companion (Sana et al. 2012), exchanging matter at some stage during their evolution, eventually forming an *accreting binary*. Of course, the evolution of such couples is related to the evolution of single stars, born within a molecular cloud, evolving in a time that depends on their mass, and then collapsing at the end of their life into a compact object—either a white dwarf, a neutron star, or a black hole (see Chapter 2). But the evolution of stellar couples is also very different from the evolution of single stars, as we will see in this book.

1.1 The Discovery of Accreting Binaries

The discovery in 1962, by a US *Aerobee* scientific sub-orbital sounding rocket carrying an X-ray instrument, of the first and brightest persistent X-ray source beyond the solar system (i.e., further than the Sun), located in the Scorpius region (Scorpio X-1, of X-ray luminosity $L_X \sim 6 \times 10^4 L_{\odot}$;[1] Giacconi et al. 1962), brought the Nobel prize of Physics to Riccardo Giacconi in 2002.[2] Shortly after this discovery, Shklovsky (1967) proposed that the origin of this powerful X-ray emission could be due to accretion process of gas from a normal star onto a neutron star, within a close binary system.[3] Sco X-1 happens to be the first accreting stellar binary, namely a neutron star accreting from a low-mass $M \sim 0.4 M_{\odot}$ companion star of 13th visual magnitude (Sandage et al. 1966): the first *low-mass X-ray binary* (see Chapter 5).

---

[1] This symbol indicates the Sun: $M_{\odot}$ is the solar mass, $L_{\odot}$ is the solar luminosity, etc.
[2] The Nobel Prize in Physics 2002 was awarded half to Riccardo Giacconi, "for pioneering contributions to astrophysics, which have led to the discovery of cosmic X-ray sources."
[3] This idea was not immediately accepted though, since it was thought that the supernova leading to the neutron star formation would break the binary system.

This discovery marked not only the dawn of X-ray astronomy, but also the advent of multi-wavelength astronomy, observing through different parts of the electromagnetic spectrum to disentangle physical processes of the universe. It also marked the beginning of a new field of study in astrophysics: the domain of compact object astrophysics. A couple of years later, in 1964, Cygnus X-1, the first and brightest X-ray source located within the Cygnus region, was discovered by another *Aerobee* rocket. The study of its optical counterpart revealed that the X-ray source forms part of a binary system, allowing to estimate the mass of the compact object to $M = 8.7 \pm 0.8 M_\odot$, much too heavy to be a neutron star (and now estimated to $M \sim 21 M_\odot$; Miller-Jones et al. 2021). Cyg X-1 thus became the first black hole *candidate*, orbiting a massive O9 supergiant companion star, from which it is accreting (see Figure 6.1), making this system one of the rare persistent *high-mass X-ray binary* in our Galaxy (see Chapter 6).

1.2 Celestial Laboratories of Extreme Physics

Accreting binaries are stellar couples hosting a compact object (white dwarf, neutron star or black hole, see Figure 1.1), at the stage of their evolution when the two components of the binary system exchange both matter and angular momentum (Chaty 2013). They happen to be among the best laboratories of fundamental physics in our universe, experiencing not only accretion of matter onto a compact object, but also relativistic ejection of matter, from hundreds of km s^{-1} up to relativistic velocities, and finally collision between ejected matter and their surrounding interstellar medium.

The phenomenon of accretion of matter is a general process, in which matter falls within the gravitational well of a compact object, and depends on many parameters, such as mass ratio, orbital separation, rotation, stellar winds, etc. (see Chapter 3).

Figure 1.1. Generic image of a typical accreting binary, consisting of a low-mass star, orbiting a compact object, which can be either a neutron star or a black hole. Reproduced with permission, Credit: Robert Hynes.

This process, releasing a substantial quantity of energy, is common in numerous classes of objects: from young forming stars to active nuclei in the heart of galaxies, via *cataclysmic variables* (see Chapter 4), low-mass and high-mass X-ray binaries, without forgetting *intermediate-mass X-ray binary* and *ultra-luminous X-ray sources* (see Chapter 7). *Semi-detached* binaries are numerous and diverse, close and easily detectable, and varying on timescales short enough to be extensively studied. The study of their variability allows astronomers to constrain parameters such as viscosity, and physical mechanisms such as oscillations, within accretion disks. With their fast evolution, taking place on a timescale accessible to human observation, accreting binaries constitute the best laboratories of fundamental physics, allowing astronomers to study high-energy processes of astrophysics, along with the behavior of accreted matter in dense environment, and the particle acceleration occurring in winds and relativistic jets. Outflows from accreting binaries can take various forms, from poorly collimated winds emanating from the star or the accretion disk, to relativistic, transient or persistent collimated jets, observed in different parts of the electromagnetic spectrum, from radio up to X-ray domains (see Chapter 8). Stellar compact objects, the densest structures within our universe, are the site of energetic phenomena, allowing astronomers to study extreme physics of ultra-dense matter, to test the theory of general relativity in strong-field gravity—within binary pulsars, or through the detection of gravitational waves—the physics of neutrinos, the Lorentz invariance (the dependence of light velocity c with the photon energy), etc.

Once both stellar companions forming the binary have collapsed into compact objects, becoming binary white dwarfs, neutron stars or black holes—or a combination of the three—they eventually merge, leading to emission of gravitational waves, such as those detected by the LIGO/Virgo collaboration in September 2015 (Abbott et al. 2016a, 2016b), nearly a century after Albert Einstein published the general theory of relativity (Einstein 1915), and predicted the existence of gravitational waves (Einstein 1918). This very first detection of gravitational waves, emanating from the merging of two stellar-mass black holes, marked the advent of multimessenger astronomy, combining both photon and gravitational wave messengers, awarding the 2017 Nobel Prize of Physics to Rainer Weiss, Barry C. Barish, and Kip S. Thorne, from the LIGO/Virgo collaboration.[4] Then, nearly two years later, in August 2017, the LIGO/Virgo collaboration witnessed the merging of a binary neutron star (Abbott et al. 2017a, 2017b, 2017c), accompanied by an intense and dense flux of neutrons, leading, through the so-called *r-process* of explosive nucleosynthesis, to the building of heavy atomic nuclei (Pian et al. 2017).

Accreting binaries also play an important role in the ecology of our Galaxy, like in all other galaxies: massive stars synthesize heavy elements during their whole life, which they disseminate in the surrounding interstellar medium at the time of their supernova event, ejecting all material contained in their outer shells, and eventually

---

[4] The Nobel Prize in Physics 2017 was awarded half to Rainer Weiss, and the other half jointly to Barry C. Barish and Kip S. Thorne, "for decisive contributions to the LIGO detector and the observation of gravitational waves."

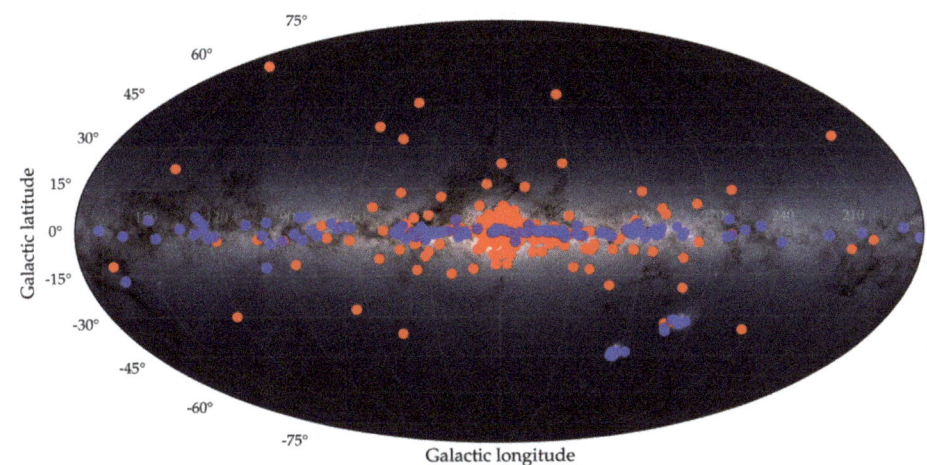

Figure 1.2. Galactic distribution, in Galactic coordinates, of accreting binaries, hosting low-mass (LMXB) and high-mass (HMXB) stars. While LMXB are mainly concentrated in the Galactic bulge, HMXB are located within the plane of our Galaxy, toward the tangential directions of the Galactic arms. All accreting binaries known to date are included, based on a catalog of 218 LMXB and 148 HMXB (Credit: Francis Fortin). The image of the Galaxy is obtained from the Gaia satellite (Credit: ESA/Gaia).

triggering additional stellar formation, by the propagation of shock waves. Their nature is intimately linked to how matter is distributed within galaxies, as revealed by their overall localization within the Milky Way (Figure 1.2): while accreting binaries hosting low-mass stars are mainly concentrated within the Galactic bulge, the ones hosting high-mass stars are located within the Galactic plane, and particularly toward tangential directions of Galactic spiral arms (Coleiro & Chaty 2013).

Accreting binaries even allow astronomers to study cosmology, since they are very luminous celestial objects, either while accreting or collapsing, or when their gravitational energy is released, making them detectable at large distances. For instance, *gamma-ray bursts* (γ-ray bursts, or GRB) allow to map the distribution of gas and *metals*[5] surrounding the galaxies, in the so-called intergalactic medium, and to constrain the epoch of reionization, which corresponds to the first era of our universe. *Type Ia* supernovae are standard candles, allowing to measure the distance, and thus the expansion rate of the universe (described by the *Hubble–Lemaitre law*), which revealed the existence of dark energy, corresponding to an accelerated expansion of our universe.

1.3 Structure of this Book

The structure of this book is as follows. We describe in Chapter 2 all astrophysical basics needed to understand the evolution of a single star, from its formation to its final endpoint. We then give in Chapter 3 all useful facts shared by the various types

---

[5] In astronomy, the term of *metals* designates all atoms heavier than hydrogen or helium.

of binary systems, and the main parameters of accretion disks, which will be needed in the whole book.

We then spend the following four chapters describing the nature, formation, and evolution of the different types of accreting stellar binaries, in which the orbital separation is small enough that both mass and angular momentum can be transferred from one object to the other. Such binaries are classified according to the true nature of the accreting object, that we call *primary*. While these classes are presented separately, in many places they share common physical processes, and we will frequently refer to one class or another, as we feel that this approach is the most appropriate to better understand these systems.

We first describe the nature, formation, and evolution of *cataclysmic variables*— binary systems hosting a white dwarf—in Chapter 4. We go on with *low-mass X-ray binaries*—a neutron star or a black hole orbiting a low-mass star—in Chapter 5. We then follow up with *high-mass X-ray binaries*—a neutron star or a black hole orbiting a massive star—in Chapter 6. We finally describe a couple of peculiar accreting binary classes in Chapter 7.

After going through all these different types of accreting binaries, we describe the ejection phenomenon taking place in most accreting sources, and sharing common properties, in Chapter 8. We then report in Chapter 9 on the various multi-wavelength observing facilities of accreting binaries, which, accompanied by heavy simulations, allow astronomers to better understand all kinds of energetic phenomena occurring in these intriguing celestial objects. Since multi-messenger astronomy is currently revolutionizing our view of the universe, we add to the list of photon-observing facilities all gravitational wave, neutrino, and cosmic ray detectors. We finally conclude this book in Chapter 10.

References

Abbott, B. P., Abbott, R., Abbott, T. D., et al. 2016a, ApJL, 818, L22

Abbott, B. P., Abbott, R., Abbott, T. D., et al. 2016b, ApJL, 826, L13

Abbott, B. P., Abbott, R., Abbott, T. D., et al. 2017a, PhRvL, 119, 161101

Abbott, B. P., Abbott, R., Abbott, T. D., et al. 2017b, ApJL, 848, L13

Abbott, B. P., Abbott, R., Abbott, T. D., et al. 2017c, ApJL, 850, L40

Chaty, S. 2013, AdSpR, 52, 2132

Coleiro, A., & Chaty, S. 2013, ApJ, 764, 185

Einstein, A. 1915, SPAW, 844

Einstein, A. 1918, SPAW, 154

Giacconi, R., Gursky, H., Paolini, F. R., & Rossi, B. B. 1962, PhRvL, 9, 439

Miller-Jones, J. C. A., Bahramian, A., Orosz, J. A., et al. 2021, Sci, 371, 1046

Pian, E., D'Avanzo, P., Benetti, S., et al. 2017, Natur, 551, 67

Sana, H., de Mink, S. E., de Koter, A., et al. 2012, Sci, 337, 444

Sandage, A., Osmer, P., Giacconi, R., et al. 1966, ApJ, 146, 316

Shklovsky, I. S. 1967, ApJL, 148, L1

Accreting Binaries
Nature, formation, and evolution
Sylvain Chaty

Chapter 2

Astrophysical Basics for Single Stars

I think there should be a law of Nature, to prevent the star from behaving in this absurd way.
—Sir Arthur Eddington, *Royal Astronomical Society, January 1935*

We describe in this chapter all astrophysical basics regarding the evolution of isolated stars, from their formation to their final endpoint, that the reader will need in the whole book. In the first part, we briefly describe the stellar formation, the main sequence and late evolution of stars of various masses. We first focus on the low-mass and intermediate-mass stars, through their red giant phase, until they finally collapse in a white dwarf. We then go through the late evolution of massive stars, up to their final collapse into a compact object, either a neutron star or a black hole. In the second part, we report on the main properties of compact objects, which will be useful throughout this book: the gravitational, magnetic, and rotational properties, before giving useful reviews, catalogs, and references.

2.1 Concise Stellar Evolution

In this chapter we recall basic facts of stellar evolution, from stellar formation to final stages, through the main sequence, the giant branch, etc., for low-mass, intermediate-mass, and high-mass stars. For more details we refer the reader to books focusing on this subject, such as Arnett (1996) and Kippenhahn et al. (2012).

doi:10.1088/2514-3433/ac595fch2

2.1.1 Stellar Formation

The whole story of this book begins inside a *molecular cloud*, made of cold molecular gas[1] and dust grains,[2] forming part of the *interstellar medium* (ISM). Molecular clouds have a typical mass in the range $M \sim 10^2 - 10^6 M_\odot$, a size from one to a few tens of parsecs,[3] an average density from a few to a few hundred particles per cm^3, and an internal temperature $T \sim 7 - 30$ K (Evans 1999).

More specifically, they can be classified in three different types, depending on their size and mass (Mac Low 2004). First, the *giant molecular clouds* (GMC), extending on tens of parsecs, with masses $M \sim 10^4 - 10^6 M_\odot$, are the birthplace of massive stars ($M \geqslant 8 M_\odot$). Second, *molecular clouds* (MC), extending on a few parsecs, with lower masses from 10^2 to $10^4 M_\odot$, are the formation site of clusters of lower mass stars. GMC and MC present similar densities $\rho \sim 50 - 500 \text{cm}^{-3}$ and temperatures $T \sim 10 - 20$ K. Third, *isolated cores*, also called *Bok globules* (Bok 1948), are small and low-mass clumps, with typical size $\leqslant 1$ pc, and mass $M \lesssim 10^2 M_\odot$, in which only a few stars form.

Molecular clouds remain stable against their own gravitational collapse, thanks to thermal pressure, centrifugal force, turbulence and/or magnetic field. This stability lasts for a few tenths of Myr, until external perturbations of their structure, due to collisions or shock waves, create sub-regions of over-densities—*clumps* or *dense cores*. This represents the first step toward the formation of a stellar core, in a process known as *fragmentation* (Bergin & Tafalla 2007). When such clumps are perturbed, or when their mass is large enough, gravitational collapse takes place, accreting most of their mass into the central object which will become the star, surrounded by an accretion disk, which will eventually form a planetary system (see Figure 2.1). Star formation was first modeled in the 1960s (Larson 1969; Penston 1969), nearly two decades before the detection of the first *protostars* allowed astronomers to bring observational constraints (Lada 1987; André et al. 1993).

The average mass of molecular clouds being much higher than a typical stellar mass, and their density much lower than a stellar density,[4] the gravitational collapse leading to stellar formation must simultaneously decrease the mass and increase the density of the stellar core. Stellar formation is thus a complex—multi-scale and non-linear—process, involving at the same time gravity, radiation, and hydrodynamics, revealed through multi-wavelength observations—various classes of young stellar objects emitting in different wavelength domains—and numerical simulations. In particular, the presence of dust grains makes molecular clouds opaque to radiation at short wavelengths (visible domain), which thus can only be observed at longer wavelengths (infrared, sub-mm, and mm domains).

---

[1] Molecular gas mainly consists of hydrogen (63%), helium (36%), and other molecules (1%) such as CO.
[2] Dust grains are composed of silicates and carbonates.
[3] A parsec (pc) is a distance unit, corresponding to the distance at which the semimajor axis of the Sun–Earth orbit is seen at an angular distance of one arcsecond: 1 pc = 3.26 light-year (l.y.).
[4] The density at the center of the Sun amounts to $\sim$160 g cm^{-3}, corresponding to 5×10^{24} particles per cm^3.

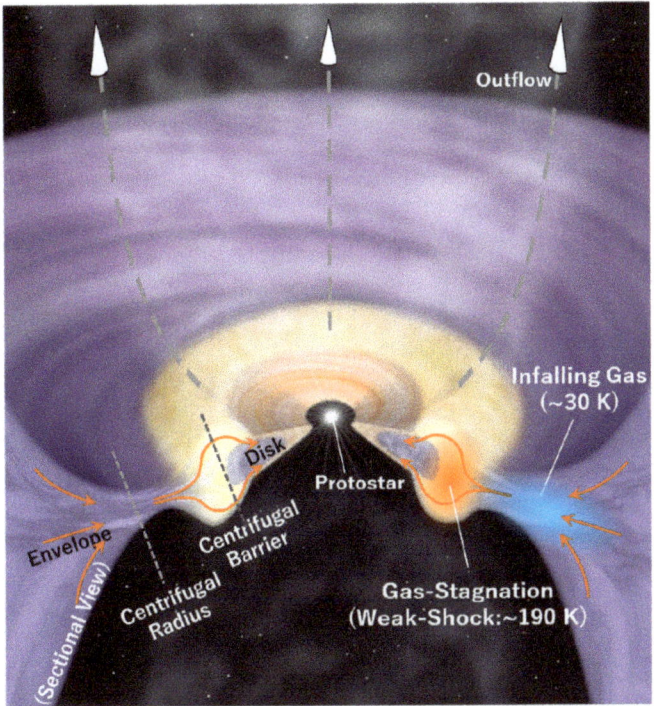

Figure 2.1. Stellar formation scenario: the central protostar is surrounded first by a dust-free region, and then by a disk with a dust sublimation front, shielding a further absorbing envelope, in which the accretion flow takes place (Credit: ALMA/RIKEN).

2.1.1.1 Scenario of Stellar Formation

Fluctuations in the density profile occur in sub-regions of the parent molecular cloud, leading to the formation of gravitationally-bound stellar cores, which are observed in two subsequent phases: *pre-stellar* and *proto-stellar* cores. Pre-stellar cores are objects in contraction, already decoupled from their environment within the cloud, but still before the formation of any stellar object. In the standard stellar formation scenario, the transition from pre-stellar to proto-stellar cores occurs through several steps, from cloud collapse, to the formation of a rotating disk, in which the accretion flow takes place, acting as a reservoir of matter infalling toward the central core, and simultaneously evacuating angular momentum through out-flows (see Figure 2.1). Both simulations and multi-wavelength observations have revealed the following steps of stellar formation, given here in a chronological order (Lada 1987; André & Montmerle 1994; André et al. 2010):

- *Pre-stellar* dense cores, collapsing within the molecular cloud, with a radius of a few au, a mass of $M \sim 0.01 - 0.1 M_\odot$, and temperatures of $T \sim 100 - 2000$ K;
- *Class 0* protostar (proto-stellar phase), with a radius of a few $R_\odot$ and a mass $M \sim 0.001 - 0.1 M_\odot$, lasting $\sim 10^4$ years and revealed in the sub-millimeter domain. Most of the mass, still contained in the envelope ($M_{env} \geqslant M_{star}$), is accreted onto the central object, with the formation of bipolar outflows,

perpendicular to the surrounding accretion disk of size $10 - 100$ au. This phase lasts for $\sim 5 \times 10^4 - 10^5$ years, until the central object has gained most of its mass;

- *Class I* protostar (proto-stellar phase), lasting $\sim 10^5$ years and revealed in the far-infrared domain: the central star has accumulated $\sim 90\%$ of its final mass, and the matter of the envelope—not accreted by the central object—is ejected through stellar winds and outflows;
- *Class II* young star or *T-Tauri star* (pre-main sequence phase), lasting $\sim 10^6$ years and revealed in the near-infrared domain: most of the envelope has disappeared, the young star is only surrounded by an optically-thick disk of size $10 - 150$ au and mass $0.005 - 0.14 M_\odot$ (André & Montmerle 1994);
- *Class III* young star or *post T-Tauri stars* (pre-main sequence phase), lasting 10^7 years and revealed in the visible domain: most of the material in the now optically thin disk has finally been accreted and even processed to form planets. Since thermal pressure cannot sustain the star against its own gravity, a final collapse takes place. The characteristic time for this contraction is the Kelvin–Helmholtz time:

$$t_{KH} = \frac{GM^2}{RL} \tag{2.1}$$

where G is the constant of gravity, and M, R, and L the mass, radius, and luminosity of the star, respectively. While for low-mass stars, this time of $\sim 10^8$ years is longer than the accretion phase, for high-mass stars this time is smaller, and the star will enter the main sequence before it has accreted all its mass. As soon as the temperature becomes high enough in the central core of the star ($T \geqslant 10^7$ K), thermonuclear reactions are triggered, burning hydrogen into helium, and the star enters the main sequence.

Before distinguishing the low-mass and high-mass regimes in this process of stellar formation, we now give the basic parameters of gravitational collapse.

2.1.1.2 Basic Parameters of Gravitational Collapse

A molecular cloud of gas of density ρ, subject to ionization at the sub-parsec ($\leqslant 1$ pc) scale, is in hydrostatic equilibrium as long as the internal pressure p counterbalances the mass M_{enc} enclosed in a radius r:

$$\frac{dp}{dr} = -\frac{G\rho(r)M_{enc}(r)}{r^2}. \tag{2.2}$$

A cloud of gas of low mass, enclosed in a small length scale, remains stable. The gravitational collapse begins to occur as soon as the gravitational energy becomes greater than all other (thermal, rotational, turbulent, and magnetic) components present within the cloud:

$$|E_{grav}| \geqslant E_{therm} + E_{rot} + E_{turb} + E_{mag} + ... \tag{2.3}$$

We define the *Jeans length* λ_J and *Jeans mass* M_J, as the minimum length and mass, respectively, for a cloud to be gravitationally bound. Once gravitational energy becomes dominant in the cloud, it begins to collapse toward stellar formation, with a length scale larger than the Jeans length:

$$\lambda_J = \frac{c_s}{\sqrt{G\rho}} \sim 0.4\text{pc}\frac{c_S}{0.2\text{km s}^{-1}}\sqrt{\frac{10^3\text{cm}^{-3}}{\rho}} \tag{2.4}$$

c_s being the sound speed in the cloud of gas. From there we derive the Jeans mass, corresponding to the mass enclosed in a sphere of radius $R_J = \frac{\lambda_J}{2}$:

$$M_J = \frac{4\pi\rho R_J^3}{3} = \frac{\pi c_s^3}{6G^{3/2}\rho^{1/2}} \sim 2M_\odot\frac{c_s}{0.2\text{km s}^{-1}}\sqrt{\frac{10^3\text{cm}^{-3}}{\rho}}. \tag{2.5}$$

The timescale, for a cloud to collapse in the absence of pressure, is given by the free fall time τ_{ff}:

$$\tau_{ff} = \sqrt{\frac{3\pi}{32G\rho}}. \tag{2.6}$$

2.1.1.3 Low-mass Regime

In the low-mass regime, the internal stellar structure is parameterized using the *equation of state* (EoS)—linking various thermodynamical parameters such as pressure, internal energy, temperature, density, etc.—of a *polytrope*, in which the pressure P only depends on the density ρ in the form:

$$P = K\rho^\gamma \tag{2.7}$$

where K is a constant, and γ a parameter that can take different values, depending on the collapsing stage, as described below. The temperature T is estimated as:

$$T \propto \rho^{\gamma-1}. \tag{2.8}$$

The various steps of stellar collapse are subsequently parameterized as:

- The collapse is first modeled as an isothermal process ($\gamma = 1$), the energy released by the gas compression being efficiently evacuated through thermal radiation, the gas being optically thin, until reaching densities of $10^{-14} - 10^{-13}$ g cm^{-3}. During this step, the Jeans mass decreases.
- Internal temperature then rises, the internal region becomes opaque to infrared radiation, and the collapse is now modeled as an adiabatic process ($\gamma = 7/5$), increasing its mass in a quasi-static contraction: this is the formation of a first proto-stellar core, of mass $M \sim 0.01M_\odot$, and radius of the order of the *astronomical unit*[5] (au).

---

[5] An astronomical unit (au) corresponds to the average distance between the Sun and the Earth: 1 au $= 149.597\,870\,700 \times 10^6$ km.

- When the central temperature reaches $T \sim 2000$ K, the molecular hydrogen H_2 starts to dissociate, in an endothermic reaction, absorbing thermal energy. Temperature decreases and thermal pressure is no longer able to support gravity, triggering a second gravitational collapse in a small central region ($\gamma = 1.1$), during which the Jeans mass decreases.
- Finally, once all the gas has become fully atomic, it forms a second hydrostatic proto-stellar core of stellar size, which continues to increase its mass in a quasi-static contraction ($\gamma = 5/3$; Larson 1969).

2.1.1.4 High-mass Regime

In the high-mass regime, massive stars result from the fusion of filamentous ridges of high density ($10^4 - 10^5$ g cm^{-3}, on a scale of a few pc), formed by accretion of gas within the molecular cloud, first predicted by theoretical models, and later confirmed by extensive far-infrared observations of the Gould Belt Survey, performed by the Herschel satellite in the $70 - 500$ μm domain (see, e.g., André et al. 2010). These observations reveal numerous filamentary structure, dense cores embedded in these filaments, up to 500 pre-stellar and 60 proto-stellar cores, and $\sim$300 unbound starless cores. They allow to constrain a stellar formation scenario in which complex networks of long and thin filaments first form within molecular clouds, before the densest filaments fragment into a number of pre-stellar cores via gravitational instability.

The whole process leading to the formation of massive stars still remains poorly understood, implying to simultaneously solve both radiative transfer and hydro-dynamical equations. Two families of model are used to study the formation of massive stars: those adapted from the low-mass model, and others trying to adopt a more dynamical approach. Some models tend to unify both aspects, for instance by applying the core-collapse model for low-mass stellar cores, and then, by continuous accretion processes, until eventually forming high-mass cores.

The initially rotating massive pre-stellar core, of size corresponding to a Jeans length $R_J \sim 0.1$ pc and mass $M_J \sim 150 M_\odot$, collapses in an accreting protostar of typical size $\sim$0.01 pc, until eventually forming a planetary system of size $\sim$0.000 1 pc. Surrounding the central protostar, we first find a dust-free region, and then the accretion disk in which the accretion flow takes place, with a dust sublimation front shielding a further absorbing envelope, with a density profile $\rho(r) \propto \frac{1}{r^{1.5}}$ (see Figure 2.1). The accretion disk reaches a typical size of $R_{\text{disk}} \sim$ 500–2000 au, and a mass of $M_{\text{disk}} > 10 M_\odot$. 2D and 3D numerical simulations reveal that, with the presence of an accretion disk substantially contributing to the total accreted mass, a forming star can reach a maximum mass of $\sim$100$M_\odot$. In the stellar formation process, we find the simultaneous presence of both disks and outflows (Bachiller 1996), showing that the accretion process is closely related to the ejection phenomenon, similar to the one observed in jet sources, such as microquasars (see Figures 8.1 and 8.2 in Chapter 8), or even supermassive black holes, based on magneto-hydrodynamical processes.

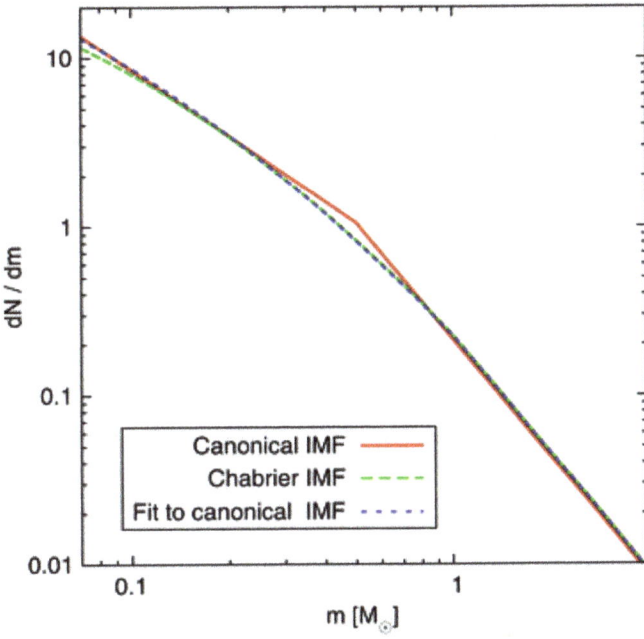

Figure 2.2. A comparison between three canonical initial mass functions (IMF), in the interval 0.07 to $4M_\odot$. The plot shows the number of stars per mass interval $\frac{dN(m)}{dm}$, versus the stellar mass m. The three IMF, indistinguishable over the whole mass interval, are nearly identical above a mass of $M \geqslant 1M_\odot$. We see from this figure that stars of mass $\sim 0.1\,M_\odot$ are ~ 50 times more numerous than solar-mass stars, depending on the IMF law. Reprinted with permission from Kroupa et al. (2013).

2.1.1.5 Initial Mass Function

The mass of newly-formed stars is given by the *initial mass function* (IMF, see Figure 2.2 for single stars), showing that stars of mass $\sim 0.1\,M_\odot$ are ~ 50 times more numerous than solar-mass stars. The mass distribution of stars heavier than the Sun is described by a power law, with a probability density function $p(M) \sim M^\alpha$, where M is the mass of the star:

$$\frac{dN}{dM} \propto M^\alpha \qquad (2.9)$$

with a slope α in the range from ~ -2.7 to -2.3, closer to $\alpha = -2.3$ in the stellar mass range $20 \leqslant M \leqslant 150 M_\odot$ (Kroupa et al. 2013), appropriate for black hole progenitors.

2.1.1.6 Struggle to Shine

From its formation within a molecular cloud, a stellar life is regulated by a permanent equilibrium between gravity F_{grav} and pressure F_{rad} interaction. A star in dynamical equilibrium is described by the *Virial* theorem, stating a relationship between its kinetic energy E_c and its potential energy E_p:

$$2E_c + E_p = 0. \tag{2.10}$$

A star, consisting of self-gravitating gas in hydrostatic equilibrium, releases energy by converting hydrogen into helium. This process of gravitational contraction, to maintain the radiative pressure, produces an increase of the mean molecular weight of the stellar core, which decreases gravitational potential energy, and as stated by the Virial theorem, simultaneously increases thermal kinetic energy, and thus its internal temperature, forcing the star to radiate even more.[6] The star then enters a *Virial cycle*, unstable on the stellar life timescale, due to negative specific heat. This cycle goes on until the stellar core becomes degenerate, since for a degenerate gas the pressure only depends on its density and not on its temperature, preventing the use of Virial theorem. The star is thus condemned to finish its life by forming a degenerate core, i.e., by collapsing into a compact object. Lifetimes of massive stars are reported in Schaller et al. (1992).

2.1.2 Main Sequence

Stars of initial mass M_{ZAMS} (ZAMS stands for Zero-Age Main Sequence) spend most of their life in the *main sequence*, in hydrostatic and thermal equilibrium, releasing energy by converting hydrogen into helium through thermonuclear reactions in their core, at a temperature $T \sim 10-15 \times 10^6$ K. As a first approximation, stars radiate like blackbodies, with a surface area of $4\pi R^2$, where R is the stellar radius. From the Stefan–Boltzmann law, the stellar luminosity L is related to its effective temperature T_{eff}:

$$L = 4\pi R^2 \sigma T_{eff}^4 \tag{2.11}$$

with σ the Stefan–Boltzmann constant. The mass–luminosity and mass–radius relations, for a typical main sequence star, are respectively:

$$\frac{L}{L_\odot} = \left(\frac{M}{M_\odot}\right)^{3.5} \text{and} \frac{R}{R_\odot} = \left(\frac{M}{M_\odot}\right)^{0.81} \tag{2.12}$$

showing that the more massive a star is, the more luminous and large it will be (Demircan & Kahraman 1991).

Grids of stellar models, reported in Schaller et al. (1992), allow us to assess the mass–age relationship of stars of mass from $M \sim 0.8$ to $\sim 120 M_\odot$ at various metallicities, and in particular the timelapse between their birth and collapse. Finally, a Hertzsprung–Russell diagram showing the whole evolution of stars of various types is reported in Figure 2.3.

---

[6] As pointed out by Tauris & van den Heuvel (2006), stars do not shine because of thermonuclear reactions occurring in their heart, but instead because of their high temperature, caused by their gravitational contraction.

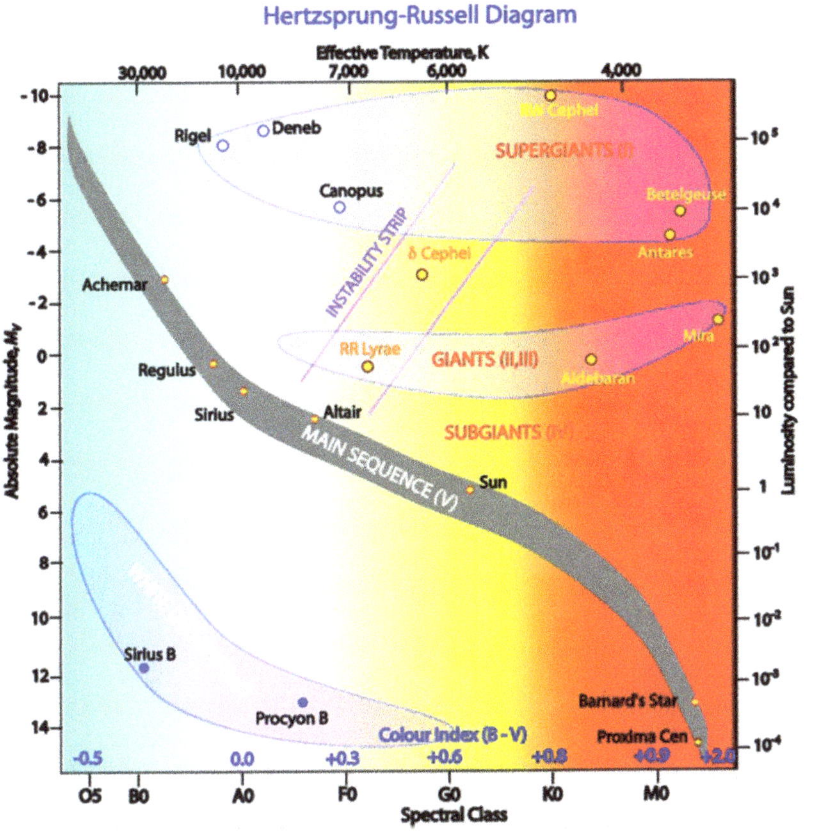

Hertzsprung-Russell Diagram

Figure 2.3. Hertzsprung–Russell diagram, named after the astronomers who devised it in the early 20th century to study stellar evolution. This diagram shows the stellar main sequence, late evolution of stars, and white dwarf localization. The y-axis shows absolute magnitude M_V (defined as $M_V = -2.5 \times \log F_{10pc}$) on the left, and bolometric luminosity compared to solar on the right, the x-axis shows effective temperature on the top, and color index (B-V) and spectral class on the bottom. Reprinted with permission from Althaus et al. (2010).

2.1.2.1 Nuclear, Thermal, and Dynamical Timescales

The total amount of atoms available for thermonuclear fusion is proportional to the stellar mass, allowing us to derive the stellar life on the main sequence phase (*nuclear timescale* τ_{MS}), revealing that a massive star has a shorter life:

$$\tau_{MS} \simeq 10\,\mathrm{Gyr}\frac{M}{M_\odot}\left(\frac{L}{L_\odot}\right)^{-1} = 10\,\mathrm{Gyr}\left(\frac{M}{M_\odot}\right)^{-2.5}. \tag{2.13}$$

Combining this τ_{MS} relationship with the initial mass function indicating the number of stars produced for each mass bin (Figure 2.2), shows that massive stars are rare in galaxies.

Another important timescale for stellar evolution is linked to thermal equilibrium, this *thermal timescale* τ_{The} corresponds to the time a star needs to emit its whole thermal energy content:

$$\tau_{\text{The}} = \frac{GM^2}{RL} = 30\text{Myr}\left(\frac{M}{M_\odot}\right)^{-2}. \tag{2.14}$$

The last timescale, related to hydrostatic equilibrium, particularly useful in the context of mass loss in accreting binaries, is the *dynamical timescale*:

$$\tau_{\text{Dyn}} = \sqrt{\frac{R^3}{GM}} = 50\text{mn}\left(\frac{R}{R_\odot}\right)^{3/2}\left(\frac{M}{M_\odot}\right)^{-1/2}. \tag{2.15}$$

2.1.3 Late Evolution of Low-mass and Intermediate-mass Stars: Toward White Dwarfs

2.1.3.1 Low-mass Stars, $M \lesssim 2.3 M_\odot$

During the phase of hydrogen shell burning, which generates the whole stellar luminosity, low-mass stars of $M \lesssim 2.3 M_\odot$ develop a small and dense degenerate core made of helium, whose mass grows until it reaches $M_{\text{He}} \sim 0.45 M_\odot$. Once they have consumed all hydrogen present in their stellar core, now made up mainly of helium, the long main-sequence phase ends, triggering the ignition of hydrogen in a shell surrounding the helium core. At this point the entire star briefly contracts, the nuclear energy release causes a sudden temperature increase up to a temperature of 10^8 K, triggering ignition of helium at the stellar core, accompanied by an *He flash*. This initiates the *red giant branch* (RGB) phase, characterized by a dense helium core surrounded by a large and convective hydrogen envelope. During this unstable phase, the star is on a helium nuclear timescale, and its stellar envelope, constituting the remaining H and He, expands, with thermonuclear fusion occurring in shells surrounding the degenerate core. The radius of the red giant is linked to the mass of its degenerate helium core, almost independent of the mass of the hydrogen-rich envelope. In the case of an accreting binary system, the core mass is linked to the orbital period of the system, up to the collapse into a white dwarf, where the mass of the resulting white dwarf is correlated to the final orbital period of the system (Tauris & van den Heuvel 2006):

$$M_{\text{WD}} = \left(\frac{P_{\text{orb}}}{b}\right)^{1/a} + c \tag{2.16}$$

where a, b, and c are parameters in the range: $a = 4.5 - 5.0$, $b = 1.0 - 1.2 \times 10^5$, and $c = 0.11 - 0.12$, only depending on the metallicity of the donor star.

2.1.3.2 Intermediate-mass Stars, $2.3 \lesssim M \lesssim 8 M_\odot$

Intermediate-mass stars, of mass between $2.3 \lesssim M \lesssim 8 M_\odot$, develop a degenerate He core during the supergiant phase, followed by a stable ignition of He. After a stable central fusion of H and He, the CO core becomes degenerate. During He shell burning, the outer radius expands again, until carbon ignites with a flash. From this point, the star becomes a red supergiant, evolving along the brief *asymptotic giant branch* (AGB). The giant and supergiant phase, and particularly massive helium stars, also called *Wolf–Rayet stars*, are characterized by intense stellar wind mass loss, continuously ejecting their outer stellar envelope, forming a surrounding nebula, until eventually leaving a naked helium core.

2.1.3.3 Formation of a White Dwarf

This stellar core eventually collapses into a cooling white dwarf, surrounded by a planetary nebula. Considering that every star of initial mass $M \leqslant 9M_\odot$ collapses into a white dwarf, this means that it is the fate of ~97% of all stars, amounting to ~50 billion of stars within the Milky Way, with a distribution similar to stars within the Galactic disk and halo. Nearly one third of white dwarfs would be hosted in a binary—or even multiple—system, with a more luminous companion of spectral type K or earlier, such as Sirius B, a white dwarf located at a distance of 2.6 pc and orbiting around Sirius A, an A-spectral type star. The population of white dwarfs in the local environment, up to a distance of 25 pc, is reported in Holberg et al. (2016). This sample of 232 white dwarfs, estimated to be complete at 68%, corresponds to a space density of white dwarfs of $4.8 \pm 0.5 \times 10^{-3}$ pc^{-3}, with a significant excess of single stars over systems containing one or more companions (74% versus 26% respectively), suggesting a loss of companions during binary system evolution. The first white dwarf, 40 Eridani, has been discovered in 1914 (Adams 1914).[7]

The composition of the newly-formed white dwarf depends on the subsequent thermonuclear fusion reactions taking place in the stellar core, thus on the mass of the stellar progenitor, before collapsing into a white dwarf. A progenitor of $M \leqslant 0.8M_\odot$ experiences H fusion only, before collapsing into a pure He white dwarf, while a progenitor of $0.8 \leqslant M \leqslant 8M_\odot$, after H and He fusion, collapses into a CO white dwarf—the most common type of white dwarf—and a progenitor of $8 \leqslant M \leqslant 9M_\odot$, after H, He, and C fusion, will collapse into a O-Ne-Mg white dwarf.

Independently of its composition, the gravity of a white dwarf is counter-balanced by the pressure created by the electron-degenerate configuration. The relatively large radius (but which can be as small as the Moon, with $R = 2140$ km; Caiazzo et al. 2021) and high temperature (with $6 \times 10^3 < T_{eff} < 3 \times 10^4$ K, by applying Stefan law for blackbodies) of white dwarfs, imply that their surface emission is detectable. 80% of white dwarfs have an A spectral type, identified by Balmer lines of neutral H. White dwarfs have a magnetic field ranging from B ~10^3 to ~10^9 G.

2.1.3.4 White Dwarf Equation of State

The existence of a localized region of white dwarfs in the Hertzsprung–Russell diagram (see Figure 2.3) is consistent with the existence of a mass–radius relationship, which is, for a typical white dwarf:

$$\frac{R}{R_\odot} = \left(\frac{M}{M_\odot}\right)^{-1/3}. \tag{2.17}$$

---

[7] The title is: "An A-type star of very low luminosity."

Using the frame of Fermi–Dirac matter distribution in a high density n environment, comparing the electron $T_{F,e}$ and nucleon $T_{F,p/n}$ Fermi temperature (m being the mass of the particle):

$$T_F \propto \frac{n^{2/3}}{m} => T_{F,e} \sim 2000 \times T_{F,p/n} \tag{2.18}$$

implies that, for a typical central temperature of a white dwarf $\sim 10^7$ K, electrons are fully degenerate, but not the nucleons:

$$T_{F,p/n} \approx 10^6 \text{K} < T_{\text{WD}} \approx 10^7 \text{K} \ll T_{F,e} \approx 2 \times 10^9 \text{K}. \tag{2.19}$$

We have already met the equation of state of a polytrope for internal stellar structure (see Equation (2.7)), which also allows us to describe the internal structure of white dwarfs, with:

$$\gamma = \frac{5}{3} \text{ for } x_F \ll 1 \text{ and } \gamma = \frac{4}{3} \text{ for } x_F \gg 1 \tag{2.20}$$

where

$$x_F = \frac{p_F}{mc} = \sqrt{\left(\frac{E_F}{mc^2}\right)^2 - 1} \tag{2.21}$$

p_F and E_F being respectively the Fermi momentum and energy.

Knowing this EoS, we can derive the maximum mass of a white dwarf, also known as *Chandrasekhar mass*, equal to $M_{\text{Chandra}} = 1.4 M_\odot$ (Chandrasekhar 1931). This maximum mass has two fundamental implications. First, only low-mass stars can become white dwarfs. Second, in case additional matter is accreted onto a white dwarf, it cannot remain stable and eventually collapses.

Finally, it is interesting to consider the cooling of white dwarf, corresponding to their final evolution stage: since no more thermonuclear reaction is occurring in the core of white dwarfs, they can only cool down with time, a long process that is called *crystallization* (see Figure 2.4; Tremblay et al. 2019). Both effective temperature and luminosity decrease, in a long time which remains difficult to accurately predict. This effect allows astronomers to derive the age of the oldest white dwarfs in our Galaxy, and thus to get a lower limit regarding the age of the Milky Way.

2.1.4 Late Evolution of Massive Stars: Toward Neutron Stars

We now describe the late evolution of massive stars, from $M \sim 9 M_\odot$, up to probably $M \sim 100 M_\odot$ (see lifetimes of massive stars in Schaller et al. 1992). These stars go through the supergiant phase, evolving through cycles of thermonuclear burning of heavier atoms, alternating with stages of exhaustion of lighter atoms, up to final stages of stellar evolution, when thermonuclear reactions are no longer possible because the massive star runs out of nuclear fuel. This occurs when all the matter of the stellar core has been transformed into iron, because thermonuclear fusion of atoms heavier than iron is an endothermic process, contrarily to fusion of lighter

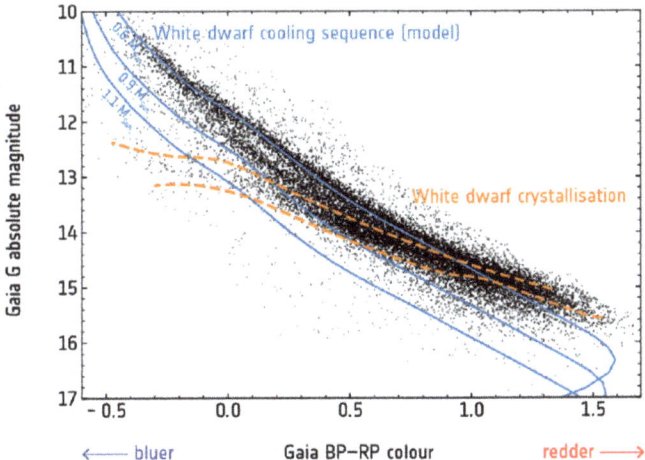

Figure 2.4. White dwarf crystallization: Hertzsprung–Russell diagram showing the absolute magnitude versus the color of ~15, 000 white dwarfs (black dots) within 100 pc, from the Gaia satellite DR2 data release. Blue lines show the cooling sequence of white dwarfs of different mass—0.6, 0.9, and $1.1 \times M_{\odot}$, respectively—predicted from theoretical models. White dwarfs of certain colors and luminosities (indicated with orange lines) pile up, due to cooling and crystallization effect of the originally hot matter inside the core. This is an observational evidence of crystallization inside white dwarfs. Reprinted with permission from Tremblay et al. (2019).

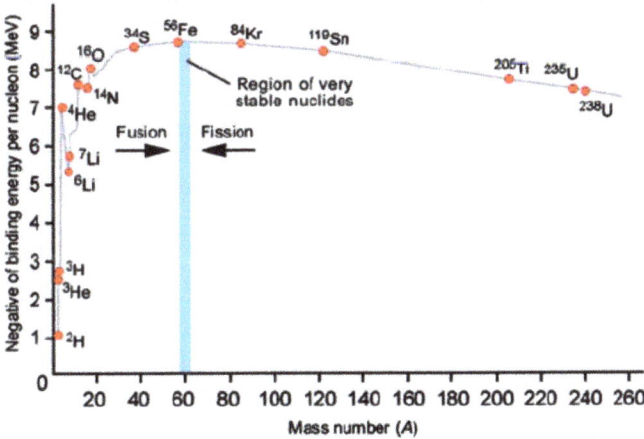

Figure 2.5. Binding energy per nucleon (MeV), versus mass number (A): fusion occurs for atoms lighter than iron, while fission occurs for atoms heavier than iron. Reprinted with permission from Lépine-Szily & Descouvemont (2012).

atoms which is exothermic (see Figure 2.5). These final phases, from carbon burning, are very-short lived (~60 yr for a $25 M_{\odot}$ star; Schaller et al. 1992), because most of the nuclear energy is released through neutrinos, which escape the stellar interior without interacting with the gas, lowering the pressure, and resulting in a runaway contraction which accelerates nuclear fusion. These massive stars are also characterized by powerful stellar winds, leading to evolved stars of final mass $8 \lesssim M \lesssim 30 M_\odot$. Once all the thermonuclear reaction cycles have occurred, the stellar iron core, more massive than the Chandrasekhar mass, collapses under its own gravitational pressure, to form a compact object, a neutron star or a black hole.

2.1.4.1 The Formation of a Neutron Star

Let us first consider the formation of a neutron star (see Figure 2.6). Before its collapse, the massive star, of radius which can reach up to a few hundreds of millions of km, consists of hydrogen, helium, and oxygen shells, surrounding an iron core, of typical mass $M = 1.4 M_\odot$ and radius $R \sim 1500$ km. The first step toward a neutron star begins when the iron core collapses, in a timescale of ~ 1 s. The massive gravitational pressure crushes most of the protons and electrons into neutrons, via the main neutralization reactions:

$$\text{inverse } \beta\text{-decay: } p \rightarrow n + e^+ + \nu_e \tag{2.22}$$

$$\text{electron capture: } p + e^- \rightarrow n + \nu_e \tag{2.23}$$

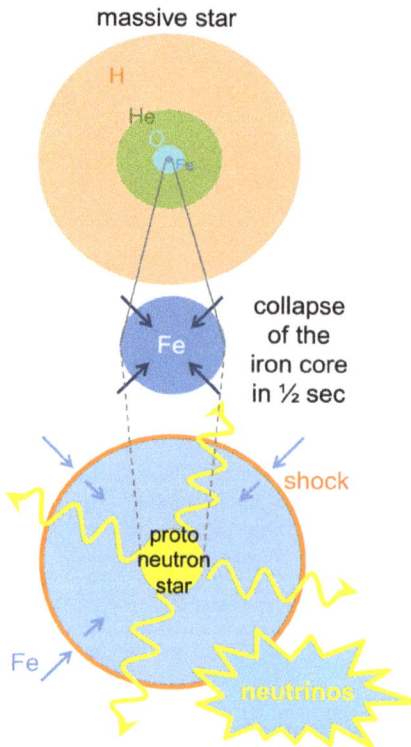

Figure 2.6. Sketch of neutron star formation: the neutrino-driven delayed explosion mechanism relies on the absorption of neutrinos by the dense post-shock gas. Reprinted with permission of the AAS from Foglizzo et al. (2015).

accompanied by photodisintegration of Fe nuclei (at $T \sim 10^9$ K):

$$^{56}\text{Fe} + 3.7\text{MeV} \rightarrow ^{56}\text{Mn} + \bar{\nu}_e. \tag{2.24}$$

This leads to the formation of a proto-neutron star, of radius $R \sim 50$ km, which, after neutrino cooling, collapses into a neutron star of radius $R \sim 12$ km, maintained by neutron degenerescence pressure at nuclear density of $\rho \sim 10^{15}$ g cm^{-3}. The gravitational energy released in this collapse, corresponding to 4×10^{53}erg $= 0.15 M_{\text{core}} c^2$, higher than the binding energy of the stellar envelope, is unavoidably accompanied by a violent explosion which ejects the outer layers with a kinetic energy of $\sim 10^{51}$erg, at a speed $\sim 10^4$km s^{-1}. This *supernova* event (of *Type II, Type Ib*, possibly associated to a short γ-ray burst, or a *macro/kilo-nova*) is due to all falling matter rebounding onto the surface of the newly-formed neutron star. The rate of Type II supernova being of nearly one per century in our Galaxy, we can extrapolate a number of nearly 10^8 to 10^9 neutron stars in the Milky Way. Only a few thousands are currently detected, mainly via their radio emission (*pulsars*, see below), or via their thermal X-ray emission, for some isolated neutron stars (Tauris et al. 2017; Vahdat et al. 2022), allowing accurate measurements of their temperature, geometry and magnetospheric properties, and also distance and age.

During the process of the supernova event, most outer shells of the star are ejected within the surrounding interstellar medium, creating a *supernova remnant* (SNR), associated to sites of acceleration of galactic cosmic rays (mostly protons) to relativistic speeds (Ackermann et al. 2013). These SNR, such as the young Cas A (Abdo et al. 2010a) and RX J1713.7-3946 (Abdo et al. 2011; Acero et al. 2017), both detected by the Fermi Space Telescope, emit in the whole electromagnetic spectrum, from radio to γ-rays.

Finally, the formation of a neutron star can be studied through analytical studies, simplified and *realistic* simulations, and also with laboratory experiments based on standing accretion shock instability (SASI), such as kitchen sink hydraulic jump, making use of physical analogies between acoustic and surface waves, shock wave and hydraulic jump, and also pressure and depth (Foglizzo et al. 2012, 2015).

2.1.4.2 Equation of State of a Neutron Star

Knowing the equation of state of a neutron star allows to reveal its interior. From its surface to its core, a neutron star is described by (see Figure 2.7):

1. an *outer crust*, a solid layer made of nuclei (ions) and electrons, and supported by electron degeneracy pressure,
2. an *inner crust*, also made of nuclei and electrons, in which superfluid neutrons leak out of nuclei at the neutron drip density $\sim 4 \times 10^{11}$ g cm^{-3}—separating the inner from the outer crust—and supported by neutron degeneracy,
3. and finally by a *core*, delimited by the crust-core boundary, where nuclei completely dissolve, at densities $\sim 2 \times 10^{14}$ g cm^{-3}. Within the core, densities even amount to several times the nuclear saturation density $\rho_{\text{sat}} = 2.8 \times 10^{14}$ g cm^{-3}, constituted at least of superfluid neutrons and superconducting protons.

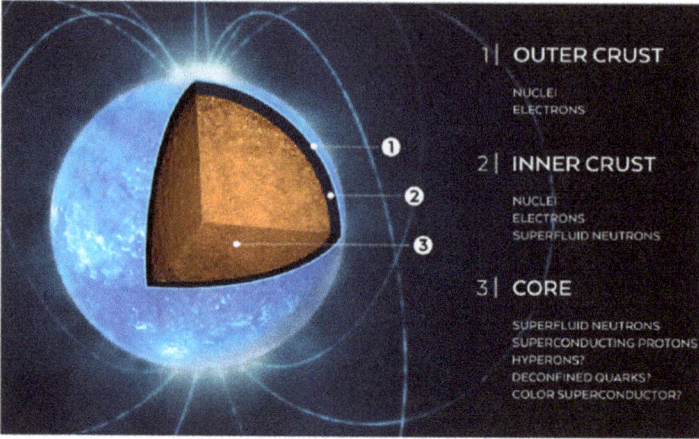

Figure 2.7. Sketch of neutron star interior. Reprinted with permission from Watts et al. (2016).

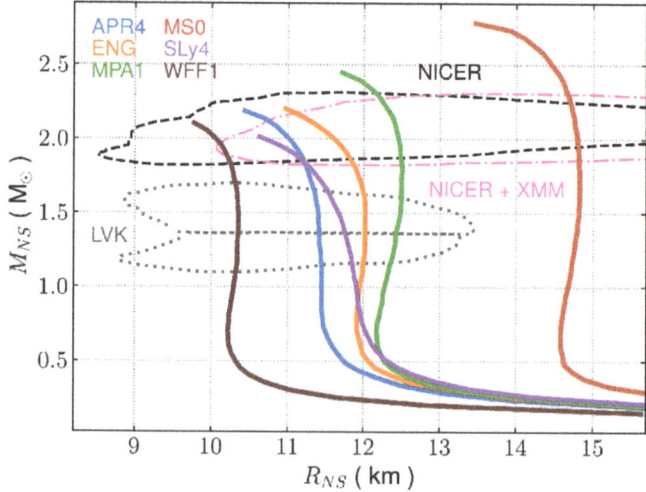

Figure 2.8. Diagram of mass–radius equation of state for a neutron star, with various models, and superimposed constraints given by NICER, XMM and LIGO–Virgo–Kagra (LVK). Reprinted with permission of the AAS from Fragione (2021).

Accurately calculating the EoS of a neutron star, which links thermodynamic variables such as temperature, pressure, and density, is a complex problem ruled by the strong nuclear force, requiring an accurate understanding of the interaction between both neutrons and remaining protons (only ∼5%) in the neutron star core. Realistic EoS of neutron stars, based on models with different microphysics, and combined with the mass–radius (M-R) relationship reported in Figure 2.8; are constrained observationally by the Neutron Star Interior Composition Explorer (NICER) and X-ray Multi-Mirror (XMM-Newton) satellites, in the range $1.2 M_\odot \leqslant M_{NS}^{low} \leqslant 1.7 M_\odot$ for low-mass NS, and $1.8 M_\odot \leqslant M_{NS}^{high} \leqslant 2.3 M_\odot$ for high-mass NS. The maximum mass of NS, theoretically

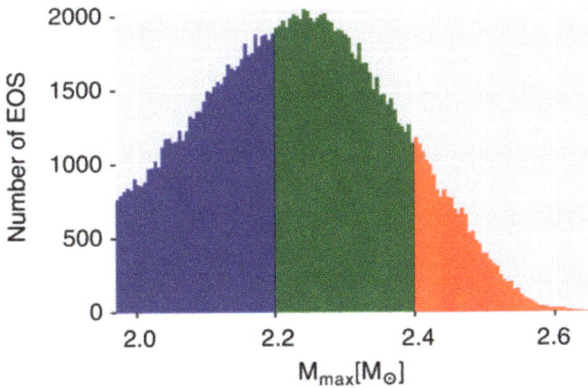

Figure 2.9. Number of EoS as a function of the maximum mass of a NS. The colors identify the subsets: $1.97 \leqslant M_{\mathrm{NS}}^{\max} \leqslant 2.20 M_{\odot}$ (blue), $2.20 \leqslant M_{\mathrm{NS}}^{\max} \leqslant 2.40 M_{\odot}$ (green), and $2.40 \leqslant M_{\mathrm{NS}}^{\max} \leqslant 2.66 M_{\odot}$ (red). Reprinted with permission from the Ferreira & Providência (2021).

confined to the maximum mass $M_{\mathrm{NS}}^{\max} \leqslant 2.6 M_{\odot}$ (Figure 2.9), is thus in agreement with the observed values, ranging from $M_{\mathrm{NS}} \sim 1.2$ up to $2.6 M_{\odot}$ (Figure 2.10). The radius of NS is less constrained, between $8.5 \leqslant R_{\mathrm{NS}} \leqslant 13.5 R_{\odot}$ by the LIGO–Virgo–Kagra collaboration, and $R_{\mathrm{NS}} \geqslant 8.5 R_{\odot}$ by NICER and XMM (Figure 2.8).

We also see from Figure 2.10 that, while all observed neutron stars within binaries are born with a canonical mass of $1.3 - 1.4 M_{\odot}$, recycled neutron stars (in low-mass X-ray binaries) may have accreted up to $\sim 1 M_{\odot}$ before eventually further collapsing into a black hole. On the high-mass end, some neutron stars might be born with higher masses, reaching $\sim 1.8 M_{\odot}$ (Özel et al. 2012; Özel & Freire 2016). An extreme case is the heaviest neutron star PSR J0740+6620, with a mass determined from electromagnetic observations in the range $2.08 \pm 0.07 M_{\odot}$. In this case, its equatorial radius $R = 13.7_{-1.5}^{+2.6}$ km has also been measured, using rotating hot spot patterns detected by NICER and XMM observations (Miller et al. 2021).

2.1.4.3 Kick Velocity of a Neutron Star

Neutron stars follow the distribution of the Galactic plane, with a width larger that the disk, due to their higher velocities: they are born with kick velocities, acquired at their formation during the supernova event, of a few hundreds of km s^{-1} (see Section 6.5.3). In this context, the pulsar wind nebula IGR J11014-6103 hosts a pulsar which likely exhibits the highest known kick velocity, between 2400 and 2900 km s^{-1}, surpassing the kick velocities of any known pulsar or even of any compact object associated with an SNR. Because the pulsar moves at such a high velocity through the interstellar medium, it creates a bow shock. The direction of motion is consistent with the pulsar being born during the event that produced the close-by supernova remnant MSH 11-61A (Tomsick et al. 2012; Halpern et al. 2014; see Figure 2.11).

2.1.4.4 Pulsars

The first pulsar was discovered in 1967, as a regular signal detected with the Interplanetary Scintillation Array in Cambridge, with a period $P_{\mathrm{spin}} = 1.337$ s

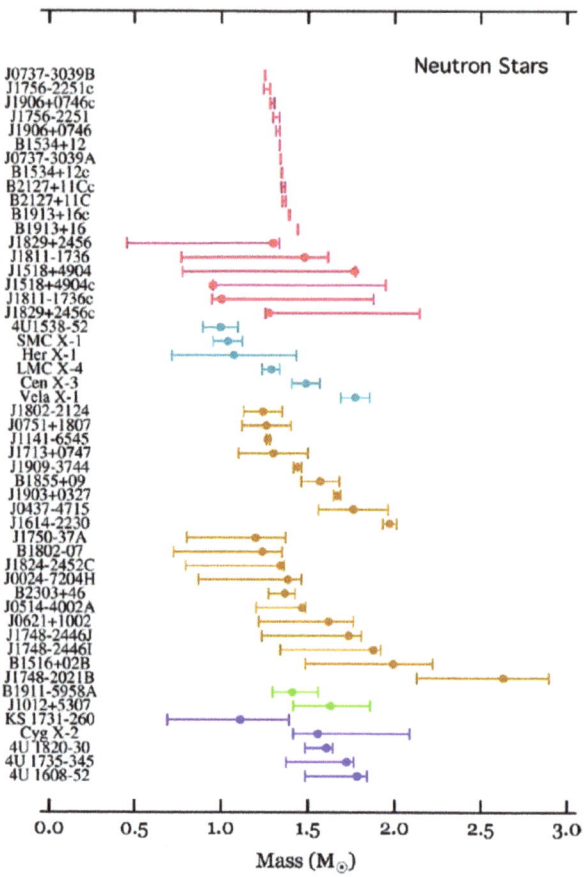

Figure 2.10. Masses of neutron stars, in various kinds of binaries: *(i)* NS in double neutron star systems (magenta); *(ii)* NS in eclipsing binaries with high-mass companions—slow pulsars—(cyan); *(iii)* NS in eclipsing binaries with white dwarf companions—recycled pulsars—(gold); *(iv)* NS with mass determined from optical observations of the white dwarf companions (green); and *(v)* NS in accreting bursters (purple). Reprinted with permission of the AAS from Özel et al. (2012).

(Hewish et al. 1968).[8] One year later, Gold (1968) interpreted radio pulsars as fast spinning and highly magnetized neutron stars, losing energy from electromagnetic radiation and emission of relativistic particles. Nearly simultaneously, the discovery of a pulsar with a short period (P_{spin} = 0.089 s) in the Vela supernova remnant (SNR, Large et al. 1968), and another one in the Crab SNR (Howard et al. 1968; P_{spin} = 0.033 s), allowed astronomers to validate the link of nature between pulsars and neutron stars.

Pulsars are neutron stars characterized by the combination of a high spin (P_{spin} ranging from 30 ms to ~25 s) and dipole magnetic field (typically of B ~ 10^{12} – 10^{13} G), also called *rotation-powered pulsars* (see Figure 2.12). Their spin period P_{spin}, spin

---

[8] First named LGM-1 (*little green man*) the source was thereafter baptized CP 1919 (*Cambridge Pulsar*).

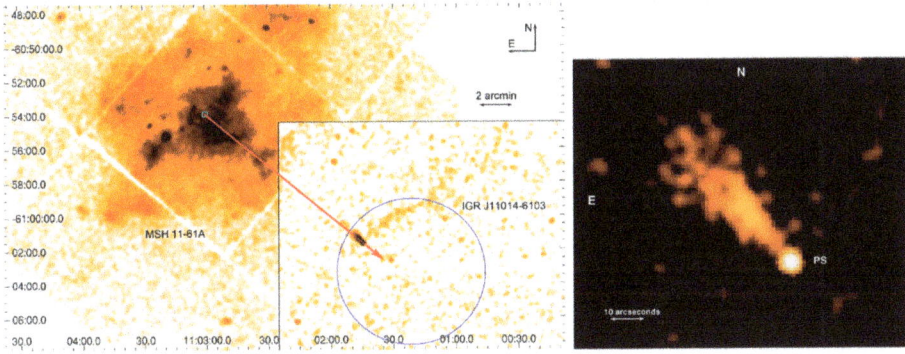

Figure 2.11. *Left*: composite image of the supernova remnant MSH 11-61A near the pulsar wind nebula IGR J11014-6103, hosting a pulsar moving at high velocity–between 2400 and 2900km s^{-1}—through the interstellar medium, creating a bow shock. The direction of motion is consistent with the pulsar born during the event which produced MSH 11-61A. The large image is from XMM-Newton (0.5–10 keV), while the inset is from Chandra–ACIS (0.3–10 keV). The blue circle is the 4.$''$3 INTEGRAL error circle. The red arrow shows the alignment of the point source (PS) with the center of MSH 11-61A, marked with a cyan square. *Right*: Chandra/ACIS 0.3–10 keV image of IGR J11014-6103, with the point source (PS) and a 4$''$ X-ray extension to the north-east (NE). Reprinted with permission of the AAS from Tomsick et al. (2012).

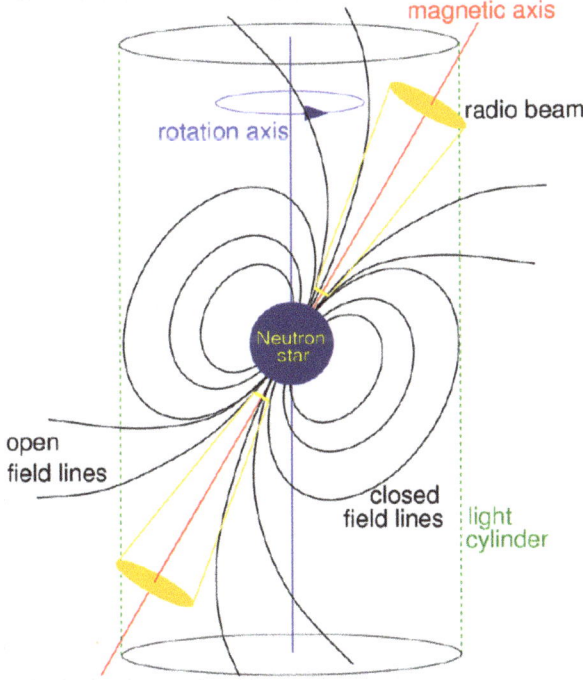

Figure 2.12. A simplified pulsar model, with the dipolar magnetosphere, the light cylinder and the radio beam emission. Reprinted with permission from Lorimer & Kramer (2004).

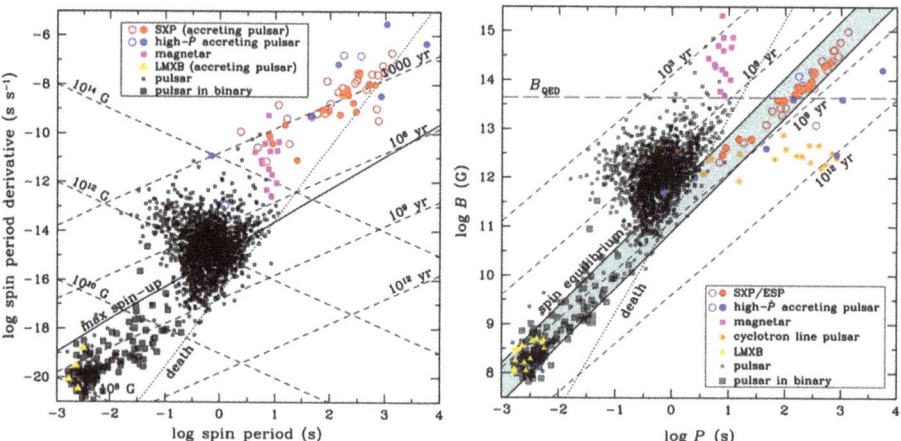

Figure 2.13. Left: Spin period derivative $\dot{P}_{spin}$ versus spin period P_{spin} of accreting pulsars, magnetars, pulsars in LMXB, pulsars in binary and isolated pulsars. Right: Magnetic field B versus spin period P_{spin} of various types of pulsars. Reprinted with permission from Ho et al. (2014).

period derivative $\dot{P}_{spin}$ and magnetic field B properties are reported in Figure 2.13; both $P - \dot{P}$ and $P - B$ diagrams allowing astronomers to distinguish the various pulsar populations (Ho et al. 2014; Wang 2016). We know a few thousands pulsars within our Galaxy and Magellanic Clouds (Manchester et al. 2005), most of them having periods P_{spin} between 0.1 s and a few seconds, and a spin period derivative $\dot{P}_{spin}$ between 10^{-17} and 10^{-13} s s^{-1}. The slowing down of pulsars, mainly due to electromagnetic radiation and particle emission, makes these (usually young) pulsars migrate to the right side of these diagrams.

Electrons relativistically accelerated along magnetic field lines create a synchrotron emission, detected in the whole electromagnetic spectrum from radio to X-rays, and even γ-rays, as shown by observations of the individual pulsar PSR J2043+2740 (Noutsos et al. 2011), and as reported by the catalog of γ-ray pulsars detected by the Fermi γ-ray space telescope (Abdo et al. 2013). Isolated pulsars interact with their remnant surrounding nebulae, creating a *pulsar wind nebulae*, commonly detected at high energies (mainly in X-rays), and sometimes in other wavelength domains (see, e.g., Curran & Chaty 2011, and references therein).

A peculiar class is populated by *magnetars* (Rea 2014; Papitto et al. 2020), also called *magnetically powered pulsars*, of which we know nearly 30 in our Galaxy and in the Magellanic Clouds, characterized by an extremely strong dipole magnetic field $B \sim 10^{14} - 10^{15}$ G, and a slow rotation period $P \sim 1 - 10$ s (see Figure 2.13 and Testa et al. 2008; Mignani et al. 2009; Rahoui et al. 2009; Esposito et al. 2010; Abdo et al. 2010b). Magnetars are likely associated to *anomalous X-ray pulsars* (AXP, Mereghetti et al. 2001), or to *soft gamma-ray repeaters* (SGR, Fuchs et al. 1999), which exhibit brief, intense and recurrent γ-ray emission, sometimes at an extraordinary high level, such as the giant burst of SGR 1806-20 (Palmer et al. 2005).

The spin-down rate $\dot{P}_{spin}$ of a pulsar, mainly depending on the magnetic field, is given by (see Figure 2.13 right):

$$\dot{P}_{\text{spin}} \propto \frac{B^2}{P_{\text{spin}}} \qquad (2.25)$$

with a characteristic decay time of magnetic field for a pulsar $\tau \sim 10^6$ yr (Bhattacharya & Srinivasan 1995). Observational data indicate that the magnetic field of a neutron star significantly decays when it is located within a binary, due to spin evolution and accretion of matter onto its surface (Bhattacharya 1995b). We estimate that only ~5% of pulsars are within binary systems, likely due to the unbound of binary systems at the supernova event, kicking out the newly-formed neutron star. Pulsars within binaries, located in the lower-left part of Figure 2.13; are characterized by short spin periods $P_{\text{spin}} \leqslant 30$ ms, small period derivative $\dot{P}_{\text{spin}} \leqslant 10^{-17}$ s s^{-1}, and weak magnetic field ($B = 10^8 - 10^{10}$ G) compared to the typical value of $10^{12} - 10^{13}$ G in isolated *canonical* pulsars. In addition, they exhibit *spin-down* timescale:

$$\tau = \frac{P}{2\dot{P}_{\text{spin}}} \sim 1 - 10 \text{ Gyr} \qquad (2.26)$$

much higher than for isolated pulsars ($\tau \leqslant 10$ Myr). These pulsars are called *spun-up, recycled*, or even *millisecond* pulsars, because an old neutron star has been spun up by transfer of mass and angular momentum via the matter accreted from its companion, process, which can last for a long time in a binary, especially in a low-mass X-ray binary (Bhattacharya 1991, 1995a).

A catalog of pulsar, including all spin-powered pulsars, AXP and SGR showing coherent pulsed emission, and excluding accretion-powered pulsars (see Chapter 5) is maintained by ATNF (Manchester et al. 2005): https://www.atnf.csiro.au/research/pulsar/psrcat.

2.1.5 Late Evolution of Massive Stars: Toward Black Holes

Black holes have been (theoretically) discovered independently, first by John Michell in 1784 (Michell 1784),[9] and then by Pierre-Simon de Laplace in 1798 (Laplace 1798).[10] Both Michell (1784) and Laplace (1798) considered a particle in a classical mechanics frame around an object of solar density, but of radius $R \geqslant 500 R_\odot$, realizing that the escape velocity of such an object would be greater than the velocity of light, so that no particle would escape, not even light. While the idea of such a body (that Laplace called *"corps obscur"*—dark body) was only mentioned in the

---

[9] Let us cite the original paragraph written on these objects by Michell (1784): "Hence, [...] if the semi-diameter of a sphere of the same density with the Sun were to exceed that of the Sun in the proportion of 500 to 1, a body falling from an infinite height toward it, would have acquired at this surface a greater velocity than that of light, and consequently, supposing light to be attracted by the same force in proportion to its inertia, with other bodies, all light emitted from such a body would be made to return toward it, by its own proper gravity."

[10] Let us now cite the original words written by Laplace (1798) on this subject: "A luminous star of the same density as the Earth, and whose diameter would be two hundred and fifty times greater than that of the Sun, would not, by virtue of its attraction, allow any of its rays to reach us; It is therefore possible that the largest luminous bodies in the Universe, by that very fact, are invisible."

book by Laplace (1798),[11] a *proof* of the possibility of its existence is given in the paper by Laplace (1799),[12] in which Laplace calculates the ratio $\frac{M}{R}$ requested to produce this effect.

The theoretical resurgence of black holes occurred in 1916, a few months only after the publication of the general relativity theory by Einstein (1915), when Karl Schwarzschild calculated the exact spherically symmetric solution of Einstein equation, with only one parameter: its mass (Schwarzschild 1916). The first analytical solution of the field equations of such a star, asymptotically shrinking to its *gravitational radius*, is derived in 1939 (Oppenheimer & Snyder 1939).[13] It was three decades later, in 1968, that John A. Wheeler baptized *black hole* such an object.[14]

The fate of massive stars is to collapse into a neutron star or a black hole, and it is now clear, both from observations and modeling, that there exists an overlap in the mass range of collapsing stars, depending on the wind strength, the quantity of matter ejected during the supernova event, and falling back on the central object, as can be seen in Figure 2.14. It appears from this figure that even massive stars of $100 M_\odot$ can collapse into neutron stars of $\sim 2 M_\odot$, losing most of their matter both through wind throughout the whole stellar life, and through ejection during the SN event.

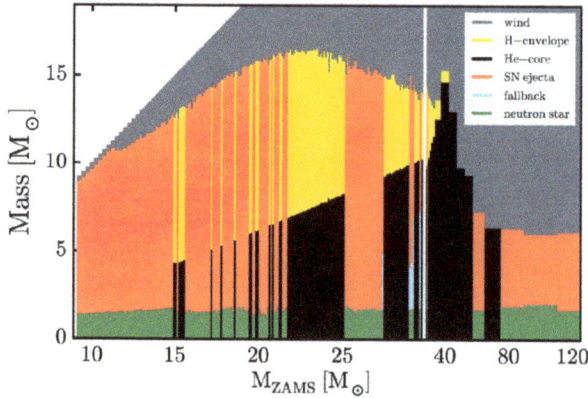

Figure 2.14. Progenitor mass versus zero-age main sequence (ZAMS) mass. There exists an overlap in the mass range of collapsing stars, depending on the wind strength, and the quantity of matter ejected during the supernova event, and falling back onto the central object. Even massive stars of $\sim 100 M_\odot$ can collapse into neutron stars of $\sim 2 M_\odot$, losing most of their matter both through wind throughout the whole stellar life, and through ejection during the SN event. Reprinted with permission of the AAS from Sukhbold et al. (2016).

---

[11] Laplace *enthusiastically* predicted: "there are therefore in celestial spaces, dark bodies as considerable, and perhaps in as great a number, as the stars."
[12] The title of Laplace (1799) is: "Proof of the proposition that the attractive force of a body can be so great that the light cannot emanate from it."
[13] The title of Oppenheimer & Snyder (1939) is: "On continued gravitational contraction."
[14] Wheeler preferred the term *black hole* to the more exact—but longer—terminology *gravitationally completely collapsed object.*

Stellar-mass black holes can be formed in two ways: either a massive star directly collapses into a black hole—without triggering any supernova explosion—or a supernova occurs, forming a proto-neutron star. In the latter case, when the energy is too low to completely unbind the stellar envelope, and a large fraction of it falls back onto the newly-formed but short-lived neutron star, it further collapses into a—delayed formed—black hole (Heger et al. 2003; Mirabel 2017).

In general, the core collapse of massive stars of $M \gtrsim 25M_\odot$ lead to the formation of black holes by fallback, and stars of $M \gtrsim 40M_\odot$ directly collapse to form black holes (see Figure 2.15 and Heger et al. 2003). Black holes are divided in three generic families: non-rotating *Schwarzschild* black holes (Schwarzschild 1916), spinning *Kerr* black holes (Kerr 1963), and charged *Newman* black holes (Newman et al. 1965). Spinning black holes are likely the most common, with black hole spins in low-mass accreting binaries covering a large range of values, usually explained by accretion after the formation of the compact object, while black hole spins in high-mass binaries having only values near the maximum, likely related to the stellar progenitor (Barack et al. 2019; Qin et al. 2019).

In the Schwarzschild metric, describing the spacetime around a static, spherical object of mass M (with $r = 0$ for a black hole), the interval ds is computed as:

$$ds^2 = -c^2\left(1 - \frac{2GM}{rc^2}\right)dt^2 = -c^2 d\tau^2 \qquad (2.27)$$

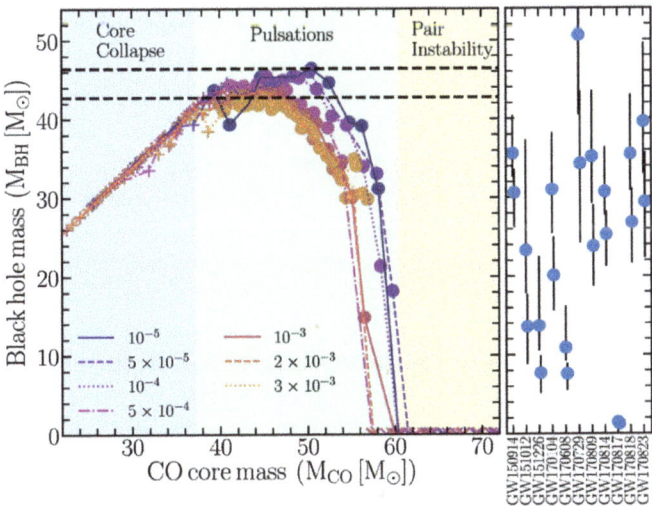

Figure 2.15. Final black hole mass versus CO core mass, for various metallicities. '+' symbols show models directly evolving toward core collapse (left/blue panel), 'circle' symbols go through at least one pulse of pulsational pair-instability supernova (middle/green panel), while 'x' symbols undergo a pair-instability supernova (right/yellow panel). Stellar metallicities, from $Z = 10^{-5}$ to $Z = 3 \times 10^{-3}$, are indicated in the lower part of the figure. Finally, mass estimates of binaries detected by LIGO–Virgo are reported in the right (Abbott et al. 2019). Reprinted with permission of the AAS from Farmer et al. (2019).

dt representing a coordinate time interval, and $d\tau$ a proper time interval, respectively. This equation leads to Einstein dilation of time:

$$\frac{dt}{d\tau} = 1 + z = \frac{1}{\sqrt{1 - \dfrac{2GM}{rc^2}}} \text{ for } r > R_S \tag{2.28}$$

where z is the gravitational redshift (see Equation (2.38), and see Section 2.2.1.4 for a definition of the Schwarzschild radius R_S).

Based on the present supernova rate in a late-type spiral galaxy like our own, of ~ 0.03 year^{-1}, and correcting for a higher rate during the first few billion years after the formation of our Galaxy, it is possible to estimate a total number of neutron stars in the Milky Way of the order of $\sim 10^9$, and a total number of stellar mass black holes of mass between 5 and $15 M_\odot$, ranging from 2×10^7 to 6×10^9 (van den Heuvel 1992; Brown & Bethe 1994).

2.1.5.1 Neutron Star–Black Hole Mass Gap

At the low end of the mass spectrum, the *neutron star–black hole mass gap* is due to an absence of any compact object in the region of mass delimited by the most massive neutron star of mass $\sim 2 M_\odot$, and the least massive black hole, such as those observed in accreting binaries, of mass $\sim 5 M_\odot$. It is not clear whether such a gap really exists, especially with the detections of gravitational waves from objects of unknown nature (heavy neutron stars of light black holes), but if it is the case, it will provide valuable information, not only on the actual physics of supernova explosion, but also on stellar binary evolution. See Chapter 6 for more details on this gap.

2.1.5.2 Various Types of Collapse and/or Supernova Events

1. *Core-collapse supernova* (CCSN): while stellar evolution models predict the final outcome of a core-collapse supernova (Foglizzo et al. 2015; Burrows & Vartanyan 2021), the relation between progenitor mass and black hole mass is not monotonic. For progenitors of He core mass M_{He} in the range $2 \lesssim M_{He} \lesssim 35 M_\odot$ (corresponding to $10 \lesssim M_{ZAMS} \lesssim 85 M_\odot$; see left/blue panel of Figure 2.15), after a hydrodynamic time of a few minutes to a few hours, fallback drives the collapsing stellar core to form either a neutron star mass, or—if it is above the maximum mass for a neutron star—a black hole, accompanied by a supernova explosion (Fryer 1999; Fryer & Kalogera 2001).

   In addition, stellar evolution models also suggest that core-collapse supernova is not the only mechanism to consider as the final fate of a massive star. An $e^- - e^+$ *pair-creation instability* process (Barkat et al. 1967; Ober et al. 1983; Bond et al. 1984), due to a substantial production of free electrons and positrons, plays a strong role in constraining the maximum mass of stellar mass black holes. Two cases involving this process can occur, according to the mass of the stellar helium core M_{He}, which can be much less than the initial mass M_{ZAMS}, depending on the strength of stellar winds, and thus on the metallicity:

2. *Pulsational pair-instability supernova* (PPISN): Massive stars with relatively small helium core mass in the range $35 \lesssim M_{He} \lesssim 65 M_\odot$ (corresponding to $85 \lesssim M_{ZAMS} \lesssim 140 M_\odot$; see middle/green panel of Figure 2.15) undergo, after the carbon fusion phase, *Pulsational Pair-Instability* (PPI), associated with mass loss of the stellar core, leading during core-collapse to explosive burning. These pulsations, through a series of contractions and expansions, are unable to unbind the star, but send shock waves into the stellar envelope. This has the effect of ejecting away many solar masses of the envelope at low velocity, until reaching a new equilibrium toward a lower stellar core mass, undergoing a long core-collapse supernova, leaving behind a remnant black hole of mass $M \lesssim 50 M_\odot$, smaller than when the PPI process is not considered (see Figure 2.16; Woosley et al. 2007; Woosley 2019).

3. *Pair-instability supernova* (PISN): For stars with helium core mass in the range $65 \lesssim M_{He} \lesssim 135 M_\odot$ (corresponding to $140 \lesssim M_{ZAMS} \lesssim 260 M_\odot$; see right/yellow panel of Figure 2.15), the central temperature reaches $\sim 10^9$ K, leading to a runaway production of $e^- - e^+$ pairs at the stellar core, which softens the EoS by reducing the internal radiation pressure, leading to a partial collapse and a runaway thermonuclear fusion of oxygen, neon, and silicon. The outcome is the entire disruption of the whole star in a single pulse, with an explosion energy of $\sim 10^{53}$ erg for a $M_{He} \sim 100 M_\odot$ progenitor, without leaving any stellar remnant (Heger et al. 2003; Woosley et al. 2007; Woosley 2017; Woosley & Heger 2021). This type of supernova only happens in massive stars of low metallicity, such as *Population III* stars of the early universe.

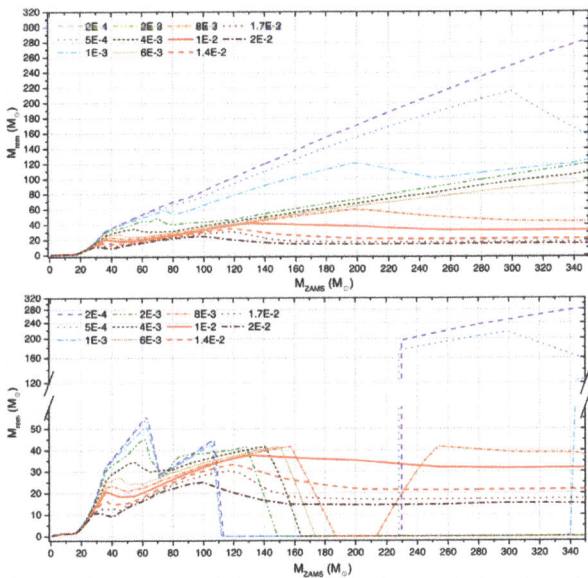

Figure 2.16. Mass of the final black hole (BH) versus zero-age main sequence (ZAMS) stellar mass, without (upper panel) and with (lower panel) PPISN and PISN respectively. The masses are computed with SEVN code, for stellar metallicities from $Z = 2 \times 10^{-4}$ to 2×10^{-2}, as indicated in the upper left. This plot shows that no remnant BH can form, from ~ 60 to $\sim 120 M_\odot$. Reprinted with permission from Spera & Mapelli (2017).

4. *Direct core-collapse* (DCC): Finally, massive progenitors with helium core mass greater than $M_{\text{Hecore}} \gtrsim 135 M_\odot$ (corresponding to $260 M_\odot \lesssim M_{\text{ZAMS}}$) experience a direct core-collapse of the large carbon–oxygen (CO) core into a black hole as massive as the helium core of the progenitor, without any supernova explosion. This is made possible because outer layers of the star are so massive that their binding energy is above a threshold that can overcome the supernova explosion, thus nothing can halt the collapse. This is the domain of the intermediate-mass black hole (IMBH), in the range $10^2 - 10^5 M_\odot$ (Heger et al. 2003; Fryer et al. 2012).

2.1.5.3 Pair-Instability Supernova Mass Gap

The $e^- - e^+$ pair-creation instability process thus prevents the formation of black holes from core-collapse for stellar progenitors in the mass range $65 \lesssim M \lesssim 135 M_\odot$ (see Figures 2.15 and 2.16), suggesting a mass gap in the black hole population, known as *pair-instability supernova mass gap*, delimited by the most massive stellar mass black hole of mass $\sim 50 M_\odot$, and the least massive intermediate-mass black hole (IMBH) of mass $\sim 135 M_\odot$ (Heger et al. 2003), although recent results suggest a maximum mass that could reach up to $\sim 161 M_\odot$ (Woosley & Heger 2021). However one has to be careful about such a mass threshold, since many uncertainties remain about the minimum and maximum threshold of this mass gap. In particular, the lower edge of the mass gap is highly uncertain, it might be as low as $M \sim 40 M_\odot$, or even above $M \sim 70 M_\odot$, depending on many parameters, such as nuclear thermonuclear $^{12}C\,(\alpha, \gamma)^{16}O$ reaction rates (Farmer et al. 2019, 2020), but also on the collapse of the residual stellar hydrogen envelope, and the impact of stellar rotation (Mapelli et al. 2020), on convection model and onset of dredge-up episodes, and even on the efficiency of accretion from companion stars and on the outcome of stellar mergers. A few events, detected in GW, fall in this category, such as GW190521[15] (see Section 6.6.2): the primary black hole falling directly within the PISN mass gap, and the remnant in the IMBH range, with masses of each component $M_1 = 85 M_\odot$ and $M_2 = 66 M_\odot$, resulting in a total merger mass of $\sim 142 M_\odot$, the most massive GW event to date (Abbott et al. 2020a, 2020b).

2.2 Fundamental Properties of Compact Objects

2.2.1 Gravitational Properties

2.2.1.1 Gravitational Potential

The *gravitational potential* $\Phi(r)$ of a self-gravitating, spherical object of mass M and radius R, is at a distance $r \geqslant R$ (with by definition $r = 0$ at its center):

$$\Phi(r) = -\frac{GM}{R}. \tag{2.29}$$

---

[15] Two other events fall in the PISN gap: GW 190426190642 and GW 200220061928.

2.2.1.2 Gravitational Potential Energy

The potential energy dE_p required to bring a shell of mass dm to infinity is:

$$dE_p = E_p(r) - E_p(\infty) = -\frac{Gm(r)}{r}dm \qquad (2.30)$$

where $m(r)$ is the mass enclosed within a radius r. Thus the *gravitational potential energy* E_{grav}, required to disassemble an object by taking the whole mass away, corresponds to a fraction of its rest mass:

$$E_{grav} = \int_0^R dE_p(r) = -\int_0^R \frac{Gm(r)}{r}dm(r) = -\alpha\frac{GM^2}{R} \qquad (2.31)$$

α being a constant.

2.2.1.3 Escape Velocity

Since a particle moving at velocity v through the gravitational field of an object possesses a constant mechanical energy per unit mass E_m:

$$E_m = \frac{1}{2}v^2 + \Phi(r) \qquad (2.32)$$

this particle can reach infinity if, and only if, the energy is positive:

$$E_m = \frac{1}{2}v_0^2 + \Phi(R) \geqslant 0. \qquad (2.33)$$

The *escape velocity* is thus the velocity v_0 corresponding to $E_m = 0$, defined as:

$$v_{esc} = \sqrt{2|\Phi(R)|} = \sqrt{\frac{2GM}{R}}. \qquad (2.34)$$

2.2.1.4 Schwarzschild Radius

Taking c as the escape velocity from a black hole, the *Schwarzschild radius* R_S, defined as:

$$R_S = \frac{2GM}{c^2} = 2.9\frac{M}{M_\odot}\text{km} \qquad (2.35)$$

corresponds to the radius at which light cannot escape if $R < R_S$, thus acting like an event horizon, R_S dividing two regions with no causal connection.

2.2.1.5 Compactness

We define the *compactness* Ξ as a dimensionless parameter, allowing to measure the gravitational potential:

$$\Xi = \frac{GM}{Rc^2} = -\frac{\Phi(R)}{c^2} = -\frac{E_{grav}}{\alpha Mc^2} = \frac{1}{2}\left(\frac{v_{esc}}{c}\right)^2 = \frac{R_S}{2R}. \qquad (2.36)$$

The mean density is:

$$\langle \rho \rangle = \frac{3M}{4\pi R^3} = 1.5 \times 10^{20} \Xi^3 \left(\frac{M}{M_\odot} \right)^{-2} \text{ kg m}^{-3}. \tag{2.37}$$

We report in Table 2.1 the mass, radius, mean density, and compactness parameter for objects of various nature. Compact objects are characterized by large compactness $\Xi \geqslant 10^{-5}$, which increase with the gravitational field, without any equivalent on Earth. Neutron stars, with a radius R close to R_S, are characterized by an escape velocity close to c, while black holes, with $R = R_S$, present the unique property to possess an escape velocity equal to c. For these two types of compact objects, hosting a large reservoir of gravitational energy, one must use a general relativistic framework.

2.2.1.6 Gravitational Redshift
From the compactness we can derive the Einstein *gravitational redshift*:

$$z \simeq \frac{1}{\sqrt{1 - 2 \times \Xi}} - 1 \tag{2.38}$$

which ranges between 0.5 and 1.5 for a neutron star, and by definition is infinite for a black hole.

2.2.1.7 Gravitational Collapse
A star of radius R_*, collapsing into a compact object of radius R, releases an energy comparable to the rest mass energy of the compact object:

$$\Delta E_{\text{collapse}} = E_{\text{grav}}(R_*) - E_{\text{grav}}(R) \simeq -E_{\text{grav}}(R) \sim \alpha \Xi M c^2. \tag{2.39}$$

Table 2.1. Compactness Parameter for Various Celestial Objects.

Object	Mass M	Radius R	Mean Density	Compactness
	$M_\odot$	km	kg m^{-3}	
Earth	3×10^{-6}	6371	5.5×10^3	7×10^{-10}
Sun	1	696340	1.4×10^3	2×10^{-6}
White dwarf	$\sim 0.1 - 1.4$	~ 7000	$10^8 - 10^9$	$2 \times 10^{-5} - 3 \times 10^{-4}$
Neutron star	$\sim 1 - 3$	~ 10	$0.5 - 5 \times 10^{18}$	$0.15 - 0.45$
Black hole	$\geqslant 3$	$2.9 \left(\frac{M}{M_\odot} \right)$	—	1^{a}
SMBH	$\sim 10^6 - 10^9$	$20\text{au} \times \frac{M}{10^9 M_\odot}$	—	1^{a}

Note.
[a] By definition (even if, when taking the radius of a black hole as the Schwarzschild radius, $\Xi = 0.5$). SMBH stands for supermassive black hole.

2.2.1.8 Accretion of Matter

A mass m, accreted from an infinite radius to a distance R—onto the surface of the compact object—converts its gravitational potential energy into kinetic energy:

$$\Delta E_k = -\Phi(R)m = \Xi mc^2. \tag{2.40}$$

We compare in Table 2.2 the energy released by the accretion of one solar mass onto the surface of various compact objects, with the energy released by thermonuclear fusion of hydrogen. Although the gravitational potential of neutron stars or black holes is higher than that of white dwarfs, allowing to release more energy by accretion process, white dwarfs release comparatively more energy by thermonuclear fusion than neutron stars.

For an object accreting matter at an accretion rate $\dot m = \frac{dm}{dt}$, we define the accretion luminosity L_{acc} as the maximum bolometric luminosity of mass m accreted at accretion rate $\dot m$, obtained in case the whole kinetic energy is radiated:

$$L_{acc} = \Xi \dot m c^2 \text{ ergs}^{-1}. \tag{2.41}$$

In the case of an object of mass M, falling from infinity down to distance R, and assuming that the whole gravitational energy is converted into radiation, L_{acc} can be rewritten as:

$$L_{acc} = \frac{GM\dot M}{R} \sim 10^{35} - 10^{38} \text{ ergs}^{-1} \tag{2.42}$$

for an active accreting binary: the gas is heated up to millions of degrees, emitting in the X-ray domain (from 0.5 to 100 keV). We then identify L_{acc} to the X-ray luminosity, mainly depending on the mass accretion rate. This relationship thus allows us to derive the mass accretion rate from the observed X-ray luminosity.

2.2.1.9 Eddington Limit

There exists an upper limit on the accreted mass flux $\dot m$, which corresponds to a maximum bolometric luminosity, called *Eddington luminosity* L_{Edd}. This luminosity is obtained by equalizing the radiation pressure on the completely ionized matter:

Table 2.2. Energy Released by Accretion onto a Compact Object of $M = 1M_\odot$, Compared to Energy Released by Thermonuclear Fusion of Hydrogen.

Nature of Compact Object	Energy Released by Accretion (mc^2)	Energy Released by Nuclear Fusion (mc^2)
White dwarf	0.000 25	0.007
Neutron star	0.2	0.007
black hole	0.1–0.42[a]	—

Note.
[a] For a maximally spinning black hole.

$$f_{\text{rad}} = \frac{L\kappa_T}{4\pi r^2 c} \tag{2.43}$$

where κ_T is the Thompson opacity, defined by $\kappa_T = \frac{\sigma T}{m_p}$, with σ_T the Thompson cross section, and m_p the proton mass, with the gravitational forces of attraction toward the surface of the compact object:

$$f_{\text{grav}} = \frac{GM}{r^2}. \tag{2.44}$$

Thus we obtain

$$L_{\text{Edd}} = \frac{4\pi GMc}{\kappa_T} \cong 10^{38}\left(\frac{M}{M_\odot}\right)\text{erg s}^{-1} \tag{2.45}$$

corresponding to a critical mass accretion rate:

$$\dot{M}_{\text{crit}} = \frac{L_{\text{Edd}}}{\eta c^2} = 3 \times 10^{-8}\left(\frac{0.06}{\eta}\right)\left(\frac{M}{M_\odot}\right)M_\odot \text{yr}^{-1} \tag{2.46}$$

with the parameter η in the range $\eta \simeq 0.06 - 0.4$.

The typical X-ray luminosities of stellar accreting binaries are in the range $10^{35} - 10^{38}$erg s^{-1}, corresponding to spherical mass accretion rate of hydrogen-rich gas onto the neutron star of $10^{-11} - 10^{-8} M_\odot$ yr^{-1}. When the mass transfer exceeds the high range value, the X-ray luminosity crosses the Eddington limit. The radiation pressure on the accreted matter then overcomes the gravitational attraction onto the compact object (neutron star or black hole). As soon as this happens, the matter accreted in excess piles up onto the compact object, forming an optically-thick cloud preventing the X-ray to escape, and limiting the X-ray luminosity. However, the mass-transfer rate may be well larger than $10^{-8} M_\odot$ yr^{-1}, even up to $10^{-4} M_\odot$ yr^{-1}, and still remain dynamically stable. In super-Eddington sources, a possibility is that the excess material is ejected in a jet, as exhibited by the jet sources, also called microquasars, such as SS 433 (see Chapter 8).

2.2.2 Magnetic Properties

Because of the conservation of the magnetic field flux ϕ_B, the magnetic field increases when a star collapses into a compact object:

$$\phi_B \propto BR^2 \rightarrow \frac{B}{B_*} \cong \left(\frac{R_*}{R}\right)^2. \tag{2.47}$$

Thus, if the magnetic field of a star of radius $R = R_\odot$ is originally $B = 100$ G, it reaches $B \sim 10^6$ G when collapsing into a white dwarf. For a star with the same original magnetic field but with original radius $R = 10R_\odot$, it reaches $B \sim 5 \times 10^{13}$ G when collapsing into a neutron star. This explains how white dwarfs and neutron stars get their large value of magnetic field at their formation. For neutron stars, the observed magnetic field ranges from 10^{12} to 10^{15} G, the objects with such high value of magnetic field being called *magnetars*.

Another property of a neutron star, that we will use throughout this book is the *Alfvén radius* r_A, which corresponds to the radius where the magnetic energy density is equal to the kinetic energy density, i.e., when magnetic stresses dominate, marking the boundary of the magnetosphere:[16]

$$r_A = \left(\frac{8\pi^2 R^{12} B^4}{GM\dot{M}^2}\right)^{1/7} = 1800 \text{ km} \times \left(\frac{R}{10 \text{ km}}\right)^{12/7} \times \left(\frac{B}{10^{12} \text{ G}}\right)^{4/7}$$
$$\times \left(\frac{M}{1.4 \, M_\odot}\right)^{-1/7} \times \left(\frac{\dot{M}}{10^{-7} \, M_\odot \text{yr}^{-1}}\right)^{-2/7}. \tag{2.48}$$

We derive from this equation that the accretion flow, onto the surface of a neutron star of radius $R \sim 10$ km and a strong magnetic field $B \sim 10^{12}$ G, typical of neutron stars within high-mass X-ray binaries, is dominated by the magnetic field.

2.2.3 Rotational Properties

Because of the conservation of the angular momentum $\vec{L}$,[17] the rotation increases when a star collapses into a compact object:

$$\vec{L} \propto \vec{\Omega} R^2 \rightarrow \frac{\Omega}{\Omega_*} \cong \left(\frac{R_*}{R}\right)^2. \tag{2.49}$$

We can estimate the increase of angular frequency when a stellar progenitor collapses in a white dwarf, or even a neutron star. If a star of radius $R = R_\odot$ has originally a rotation period of $\sim$28 days like the Sun, it decreases to a rotation period of $\sim$4 min when collapsing into a white dwarf. For a star with the same rotation period, but with original radius $R = 10R_\odot$, it reaches to a rotation period of the order of the millisecond when collapsing into a neutron star. This explains why white dwarfs and neutron stars become fast rotators at the time of their formation. For white dwarfs, the observed rotation period is from hours to days, and for neutron stars, the observed rotation period ranges from a few milliseconds to a few tens of seconds.

We also note here the useful determination of the critical rotation velocity of a star, at the limit of the breakup. It is a balance between the gravitational force and the centrifugal force, respectively written as:

$$f_{\text{grav}} = \frac{GM}{R^2} \leftrightarrow f_{\text{cent}} = \Omega^2 R, \text{ thus:} \tag{2.50}$$

---

[16] this will be useful when considering accretion on an accreting pulsar, see Chapter 6.
[17] The angular momentum $\vec{L}$ is a vector, not to be confounded with the luminosity L, a scalar.

$$\Omega_{\max} = \Omega_K = \sqrt{\frac{GM}{R^3}} = \sqrt{\frac{4\pi G\langle\rho\rangle}{3}}. \qquad (2.51)$$

To summarize, we can combine magnetic field and rotation properties for a white dwarf and a neutron star, respectively:

$$\frac{\Omega}{\Omega_*} \sim \frac{B}{B_*} \sim \left(\frac{R_*}{R}\right)^2 \sim 10^4 \left(\frac{R_*}{R_\odot}\right)^2 \left(\frac{R}{7000 \text{ km}}\right)^2 \text{for a white dwarf} \qquad (2.52)$$

$$\frac{\Omega}{\Omega_*} \sim \frac{B}{B_*} \sim \left(\frac{R_*}{R}\right)^2 \sim 5 \times 10^9 \left(\frac{R_*}{R_\odot}\right)^2 \left(\frac{R}{10 \text{ km}}\right)^2 \text{for a neutron star.} \qquad (2.53)$$

2.3 Reviews, Catalogs and References

2.3.1 Reviews

- Reviews on stellar evolution: Arnett (1996); Kippenhahn et al. (2012);
- Reviews on stellar formation: Evans (1999); Lada & Lada (2003); Lada (2005);
- Review on stellar evolution: Scilla (2016);
- Review on neutron star theory: Heiselberg & Pandharipande (2000);
- Review on black holes: Barack et al. (2019);
- Review on pulsars and magnetars: Papitto et al. (2020).

2.3.2 Catalogs

- Catalog of 232 white dwarfs in the local environment, up to a distance of 25 pc: Holberg et al. (2016);
- Catalog of pulsar, including all spin-powered pulsars, AXP and SGR showing coherent pulsed emission, and excluding accretion-powered pulsars, maintained by ATNF: Manchester et al. (2005).

References

Abbott, B. P., Abbott, R., Abbott, T. D., et al. 2019, PhRvX, 9, 011001

Abbott, R., Abbott, T. D., Abraham, S., et al. 2020a, PhRvL, 125, 101102

Abbott, R., Abbott, T. D., Abraham, S., et al. 2020b, ApJL, 900, L13

Abdo, A. A., Ackermann, M., Ajello, M., et al. 2010a, ApJL, 710, L92

Abdo, A. A., Ackermann, M., Ajello, M., et al. 2011, ApJ, 734, 28

Abdo, A. A., Ackermann, M., Ajello, M., et al. 2010b, ApJL, 725, L73

Abdo, A. A., Ajello, M., Allafort, A., et al. 2013, ApJS, 208, 17

Acero, F., Aloisio, R., Amans, J., et al. 2017, ApJ, 840, 74

Ackermann, M., Ajello, M., Allafort, A., et al. 2013, Sci, 339, 807

Adams, W. S. 1914, PASP, 26, 198

Althaus, L. G., Córsico, A. H., Isern, J., & García-Berro, E. 2010, A&ARv, 18, 471

André, P., Men'shchikov, A., Bontemps, S., et al. 2010, A&A, 518, L102

André, P., & Montmerle, T. 1994, ApJ, 420, 837

André, P., Ward-Thompson, D., & Barsony, M. 1993, ApJ, 406, 122

Arnett, D. 1996, Supernovae and Nucleosynthesis: An Investigation of the History of Matter from the Big Bang to the Present (Princeton, NJ: Princeton Univ. Press)

Bachiller, R. 1996, ARA&A, 34, 111

Barack, L., Cardoso, V., Nissanke, S., et al. 2019, CQGra, 36, 143001

Barkat, Z., Rakavy, G., & Sack, N. 1967, PhRvL, 18, 379

Bergin, E. A., & Tafalla, M. 2007, ARA&A, 45, 339

Bhattacharya, D. 1995a, in X-ray Binaries, ed. W. H. G. Lewin, J. van Paradijs, & E. P. J. van den Heuvel (Cambridge: Cambridge Univ. Press)

Bhattacharya, D. 1995b, JApA, 16, 217

Bhattacharya, D., & Srinivasan, G. 1995, in X-ray Binaries, ed. W. H. G. Lewin, J. van Paradijs, & E. P. J. van den Heuvel (Cambridge: Cambridge Univ. Press)

Bhattacharya, D., & van den Heuvel, E. P. 1991, PhR, 203, 1

Bok, B. J. 1948, in Dimensions and Masses of Dark Nebulae (Cambridge, MA: Harvard College Observatory), 53

Bond, J. R., Arnett, W. D., & Carr, B. J. 1984, ApJ, 280, 825

Brown, G. E., & Bethe, H. A. 1994, ApJ, 423, 659

Burrows, A., & Vartanyan, D. 2021, Natur, 589, 29

Caiazzo, I., Burdge, K. B., Fuller, J., et al. 2021, Natur, 595, 39

Chandrasekhar, S. 1931, ApJ, 74, 81

Curran, P. A., Chaty, S., Zurita Heras, J. A., & Coleiro, A. 2011, A&A, 534, A48

Demircan, O., & Kahraman, G. 1991, Ap&SS, 181, 313

Einstein, A. 1915, SPAW, 844

Esposito, P., Israel, G. L., Turolla, R., et al. 2010, MNRAS, 405, 1787

Evans, N. J. II 1999, ARA&A, 37, 311

Farmer, R., Renzo, M., de Mink, S. E., Fishbach, M., & Justham, S. 2020, ApJL, 902, L36

Farmer, R., Renzo, M., de Mink, S. E., Marchant, P., & Justham, S. 2019, ApJ, 887, 53

Ferreira, M., & Providência, C. 2021, PhRvD, 104, 063006

Foglizzo, T., Kazeroni, R., Guilet, J., et al. 2015, PASA, 32, e009

Foglizzo, T., Masset, F., Guilet, J., & Durand, G. 2012, PhRvL, 108, 051103

Fragione, G. 2021, ApJL, 923, L2

Fryer, C. L. 1999, ApJ, 522, 413

Fryer, C. L., Belczynski, K., Wiktorowicz, G., et al. 2012, ApJ, 749, 91

Fryer, C. L., & Kalogera, V. 2001, ApJ, 554, 548

Fuchs, Y., Mirabel, F., Chaty, S., et al. 1999, A&A, 350, 891

Gold, T. 1968, Natur, 218, 731

Halpern, J. P., Tomsick, J. A., Gotthelf, E. V., et al. 2014, ApJL, 795, L27

Heger, A., Fryer, C. L., Woosley, S. E., Langer, N., & Hartmann, D. H. 2003, ApJ, 591, 288

Heiselberg, H., & Pandharipande, V. 2000, ARNPS, 50, 481

Hewish, A., Bell, S. J., Pilkington, J. D. H., Scott, P. F., & Collins, R. A. 1968, Natur, 217, 709

Ho, W. C. G., Klus, H., Coe, M. J., & Andersson, N. 2014, MNRAS, 437, 3664

Holberg, J. B., Oswalt, T. D., Sion, E. M., & McCook, G. P. 2016, MNRAS, 462, 2295

Howard, W. E., Staelin, D. H., & Reifenstein, E. C. 1968, IAU Circ. 2110, 2

Kerr, R. P. 1963, PhRvL, 11, 237

Kippenhahn, R., Weigert, A., & Weiss, A. 2012, Stellar Structure & Evolution (Berlin: Springer)

Kroupa, P., Weidner, C., Pflamm-Altenburg, J., et al. 2013, in Planets, Stars and Stellar Systems, ed. T. D. Oswalt, & G. Gilmore (Berlin: Springer), 115

Lada, C. J. 1987, in IAU Symp. 115, Star Forming Regions, ed. M. Peimbert, & J. Jugaku (Cambridge: Cambridge Univ. Press), 1

Lada, C. J. 2005, PThPS, 158, 1

Lada, C. J., & Lada, E. A. 2003, ARA&A, 41, 57

Laplace, P. S. 1798, Exposition du Système du Monde (2,; Paris: Imprimerie du Cercle-Social)

Laplace, P. S. 1799, Allgemeine Geographische Ephemeriden, AIIGE, 4, 1

Large, M. I., Vaughan, A. E., & Mills, B. Y. 1968, Natur, 220, 340

Larson, R. B. 1969, MNRAS, 145, 271

Lépine-Szily, A., & Descouvemont, P. 2012, IJAsB, 11, 243

Lorimer, D. R., & Kramer, M. 2004, Handbook of Pulsar Astronomy (Cambridge: Cambridge Univ. Press)

Mac Low, M. M. 2004, in The Dense Interstellar Medium in Galaxies, ed. S. Pfalzner, C. Kramer, C. Staubmeier, & A. Heithausen (Berlin: Springer), 379

Manchester, R. N., Hobbs, G. B., Teoh, A., & Hobbs, M. 2005, AJ, 129, 1993

Mapelli, M., Spera, M., Montanari, E., et al. 2020, ApJ, 888, 76

Mereghetti, S., Mignani, R. P., Covino, S., et al. 2001, MNRAS, 321, 143

Michell, J. 1784, RSPT, 74, 35

Mignani, R. P., Rea, N., Testa, V., et al. 2009, A&A, 497, 451

Miller, M. C., Lamb, F. K., Dittmann, A. J., et al. 2021, ApJL, 918, L28

Mirabel, F. 2017, NewAR, 78, 1

Newman, E. T., Couch, E., Chinnapared, K., et al. 1965, JMP, 6, 918

Noutsos, A., Abdo, A. A., Ackermann, M., et al. 2011, ApJ, 728, 77

Ober, W. W., El Eid, M. F., & Fricke, K. J. 1983, A&A, 119, 61

Oppenheimer, J. R., & Snyder, H. 1939, PhRv, 56, 455

Özel, F., & Freire, P. 2016, ARA&A, 54, 401

Özel, F., Psaltis, D., Narayan, R., & Santos Villarreal, A. 2012, ApJ, 757, 55

Palmer, D. M., Barthelmy, S., Gehrels, N., et al. 2005, Natur, 434, 1107

Papitto, A., Falanga, M., Hermsen, W., et al. 2020, NewAR, 91, 101544

Penston, M. V. 1969, MNRAS, 144, 425

Qin, Y., Marchant, P., Fragos, T., Meynet, G., & Kalogera, V. 2019, ApJL, 870, L18

Rahoui, F., Chaty, S., & Lagage, P. O. 2009, A&A, 493, 119

Rea, N. 2014, AN, 335, 329

Schaller, G., Schaerer, D., Meynet, G., & Maeder, A. 1992, A&AS, 96, 269

Schwarzschild, K. 1916, AbhKP, 1916, 189

Scilla, D. 2016, JPhCS, 703, 012002

Spera, M., & Mapelli, M. 2017, MNRAS, 470, 4739

Sukhbold, T., Ertl, T., Woosley, S. E., Brown, J. M., & Janka, H. T. 2016, ApJ, 821, 38

Tauris, T. M., Kramer, M., Freire, P. C. C., et al. 2017, ApJ, 846, 170

Tauris, T. M., & van den Heuvel, E. P. J. 2006, in Cambridge Astrophysics Ser. 39, Compact Stellar X-ray Sources, ed. W. Lewin, & M. van der Klis (Cambridge: Cambridge Univ. Press), 623

Testa, V., Rea, N., Mignani, R. P., et al. 2008, A&A, 482, 607

Tomsick, J. A., Bodaghee, A., Rodriguez, J., et al. 2012, ApJL, 750, L39

Tremblay, P-E., Fontaine, G., Fusillo, N. P. G., et al. 2019, Natur, 565, 202

Vahdat, A., Posselt, B., Santangelo, A., & Pavlov, G. G. 2022, A&A, 658, A95

van den Heuvel, E. P. J. 1992, in Environment Observation and Climate Modelling Through International Space Projects, 29

Wang, J. 2016, AdAst, 2016, 3424565

Watts, A. L., Andersson, N., Chakrabarty, D., et al. 2016, RvMP, 88, 021001

Woosley, S. E. 2017, ApJ, 836, 244

Woosley, S. E. 2019, ApJ, 878, 49

Woosley, S. E., Blinnikov, S., & Heger, A. 2007, Natur, 450, 390

Woosley, S. E., & Heger, A. 2021, ApJL, 912, L31

Chapter 3

Astrophysical Basics for Binaries

Encore une fois, je me laisse aller à faire des étoiles trop grandes.
Once again I'm allowing myself to do stars too big.
—Vincent van Gogh, *Lettre à Emile Bernard, 26 novembre 1889*

We describe in this chapter astrophysical basics needed to understand binary systems, from their formation to their final endpoint, that will be needed through the whole book. We first report on how to form a binary, and how to define its general properties, through the fundamental binary parameters, such as dynamical mass function, orbital energy, and angular momentum. We then introduce the Roche geometry, with gravitational equipotentials, and how this geometry constrains the transfer of both mass and angular momentum, passing from one component to the other. After this introduction, we describe the common envelope evolution, a crucial phase in the course of the binary evolution. We then mention the main radiation mechanisms, distinguishing thermal and non-thermal processes. We finish this chapter by giving details on different types of accretion disks, geometrically thin (thick) and optically thick (thin) accretion disks, how to model them, and how to image them, via the astrotomography. We finish by giving useful reviews and references.

3.1 Formation of a Binary

First, let us mention that the two individual stars forming a binary system, either of low or high mass, follow the same formation channel than isolated stars, depending on the specificity of the interstellar cloud where they are born, such as its composition. A theory of stellar formation must thus take into account the formation of binary and even multiple systems, reproducing parameters such as orbital separation (on a wide range, from day to year timescale), mass ratio, and eccentricity. Such a system can form either from the fragmentation and gravitational collapse, within a molecular cloud, of a pair (or more) of stars born close-by, or by dynamical capture/exchange within a dense environment, such as a globular, stellar, or nuclear cluster. The current census is that more than 70% of massive stars have experienced binarity during their stellar life, through exchange of matter or angular momentum (Sana et al. 2012).

The process of *fragmentation* is likely the best mechanism in order to form most binary and multiple systems, occurring when an object, not maintained in equilibrium by internal pressure, splits into distinct bodies, while dynamically evolving in a gravitational collapse. For this fragmentation process to occur, both objects must be individually gravitationally stable, thus the available mass M_{tot} must be greater than twice the Jeans mass M_J (see Section 2.1.1). This occurs preferentially when M_J decreases during or after the two collapsing phases, toward the formation of a proto-stellar core. The length scale of fragmentation is of the order of Jeans radius R_J, since closer objects cannot form separately and instead merge together. Thus, for fragmentation to form binaries with orbital separation less than 1 au, densities greater than $\rho \geqslant 10^{-10}$ g cm^{-3} are required, corresponding to fragments of Jeans mass $M_{frag} \leqslant 0.01 M_\odot$, such that close binaries formed by fragmentation necessarily have to accrete most of their final mass after this process. Such binaries are in general characterized by high eccentricity, with mass ratio lying in the range $0.1 \leqslant q = M_2/M_1 \leqslant 1.0$. For more details on binary formation, see the review by Kratter (2011).

3.2 General Properties of a Binary

3.2.1 Binary Parameters

We consider here a binary system consisting of two points of mass M_1 and M_2. We define M_1 as the primary, i.e., the compact object (also called accretor when mass transfer occurs), M_2 as the secondary, i.e., the companion star (or donor star when mass transfer occurs) and $M = M_1 + M_2$ as the total mass of the system. In the whole book, we will use the following parameters:

- the *mass ratio* $q = M_2/M_1$; $q < 1$ for *low-mass X-ray binaries* (LMXB) hosting a solar-mass companion star orbiting a neutron star or a black hole of mass $M \geqslant 1 M_\odot$, and $q \geqslant 1$ for *high-mass X-ray binaries* (HMXB) hosting a massive star $M \gtrsim 10 M_\odot$ and a compact object;
- the *semi-major axis* a (also called *orbital separation*), and the *semi-minor axis* b (see Figure 3.1) are the main parameters of the ellipse defined by the equation $x^2/a^2 + y^2/b^2 = 1$. These parameters are linked by the relation $a^2 = b^2 + c^2$, where c is the ellipse focus abscissa. a is linked to the orbital period P by the third Kepler law:

$$\frac{a^3}{P^2} = \frac{G(M_1 + M_2)}{4\pi^2} = \left(\frac{215}{365}\right)^2 (M_1 + M_2) \tag{3.1}$$

with a in $R_\odot$, P in days, M_1 and M_2 in $M_\odot$ (for a given period, a double mass leads to an increase of the semi-major axis by 26% only);
- the *eccentricity* $e = c/a$. For stellar binaries located on the main sequence, q and P remain practically unchanged since the formation of the binary, after

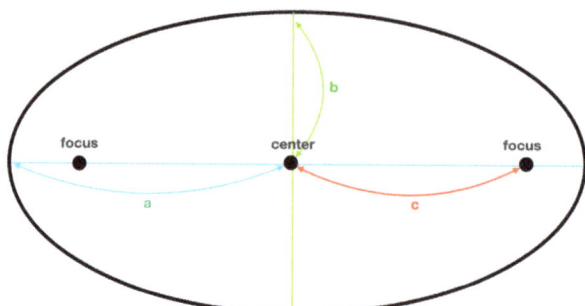

Figure 3.1. Schema of ellipse with semi-major axis a, semi-minor axis b, and focus abscissa c indicated. Reproduced with permission. Credit: Sylvain Chaty.

the dissipation of the proto-stellar nebula. In the case of tight binaries, the eccentricity decreases with time, because of circularization of orbits due to tidal effects leading to energy dissipation, with a timescale depending on the orbital period, which can be as short as $\sim 10^4$ years for a giant star filling its Roche lobe (Zahn 1977). On the opposite, for wide binaries mean eccentricity increases with orbital period: while the majority of systems with $P < 10$ d have circular orbits ($e \sim 0$), systems with longer orbital periods usually exhibit eccentric orbits. The rare long period and low eccentricity systems mainly host white dwarfs, which have had time to circularize during their red giant phase.

3.2.2 Dynamical Mass Function of a Binary

Because of their binarity, binary systems are characterized by a signal modulated by the Doppler effect at their orbital period P, with inclination[1] i. The observation of lines emitted by at least one component of the system thus allows the observer to determine a combination of primary and secondary masses M_1 and M_2:

$$f(M) = \frac{(M_2 \sin(i))^3}{(M_1 + M_2)^2} = \frac{4\pi^2}{G} \frac{(a\sin(i))^3}{P^2}. \tag{3.2}$$

In the case where lines are detected for both components, then both masses can be determined. In addition, if jets are observed, usually perpendicular to the orbital plane, the uncertainty of the inclination i can be removed. While it is possible to determine masses dynamically both for neutron stars and black holes (as reported in Figure 3.2, Kaper et al. 2006), they are usually more accurate for neutron stars within binary pulsar systems (Kiziltan et al. 2013).

The high-mass X-ray binary Cyg X-3 has been intensively studied through orbital-phase-resolved infrared spectra, both in outburst and quiescence, allowing astronomers to derive its dynamical mass function, with a mass range for the Wolf–Rayet He star within $5 \leqslant M \leqslant 11 M_\odot$, orbiting a compact object—either a neutron star or a black hole—and to produce Doppler tomograms (Hanson et al. 2000; see Figure 3.8).

---

[1] The inclination i is measured by convention as the angle between the perpendicular to the orbital plane of the binary, and the line of sight (from Earth to the binary).

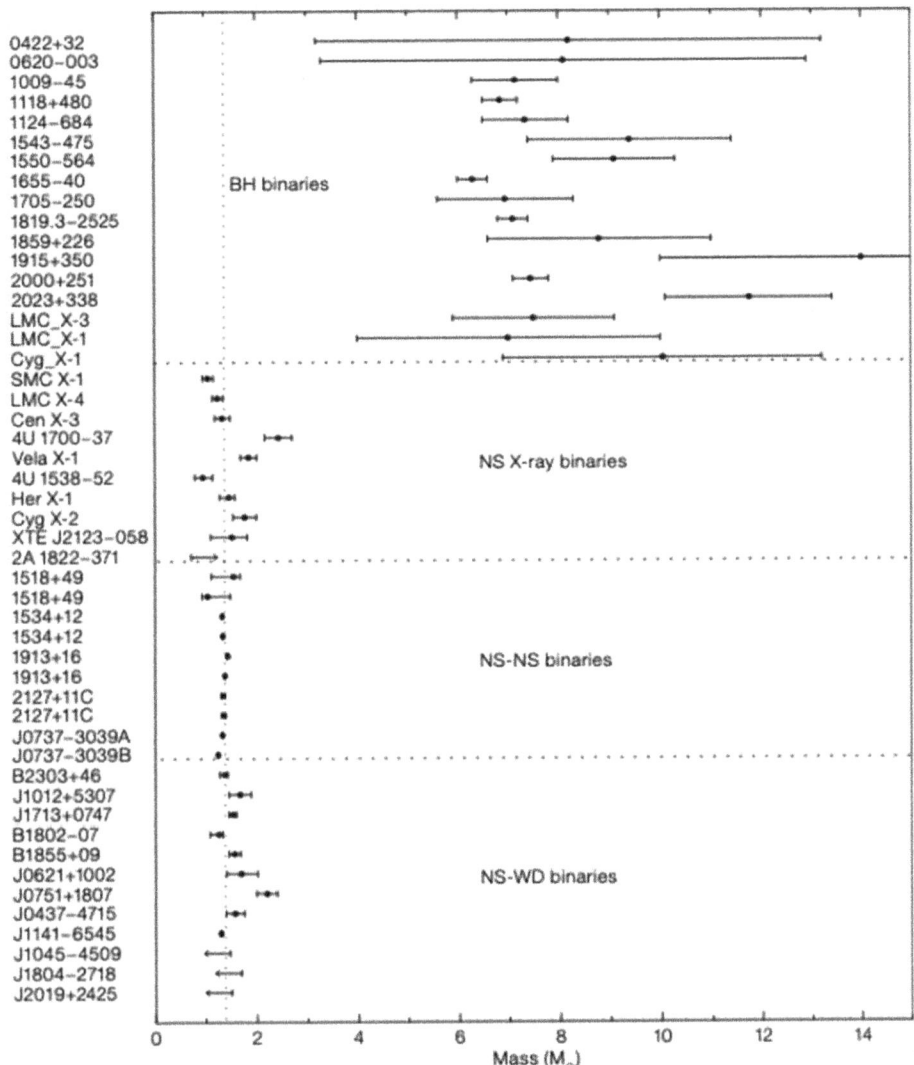

Figure 3.2. Masses determined dynamically for black holes (*top*) and neutron stars (*bottom*). Neutron stars, both within binary radio pulsars (NS+NS and NS+WD binaries), and within X-ray pulsars (NS X-ray binaries) occupy a narrow mass range near $1.35 M_\odot$, with the exception of a few systems with $M_{NS} \sim 2M_\odot$. Black holes show a wider spread, due to different formation mechanism. NS and BH masses come from Stairs (2006) and McClintock & Remillard (2006). Credit: ESO, reprinted with permission from Kaper et al. (2006).

3.2.3 Orbital Energy and Angular Momentum of a Binary

The orbital energy of a binary system is given by:

$$E_{\text{orb}} = -\frac{GM_1 M_2}{2a} = -\frac{GM_1 M_2}{a} + \frac{1}{2}\mu v_{\text{rel}}^2 \qquad (3.3)$$

where $\mu = \frac{M_1 M_2}{M_1 + M_2}$ is the reduced mass of the system, and $v_{rel} = \sqrt{\frac{G(M_1 + M_2)}{a}}$ is the relative velocity of both stars in a circularized binary.

The orbital angular momentum of a binary system is given by:

$$J_{orb} = \mu \Omega a^2 \sqrt{1 - e^2} \qquad (3.4)$$

where Ω is the angular velocity.

3.3 Roche Geometry

We now describe the geometry of a binary system, using the *Roche potential* and *Roche lobe geometry*, with the three alternatives: detached, semi-detached, and contact binaries. In tight binaries, tidal effects become important, leading to circularization of the orbits in a short timescale, spin-up of stellar rotational velocity, and synchronization of stellar rotation with the orbital motion.

3.3.1 Roche Equipotentials

We assume here that the two components of the binary system are point-like masses of mass M_1 and M_2, located respectively at the positions r_1 and r_2, r being the distance to the center of mass. We first consider a circular orbit (a good approximation, with tidal forces quickly circularizing the orbit; Zahn 1977). The orbital separation is given by Kepler's third law:

$$a = 3.53 \times 10^8 (M_1 + M_2)^{1/3} P^{2/3} \, \text{m} \qquad (3.5)$$

where P is the orbital period measured in hours. In the co-rotating frame of the binary system, the conservation equation of the gas momentum is:

$$\frac{\partial v}{\partial t} + (v. \, \nabla)v = -\nabla \Phi - 2\Omega \times v - \frac{1}{\rho} \nabla P \qquad (3.6)$$

where v is the gas velocity, ρ its density, P its pressure, Ω the orbital angular velocity of the system:

$$\Omega = \sqrt{\frac{GM}{a^3}} \qquad (3.7)$$

and Φ is the effective gravitational potential felt by a particle in a binary system, as the sum of the gravitational potential of each stellar mass, and the centrifugal potential due to the orbital motion of the binary:

$$\Phi = -\frac{GM_1}{|r - r_1|} - \frac{GM_2}{|r - r_2|} - \frac{(\Omega \times r^2)^2}{2}. \qquad (3.8)$$

At equilibrium, with a velocity $v = 0$, the conservation equation of the gas momentum (Equation (3.6)) becomes:

$$\nabla\Phi = -\frac{1}{\rho}\nabla P. \tag{3.9}$$

The pressure thus remains constant along Φ equipotentials, and the gas state equation implies that the density and temperature also remain constant on these equipotentials. Figure 3.3 illustrates the shape of the equipotential surfaces in the *Roche potential*, a comoving frame representing the potential well of each star in the orbital plane, and at large distance the potential due to the centrifugal force, forming a paraboloid. These Roche equipotential contours passing through L_1 roughly represent possible stellar surfaces, defining the so-called *pear-shaped* Roche lobe.

The Roche potential presents a few points where the potential gradient is null, these can be extrema or saddle points, which are points where equipotentials cross each other. These extrema, called *Lagrangian* points, given by the condition $\nabla\Phi = 0$, are located in the orbital plane. The Lagrangian points are numbered by increasing energy:

$$\text{Energy}(L_1) < \cdots < \text{Energy}(L_i) < \cdots < \text{Energy}(L_5). \tag{3.10}$$

For each value of the energy, there exists two or three distinct equipotentials, one or two close to the binary system, and one at the exterior. The three points L_1, L_2, and L_3 are three solutions to this condition, representing unstable equilibrium

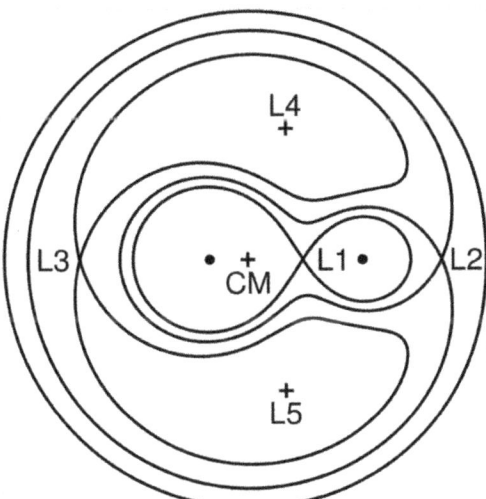

Figure 3.3. Contours of Roche equipotentials within the orbital plane, with indication of the center of mass (CM) and five Lagrangian points. The equipotential contour passing through L_1 approximately represents possible stellar surfaces, with the apparent *pear-shaped* Roche lobes. The most massive star occupies the larger Roche lobe. Thus, for a CV or an LMXB, the smaller lobe would be filled by the secondary star, while the primary (white dwarf, neutron star, or black hole) would fill the larger lobe. In contrast, for an HMXB the massive secondary star would fill the larger lobe, while the primary (neutron star or black hole) would fill the smaller lobe. L_1 is a critical saddle point of the potential, located right in between the two point masses, where mass transfer occurs. Reprinted with permission from Wikipedia, licensed under the Creative Commons Attribution-Share Alike 4.0 International license.

points. The two points L_4 and L_5 are two other solutions, also unstable, though the Coriolis forces make them stable.[2]

Let us choose a reference frame in which the origin is the center of mass of the system, and the two masses M_1 and M_2 are located on an x-axis, respectively at the positions $x_1 = \frac{-1}{1+q}$ and $x_2 = \frac{q}{1+q}$, such that the distance between the two point masses is normalized by the orbital separation: $x_2 - x_1 = 1$. The position of L_1 point, the most important to consider in the context of binary systems, because it is through this point that matter will pass from one component to the other, can be approximated by the following formula (Plavec & Kratochvil 1964):

$$x_{L_1} - x_1 = 0.5 - 0.227 \log(q) \tag{3.11}$$

for which the particular case of equal mass $M_1 = M_2$ gives $x_{L_1} - x_1 = 0.5$, with an L_1 point located at equidistance between M_1 and M_2.

3.3.2 Transfer of Mass and Angular Momentum

As seen in Figure 3.3; the equipotential *pear-shaped* surfaces, also called *Roche lobes*, approximately represent the stellar surfaces. The Roche geometry leads to three distinct possibilities: *detached*, *semi-detached*, or *contact binaries* (see Figure 3.4).

3.3.2.1 Detached Binaries
Binaries are detached when the two stars are each well within an equipotential, within their Roche lobe, separated without mass transfer, with limited deformation due to binarity. We will see later that post common-envelope phase binaries and HMXB hosting Be stars are good examples of such detached systems.

3.3.2.2 Semi-detached Binaries
As soon as one of the star of the binary system, initially born with a radius smaller than its Roche lobe, later evolves, its envelope expands and its radius increases, until it reaches the Lagrangian point L_1. The star now begins to fill its Roche lobe, initiating *Roche-lobe overflow* (RLO), and the slightest instability at L_1 will initiate a mass transfer by sending material from the secondary, over the saddle point, to be

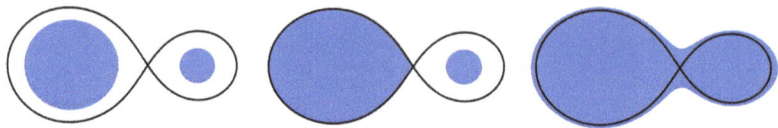

Figure 3.4. Schema of the Roche equipotential of a binary system in the orbital plane, with a mass ratio $q = 3$, where are shown from left to right the three different cases: detached, semi-detached, and contact binary. Reprinted with permission from Wikipedia, licensed under the Creative Commons Attribution-Share Alike 4.0 International license.

---

[2] This is why some Trojan asteroids are located at L_4 and L_5 Lagrangian points of the Sun–Jupiter system.

accreted onto the primary. This is the case for cataclysmic variables, low-mass X-ray binaries, some high-mass X-ray binaries such as Cyg X-1, and the massive and short-period *Algol* binaries (Sen et al. 2022).

The accretion flow of matter possesses some angular momentum, and thus does not fall directly onto the surface of the accretor, but instead goes orbiting around it, this accumulation of matter leading to dissipation and formation of an accretion disk, with gas spiraling inwards through the disk. For conservation of angular momentum, a small amount of gas also needs to be carried away, via winds, outflows or even jets. The maximum disk radius, allowed by tidal interaction, reaches up to ~90% of the Roche lobe radius, meaning that the disk size is much bigger than a comparatively small compact object such as a white dwarf, and even more for a neutron star or a black hole. The Roche lobe radius only depends on the orbital separation of the system a and the mass ratio q (Eggleton 1983):

$$\frac{R_L}{a} = \frac{0.49q^{2/3}}{0.6q^{2/3} + ln(1 + q^{1/3})}.$$ (3.12)

The process of mass transfer has an impact on the orbit of the binary system, depending on the mass ratio. For a conservative system, in which both the total mass and angular momentum are conserved, there are two alternatives, that we describe now.

1. First, if we consider a mass transfer from the less massive to the more massive component, the mass which is transferred is moving closer to the center of mass of the system, decreasing the angular momentum. To conserve the angular momentum, the orbital separation has to increase. As a consequence, the less massive star does not fill its Roche lobe any more, stopping the mass transfer (unless its radius increases, for instance in case of late evolution toward red giant). This is a stable mass transfer on a long timescale, such as seen in CV or LMXB, for instance when $q \leqslant 0.5$. From Kepler's third law, we derive that R_2 only depends on the mass M_2 and the orbital period of the system (but not on the mass M_1):

$$\frac{R_2}{R_\odot} = 0.23\frac{M_2}{M_\odot}^{1/3} P_{hr}^{2/3}.$$ (3.13)

In addition, the mean density of the secondary star (in the main sequence) only depends on the orbital period of the system (and not on its mass), with a typical value, for an orbital period of a few hours:

$$\rho_2 = \frac{M_2}{\frac{4}{3}\pi R_2^3} = \frac{0.11}{P_{hr}^2}\text{kgcm}^{-3}.$$ (3.14)

2. Second, if we consider a mass transfer from the more massive to the less massive component, the transferred mass is moving away from the center of mass of the system, increasing the angular momentum. To conserve the

angular momentum, the orbital separation has to decrease, and as a consequence, the most massive star overfills even more its Roche lobe, increasing the mass transfer rate. We thus reach an unstable mass transfer, on a dynamical timescale, seen in some CV.

Introducing Roche geometry in models of internal stellar structure only brings low corrections, compared to spherical geometry models describing the structure of isolated stars. This is the reason why it is generally assumed that a star filling its Roche lobe within a semi-detached binary can be modeled by a sphere of the same volume than the lobe. However, the tidal deformations of the secondary, constrained by the Roche potential, are detectable by ellipsoidal variations observed in CV and LMXB in quiescence, allowing us to constrain the mass of the compact object. This is, along with the spectroscopic determination of the dynamical mass function (see Section 5.1.4), an unambiguous proof of the existence of black holes in accreting stellar binaries. Once RLO is initiated, it goes on until the secondary has lost its entire hydrogen-rich envelope (which can represent up to ~70% of its mass), and then no longer fills its Roche lobe.

As originally proposed by Kippenhahn & Weigert (1967), we can define two different types of RLO, called *case A* and *case B*, to which was later added *case C* (see for instance Tauris & van den Heuvel 2006):

- *Case A*: the binary system is so tight that the secondary star begins to fill its Roche lobe when it is still burning hydrogen in its core (thus during the main sequence phase);
- *Case B*: the binary system is wider than in previous case, so that the secondary star begins to fill its Roche lobe after the end of hydrogen burning in its core, but before helium ignition in the surrounding shell (thus during the red giant phase);
- *Case C*: the binary system is even wider than in the previous case (orbital periods up to ~10 years), so that the secondary star begins to fill its Roche lobe during or even after helium shell burning (thus during the asymptotic giant branch).

3.3.2.3 Contact Binaries

If mass transfer goes on within the binary system, the second star, accreting matter from the donor, increases its volume, until it also begins to overfill its Roche lobe. The system then becomes a contact binary, and with expansion going on, it is accompanied by the formation of a common envelope. We describe below the outcome of this phase.

3.4 Common Envelope Evolution

A fundamental phase in the overall evolution of stellar accreting binaries, that we will encounter throughout this whole book, is when the binary passes through the *common envelope* (CE) phase, initially invoked to explain how two evolved massive stars, initially too big to fit within their current orbital separation, eventually fit together within a tight binary system. In the context of this scenario, initially proposed by Paczynski (1976), both initial stars are born within a relatively wide binary, which allowed each star to expand during its evolution. When one of the stars expands enough that its atmosphere begins to enshroud the whole binary system, the compact object is captured, forced to orbit within the envelope of the evolved star. The common envelope phase is then triggered, accompanied by a dynamically unstable mass transfer (Ivanova et al. 2013). During this phase, a large frictional drag force is produced by the motion of one of the component of the binary, through the envelope of the evolved star. This drag force allows to reduce orbital angular momentum from the binary, leading to spiral-in of binary components and shrinking of the orbit, depositing energy within the envelope, often concluding the whole process by the ejection of the envelope. This phase allows to explain how precursors of compact object within a binary system had in the course of their previous evolution a radius much larger than the current orbital separation of the system. This phase is however full of uncertainties, mainly because a successful envelope ejection may occur on a dynamical timescale (Podsiadlowski 2001)—which can be short, in the range $10^2 - 10^3$ years—involving multi-dimensional hydrodynamical calculations (Fragos et al. 2019; García et al. 2021).

We can estimate the reduction of the orbital separation, by equalizing the binding energy of the envelope around the (sub-)giant donor, to the difference in orbital energy, before and after the CE phase. Following the so-called *energy formalism* (Webbink 2008), the main energy source needed to eject the stellar envelope is provided by the orbital energy reservoir and thus, by the spiral-in of the companion. Changes in these two quantities are linked through a free parameter called α_{CE}, the CE efficiency representing the fraction of orbital energy deposited as kinetic energy of the envelope components:

$$\Delta E_{env} = \alpha_{CE} \Delta E_{orb} \qquad (3.15)$$

where ΔE_{env} is the change in the binding energy of the donor star envelope, while ΔE_{orb} represents the orbital energy released through the in-spiral. Thus $0 < \alpha_{CE} < 1$ describes the efficiency to eject the envelope, by converting orbital energy into kinetic energy, eventually leading to the ejection of the envelope. The total binding energy of the envelope to the core, E_{env}, is given by:

$$E_{env} = -\int_{M_{core}}^{M_{donor}} \left(\frac{GM(r)}{r} + \alpha_{th} U \right) dm \qquad (3.16)$$

in which the first term is the gravitational (binding) energy, and the second is the specific internal thermodynamic energy of the envelope (Ivanova et al. 2013; Kruckow et al. 2016). There are two possible global outcomes of this CE phase: on one hand, if there is not enough orbital energy available to eject the envelope, both stellar components eventually merge at the end of this phase, becoming a *Thorne–Zytkow object* (TZO, Thorne & Zytkow 1975). On the other hand, in case the envelope is ejected, the binary system survives this critical phase of its evolution, and the global result, due to spiral-in, is a drastic reduction of its orbital separation by a factor of ~ 100 (see Figure 3.5). We will describe in more details in Section 6.5.1 how high-mass X-ray binary systems go through the common envelope phase, in the course of their evolution (in particular, see Figure 6.14).

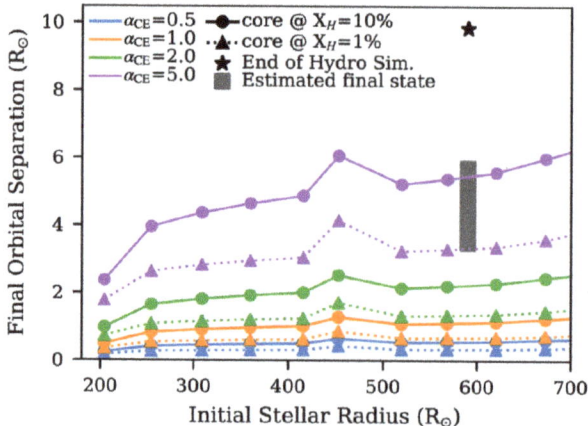

Figure 3.5. Final orbital separation (in $R_\odot$), after the common envelope phase, versus the initial stellar radius (in $R_\odot$), comparing various α_{CE} prescriptions given in the upper left part of the plot. The estimated final orbital separation is denoted by the gray rectangle, showing a reduction of the orbital separation by a factor of ~ 100. Reprinted with permission of the AAS from Fragos et al. (2019).

3.5 Radiation Mechanisms

We briefly mention here the main radiation mechanisms that will be needed throughout this book. For more details, the reader can refer to Rybicki & Lightman (1979), an extensive review on radiative processes in astrophysics.

3.5.1 Thermal Processes

The main thermal process that we will encounter in this book is the *blackbody* radiation, a thermal electromagnetic radiation emitted by an opaque and non-reflective body, which happens to be in thermodynamic equilibrium within its environment. It emits a specific spectrum on the frequency ν, depending only on its temperature T, assumed to be uniform and constant. The spectral radiance $B_\nu(\nu, T)$ of this spectrum is defined by the *Planck law* of blackbody radiation:

$$B_\nu(\nu, T) = \frac{2h\nu^3}{c^2} \frac{1}{e^{h\nu/kT} - 1} \tag{3.17}$$

where h is the Planck constant, c the speed of light, and k the Boltzmann constant.

From this law, we can derive, by integrating $B_\nu(\nu, T)$ over the frequency range, the total energy radiated per unit surface of a blackbody:

$$j^\star = \sigma T^4 \tag{3.18}$$

where σ is the Stefan–Boltzmann constant.

The total power radiated from a blackbody, taking into account its surface area A, and $\epsilon = 1$ in case of a perfect blackbody,[3] is thus:

---

[3] We speak instead of a graybody in case $\epsilon \leqslant 1$.

$$P = A\epsilon\sigma T^4. \tag{3.19}$$

Equations (3.18) and (3.19), in which $j^\star$ and P are directly proportional to the fourth power of the blackbody temperature T, are known as the *Stefan–Boltzmann law*.

3.5.2 Non Thermal Processes

3.5.2.1 Compton and Inverse Compton Effect

The *Compton effect* is the scattering (by an angle θ) of a photon of energy E_γ, after an interaction with a charged particle, usually an electron (of mass m_e), resulting in a decrease in energy of the (X- or γ-ray) photon, part of its energy being transferred to the electron. The resulting energy, after interaction, is:

$$E_{\gamma'} = \frac{E_\gamma}{1 + (E_\gamma/m_e c^2)(1 - \cos\theta)}. \tag{3.20}$$

Inverse Compton effect is the opposite case, in which a charged particle (an electron) transfers part of its energy to a photon. This effect is more common in astrophysics, especially when studying the X-ray radiation emitted in the surroundings of an accretion disk. In this case, relativistic *non-thermal* electrons from the corona surrounding the black hole transfer part of their energy to low-energy *thermal* photons emitted by the accretion disk, thus creating a power-law component in X-ray spectra (keV domain) of accreting black holes, as described in Chapter 5.

3.5.2.2 Bremsstrahlung

Bremsstrahlung (braking radiation) is the electromagnetic radiation produced by the deceleration of a charged particle (typically an electron), when deflected by another charged particle (such as an atomic nucleus). The kinetic energy from the incident particle is converted into radiation, through emission of photons. This Bremsstrahlung effect occurs mainly in high temperature processes, thus mainly related to X-ray emission ($\geqslant$10 keV).

Bremsstrahlung emitted from plasma is sometimes referred to as *free–free* radiation, since it is created by free electrons, which remain free after the emission of a photon (contrary to the opposite *bound–bound* radiation, and to the intermediate *free–bound* radiation).

3.5.2.3 Synchrotron Radiation

Synchrotron radiation is the relativistic version of the cyclotron radiation: any charge moving inside a magnetic field B emit a braking radiation due to the Abraham–Lorentz force. In the non-relativistic case, the electron gyration frequency is:

$$\nu_0 = \frac{eB_\perp}{2\pi m_e} \tag{3.21}$$

where e is the electron charge, and $B_\perp$ the component of B perpendicular to the electron trajectory.

When the Lorentz factor $\gamma = \frac{1}{\sqrt{1-\beta^2}} > 1$ (where $\beta = \frac{v}{c}$), the relativistic aberration becomes important, focusing the emission along the direction of movement, with a semi-angle of $\frac{1}{\gamma}$. Since the relativistically accelerated electrons spiral along the local axis of the magnetic field B, the observer detects the radiation only when the emission cone coincides with the line of sight, during a time interval Δt:

$$\Delta t \sim \frac{1}{2\pi\nu_0\gamma^2}. \tag{3.22}$$

This process has the particularity to generate high frequency harmonics. $\nu_c = \frac{3}{2}\gamma^2\nu_0$ is the critical frequency among which half of the energy is radiated, the emission peaking at $0.29\nu_c$ (Rybicki & Lightman 1979). For electrons following a power-law energy distribution of index p:

$$N(E)dE = E^{-p}dE \tag{3.23}$$

their radiation is also described by a power-law:

$$S(\nu) \sim \nu^{-(p-1)^2}. \tag{3.24}$$

Flat radio spectra observed in astrophysical jets result from the sum of synchrotron power-law emission, coming from different parts of the jet in the stationary case, in which plasma is injected from the central source at constant velocity (Blandford & Königl 1979; Hjellming & Johnston 1988). Variable radio emission thus suggests discrete ejections of plasma bubbles expanding adiabatically (see Figure 8.6). Along a conical jet, the radius r of a plasma bubble linearly increases with time ($r \sim t$), while electron energy decreases as:

$$E = \gamma m_e c^2 = 3kT \sim t^{-1} \tag{3.25}$$

and magnetic field decreases as:

$$B \sim R^{-2} \sim t^{-2}. \tag{3.26}$$

For an electronic plasma bubble described by a power-law energy distribution, the time of maximum emission is:

$$t_\lambda^{\max} \sim \lambda^{(p+4)(4p+6)} \tag{3.27}$$

and the maximum flux is:

$$F_\lambda^{\max} \sim \lambda^{-(7p+3)/(4p+6)}. \tag{3.28}$$

Thus, when a hot plasma bubble expands, the emission peak is displaced toward longer wavelengths, while its intensity decreases (van der Laan 1966; see an example in Figure 8.7).

3.6 Accretion Disk

We distinguish geometrically thin and optically thick accretion disks—radiatively efficient —from geometrically thick and optically thin accretion disks—radiatively inefficient.

3.6.1 Geometrically Thin and Optically Thick Accretion Disks

Standard geometrically thin and optically thick accretion disks are characterized by $H \ll R$, where H is the disk height and R its radius (see Figure 3.6). The accretion rate in this type of accretion disks is constrained between:

$$10^{-2} \leqslant \frac{\dot{M}}{\dot{M}_{\mathrm{crit}}} \leqslant 1. \tag{3.29}$$

We assume a hydrostatic equilibrium in the vertical direction (perpendicular to the accretion disk), so that the vertical velocity $v_z = 0$. In such an accretion disk, the orbits are circular, and the tangential rotation velocity v_ϕ of the matter orbiting the compact object is close to Keplerian velocity, which is much greater than a radially-falling velocity:

$$v_\phi \simeq \sqrt{\frac{GM}{R}} = R \times \Omega \tag{3.30}$$

with an orbital period:

$$P_{\mathrm{orb}} = \frac{2\pi R}{v_\phi} = \frac{2\pi}{\Omega} \tag{3.31}$$

and the dynamical timescale τ_{dyn} is thus:

$$\tau_{\mathrm{dyn}} = \sqrt{\frac{r_k^3}{GM}} \tag{3.32}$$

with r_k the Keplerian radius.

Without viscosity, matter in the disk would orbit the compact object indefinitely, without being accreted. These disks are called accretion disks, because viscosity is

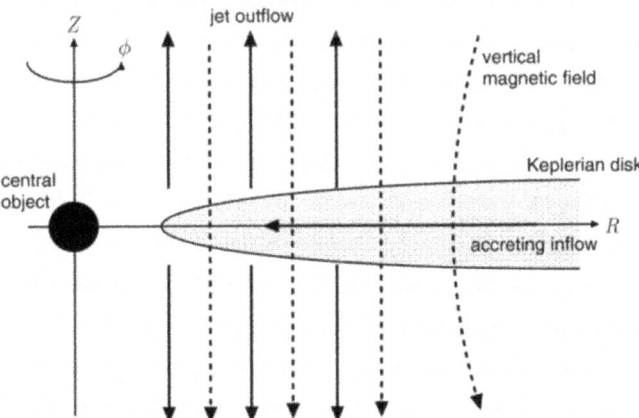

Figure 3.6. Schema of an accretion disk. Reprinted with permission of the AAS from Shiraishi et al. (2009).

present. Viscosity is efficient at slowing down matter inwards within an accretion disk, and in parallel at transporting angular momentum outwards, orthogonally to the gas, because of differential rotation between orbits implied by Keplerian rotation velocity. Angular momentum transfer within the disk results from a magneto-rotational instability (Balbus 2003). The origin of viscosity is still unknown: probably not molecular (i.e., created by collision of atoms or ions composing the gas within the disk), but more likely turbulent, magnetic, or non-local (such as spiral waves propagating in the disks), or even due to a mixture of these causes. We point out an alternative study to characterize anomalous transport in accretion disks in Greenhough et al. (2002, 2003).

This process of accretion via viscosity, called *shear viscosity*, develops between orbits with the following viscous timescale t_{visc}:

$$t_{\mathrm{visc}}(R) \simeq \frac{R^2}{\nu} \tag{3.33}$$

where $\nu = \alpha c_S H$ is the kinematic viscosity coefficient, $\alpha < 1$ is the standard viscosity parameter, so-called *α-prescription coefficient* of Shakura & Sunyaev (1973), and c_S the sound speed within the accretion disk. The shear viscosity leads to a radial velocity component, for $R \gg R_{\mathrm{in}}$, the inner radius of accretion disk:

$$v_R \simeq -\frac{R}{t_{\mathrm{visc}}(R)} \simeq -\frac{\nu}{R} \simeq -\alpha c_S \frac{H}{R} \ll c_S. \tag{3.34}$$

Using the vertical structure of the disk, we can show that, in the case of a geometrically thin disk ($H \ll R$):

$$c_S \ll \left(\frac{GM}{R}\right)^{1/2} \tag{3.35}$$

implying that the accretion disk is a highly supersonic medium.

3.6.1.1 Innermost Stable Circular Orbit

The general relativity predicts the existence of a minimal distance of stable circular orbit, called last stable circular orbit, or *innermost stable circular orbit* (ISCO), meaning that no accretion disk exists at $R < R_{\mathrm{ISCO}}$, the matter falling in free fall toward the compact object (which can be either a neutron star or a black hole). The ISCO radius is:

$$R_{\mathrm{ISCO}} = \frac{6GM}{c^2} = 12.5\frac{M}{1.4M_\odot}\mathrm{km} \tag{3.36}$$

corresponding to $3R_S$ for a Schwarzschild black hole. The ISCO Keplerian frequency is:

$$\nu_{\mathrm{ISCO}} = \frac{1580}{1.4M_\odot}\mathrm{Hz}. \tag{3.37}$$

3.6.2 Geometrically Thick and Optically Thin Accretion Disks

Geometrically thick and optically thin accretion disks are described in the frame of the *advection-dominated accretion flow* (ADAF), based on the fact that the optically thin gas radiates inefficiently, so that most of the viscously generated energy is advected radially (Narayan & Yi 1994). These accretion solutions are present at relatively low-mass accretion rates (Narayan 1996):

$$\frac{\dot{M}}{\dot{M}_{\text{Edd}}} \leqslant 0.3\alpha^2 \tag{3.38}$$

with α the standard viscosity parameter.

These equilibrium accretion flows, dominated by $e^+ - e^-$ pairs, present a two-temperature structure, where ions are nearly virial, and electrons are relativistic, with $T_e \sim 10^9 - 10^{10}$ K. Because of this high temperature, the flow exhibits a nearly spherical morphology, usually called *corona*, above the thin disk (Esin 1999). These accretion solutions are based on anti-correlated accretion rate $\dot{M}$ and transition radius R_{tr} between the inner accretion disk and the ADAF, with R_{tr} decreasing when $\dot{M}$ increases (Esin et al. 2001). We will mention in more detail such accretion solutions in Chapter 5, since they can reproduce the spectra of low-luminosity accreting black holes in the low state, and also the transition from low to high state, at a critical mass accretion rate (Narayan & Yi 1995).

3.6.3 Modeling Accretion Disks

Modeling the thermal part of the X-ray spectrum allows us to estimate the total luminosity of the disk L_d, the accretion rate $\dot{M}$, and the value of the temperature T_{in} at the location of the inner radius R_{in} of the accretion disk. Assuming that the disk radiates in all points as a blackbody, following the *Stefan–Boltzmann law*, the energy emitted by a disk ring delimited between radii r and $r + \Delta r$ is:

$$E(r, r + \Delta r) = 4\pi r \Delta r \sigma T^4 \sim r^{-2}\Delta r \tag{3.39}$$

since the radiated energy corresponds to gravitational energy obtained from $r + \Delta r$ and r. The disk effective temperature thus varies like $T_{\text{eff}} \propto r^{-3/4}$. This is described by the expression *multi-color blackbody*.

The emission spectrum, integrated on the whole disk blackbody, is thus:

$$f_d(E) \sim \int_{R_{\text{in}}}^{R_{\text{out}}} 2\pi r B(E, T) \, dr \tag{3.40}$$

where R_{in} and R_{out} are the inner and outer radii, and T_{in} and T_{out} are the temperatures at inner and outer radii of the accretion disk.

Assuming $R_{\text{out}} \gg R_{\text{in}}$, the total luminosity of the disk can be written as:

$$L_d = \int_{R_{\text{in}}}^{R_{\text{out}}} 4\pi r \sigma T^4(r) dr = 4\pi R_{\text{in}}^2 \sigma T_{\text{in}}^4 \tag{3.41}$$

the shape of the X-ray spectrum thus giving access to T_{in}. Knowing the distance of the source and the disk inclination, we can easily derive R_{in}, as well as the accretion rate $\dot{M}$:

$$\dot{M} = \frac{8\pi R_{in}^3 \sigma T_{in}^4}{3GM} \qquad (3.42)$$

with M being the mass of the compact object.

Thus the temperature profile of the accretion disk is, for $R \gg R_{in}$:

$$\left(\frac{T(R)}{T_{in}}\right)^4 = \left(\frac{R}{R_{in}}\right)^{-3} \qquad (3.43)$$

where

$$T_{in}^4 = \left(\frac{3GM\dot{M}}{8\pi R_{in}^3 \sigma}\right). \qquad (3.44)$$

From this last formula, we can infer that accretion disks around white dwarfs emit in the visible domain, while accretion disks surrounding neutron stars and black holes emit mainly in the X-ray domain (see an example of such a modeling of accretion disk in Figure 5.14; Chaty et al. 2003).

3.6.4 Astrotomography, or How to Image Accretion Disks

Accretion disks in binaries cannot be spatially resolved yet, even using large interferometers: at a distance of 1 kpc, the typical separation of the two components of the binary system, of $\sim 10^9$ m, corresponds to an angle of $<10\mu$ arcsecond, and in addition we are usually looking for features ~ 100 times smaller. Fortunately, there exists now a few methods to allow the observer to indirectly reconstruct the image of an accretion disk with accuracy, these methods being regrouped under the generic term of astrotomography. These techniques, such as eclipse mapping, Doppler tomography and Roche tomography, allow observers to map emission patterns in the accretion flow and in the secondary star, by inverting the line profiles, and deducing information about the intensity and the velocity field. They use information on the varying geometry of the stars, as they orbit within the binary system, their geometry being constrained by the Roche geometry in the semi-detached scenario. They allowed astronomers to discover the spiral structure during outburst in accretion disks within binaries of various nature, hosting white dwarf (see, e.g., Figure 3.7 in the case of a cataclysmic variable) and neutron star or black hole (see, e.g., Figure 3.8 for a high-mass X-ray binary). We review here the main specificities of these methods, and we refer the interested reader to Boffin et al. (2001) for additional details.

3.6.4.1 Eclipse Mapping

The first method, *eclipse mapping*, consists in computing the brightness of each component of the binary system (the accretion disk, the hot spot, and the companion

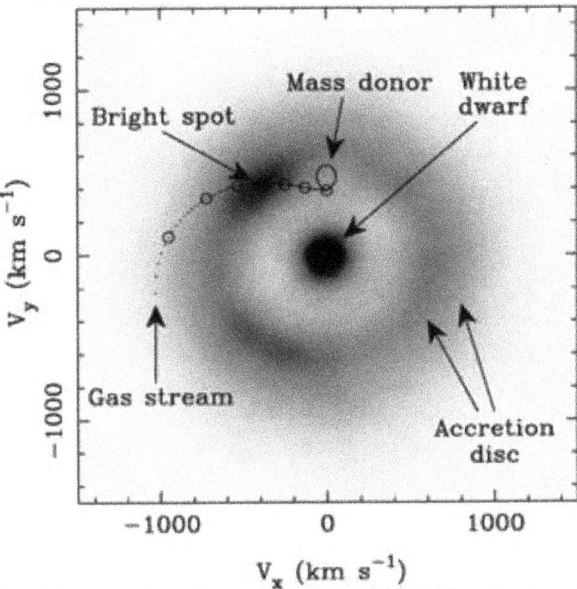

Figure 3.7. Doppler map of the cataclysmic variable CE 315, this system has an extreme mass ratio, thus the white dwarf is almost at the center of mass at [0, 0]. The accretion disk is seen as a spiral structure surrounding the white dwarf. Reprinted with permission from Marsh (2005).

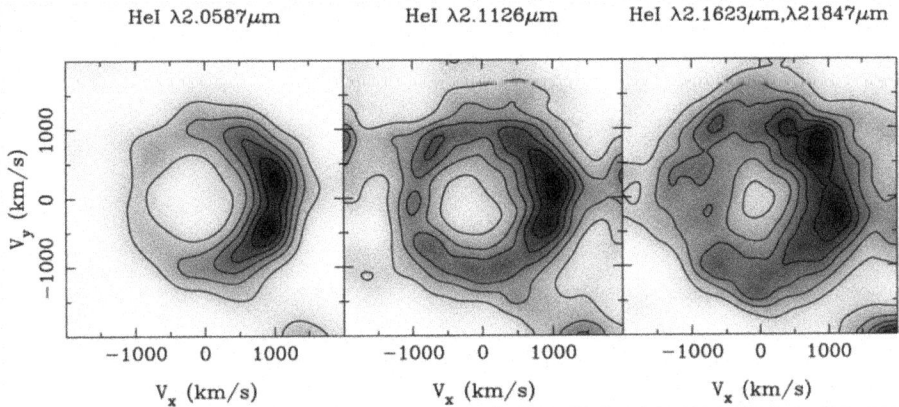

Figure 3.8. Doppler tomograms of the four He I lines of the high-mass X-ray binary Cyg X-3, during its 1997 June outburst, derived from infrared spectra of the companion Wolf–Rayet He star. A ring structure is clearly visible, similar to the tomographic signature of an accretion disk, and consistent with a disk-wind geometry, rotating at velocities of the order of $1000 \, \mathrm{km \, s^{-1}}$. Reprinted with permission of the AAS from Hanson et al. (2000).

star), when the secondary passes in front of the accretion disk, causing a partial or even total eclipse of the accretion disk. The brightness variations of the system can be reconstructed from the shape of the eclipse profile. However, the system emits in 2D, while the eclipse profile, as seen by the observer, is in 1D, therefore the solution to the reconstruction of the lightcurve is not unique, and one has to use the concept of image entropy. This technique, which can be extended from photometry to spectroscopy, allows the observer to determine how the effective temperature varies within the disk of transient systems, during their quiescent and flaring phases. This technique, involving an eclipse, only works for systems which are viewed almost edge-on.

3.6.4.2 Doppler Tomography

The second technique, *Doppler tomography*, uses the kinematic (velocity field) information given by the variations in the emission line profiles, with respect to the orbital phase of the binary system. Since line profile is largely dominated by the Doppler effect of each emitting part of the disk, the observed emission lines coming from a disk are often broad, due to large range of orbital Keplerian velocities within the disk ($\sim 10^3 km\ s^{-1}$), and exhibit two characteristic peaks. This technique, which presents the advantage of mapping the whole orbit, and not only the eclipse like the previous one, allowed observers to reveal the existence of asymmetric structures within the accretion disk of some CV, which can be interpreted as spiral waves. This technique also allows astronomers to detect the secondary, via narrow line profiles in absorption ($\sim 100 km\ s^{-1}$).

3.6.4.3 Roche Tomography

The third technique, *Roche tomography*, aims at including the secondary star, by taking into account the Roche geometry, which constrains the shape of the secondary inside its Roche lobe, as an effect of the tidal distortion from the primary. However, one also has to take into account the irradiation of the secondary by the disk and the compact object itself, and also the emission of the hot spot, where the matter coming from the secondary hits the accretion disk. This irradiation, by ionizing the neutral atoms that are used to map the radial velocity to build the tomogram, causes an artificial displacement of the emission coming from the secondary, toward the external part, the one that is not irradiated by both the disk and the primary. This technique also allows to reveal *starspots* on the surface of the secondary star.

3.7 Reviews and References

3.7.1 Reviews

- Review on binary formation: Kratter (2011);
- Review on various kinds of binary systems: Tauris & van den Heuvel (2006);
- Review on the Common Envelope Phase: Ivanova et al. (2013);
- Review on the radiative processes in astrophysics: Rybicki & Lightman (1979);
- Review on ADAF solutions: Esin et al. (2001);
- Review on astrotomography: Boffin et al. (2001).

References

Balbus, S. A. 2003, ARA&A, 41, 555

Blandford, R. D., & Königl, A. 1979, ApJ, 232, 34

Boffin, H. M. J., Steeghs, D., & Cuypers, J. 2001, in Astrotomography (Berlin: Springer)

Chaty, S., Haswell, C. A., Malzac, J., et al. 2003, MNRAS, 346, 689

Eggleton, P. P. 1983, ApJ, 268, 368

Esin, A. A. 1999, ApJ, 517, 381

Esin, A. A., McClintock, J. E., Drake, J. J., et al. 2001, ApJ, 555, 483

Fragos, T., Andrews, J. J., Ramirez-Ruiz, E., et al. 2019, ApJL, 883, L45

García, F., Simaz Bunzel, A., Chaty, S., Porter, E., & Chassande-Mottin, E. 2021, A&A, 649, A114

Greenhough, J., Chapman, S. C., Chaty, S., Dendy, R. O., & Rowlands, G. 2002, A&A, 385, 693

Greenhough, J., Chapman, S. C., Chaty, S., Dendy, R. O., & Rowlands, G. 2003, MNRAS, 340, 851

Hanson, M. M., Still, M. D., & Fender, R. P. 2000, ApJ, 541, 308

Hjellming, R. M., & Johnston, K. J. 1988, ApJ, 328, 600

Ivanova, N., Justham, S., Chen, X., et al. 2013, A&ARv, 21, 59

Kaper, L., van der Meer, A., van Kerkwijk, M., & van den Heuvel, E. 2006, Msngr, 126, 27

Kippenhahn, R., & Weigert, A. 1967, ZA, 65, 251

Kiziltan, B., Kottas, A., De Yoreo, M., & Thorsett, S. E. 2013, ApJ, 778, 66

Kratter, K. M. 2011, in ASP Conf. Ser. 447, Evolution of Compact Binaries, ed. L. Schmidtobreick, M. R. Schreiber, & C. Tappert (San Francisco, CA: ASP), 47

Kruckow, M. U., Tauris, T. M., Langer, N., et al. 2016, A&A, 596, A58

Marsh, T. R. 2005, Ap&SS, 296, 403

McClintock, J. E., & Remillard, R. A. 2006, in Cambridge Astrophysics Ser. 39, Compact Stellar X-Ray Sources, ed. W. Lewin, & M. van der Klis (Cambridge: Cambridge Univ. Press), 157

Narayan, R. 1996, ApJ, 462, 136

Narayan, R., & Yi, I. 1994, ApJ, 428, L13

Narayan, R., & Yi, I. 1995, ApJ, 452, 710

Paczynski, B. 1976, in IAU Symp. 73, Structure and Evolution of Close Binary Systems, ed. P. Eggleton, S. Mitton, & J. Whelan (Dordrecht: Reidel), 75

Plavec, M., & Kratochvil, P. 1964, BAICz, 15, 165

Podsiadlowski, P. 2001, in ASP Conf. Ser. 229, Evolution of Binary and Multiple Star Systems, ed. P. Podsiadlowski, S. Rappaport, A. R. King, F. D'Antona, & L. Burderi (San Francisco, CA: ASP), 239

Rybicki, G. B., & Lightman, A. P. 1979, Radiative Processes in Astrophysics (New York: Wiley-Interscience)

Sana, H., de Mink, S. E., de Koter, A., et al. 2012, Sci., 337, 444

Sen, K., Langer, N., Marchant, P., et al. 2022, A&A, 659, A98

Shakura, N. I., & Sunyaev, R. A. 1973, A&A, 24, 337

Shiraishi, J., Yoshida, Z., & Furukawa, M. 2009, ApJ, 697, 100

Stairs, I. H. 2006, JPhG, 32, S259

Tauris, T. M., & van den Heuvel, E. P. J. 2006, in Cambridge Astrophysics Ser. 39, Compact Stellar X-Ray Sources, ed. W. Lewin, & M. van der Klis (Cambridge: Cambridge Univ. Press), 623

Thorne, K. S., & Zytkow, A. N. 1975, ApJL, 199, L19

van der Laan, H. 1966, Natur, 211, 1131

Webbink, R. F. 2008, in Short-Period Binary Stars: Observations, Analyses, and Results, ed. E. F. Milone, D. A. Leahy, & D. W. Hobill (Berlin: Springer), 233

Zahn, J. P. 1977, A&A, 57, 383

Accreting Binaries
Nature, formation, and evolution
Sylvain Chaty

Chapter 4

Cataclysmic Variables (CV)

we are locked together
like a binary star system
he comes sometimes and vanishes
suddenly into nothingness

—Laura Stars, *Locked Together, 2016*

We describe in this chapter the nature, formation, and evolution of cataclysmic variables, which are binary systems hosting a main sequence star orbiting an accreting white dwarf. We first review the nature of these systems, then give the various types of cataclysmic variables, describing novae, magnetic systems, super-soft X-ray sources, AM CVn, binary white dwarfs. We then go through the formation, evolution, and final fate of a cataclysmic variable, in the context of isolated binary evolution, focusing at the end on the distribution of orbital period. We finish this chapter by giving useful links to reviews, catalogs and references.

4.1 Nature of Cataclysmic Variables

Cataclysmic variables (CV) are binary systems hosting a low-mass main sequence star (the secondary), orbiting an accreting white dwarf (the primary). The secondary star, usually a K-M spectral type, fills its Roche lobe (semi-detached system, see Roche lobe geometry in Section 3.3), and transfers its mass onto a CO or ONe white dwarf (see Figure 4.1). The hydrogen-rich gas leaves the secondary via the Lagrange point L_1, forming a highly collimated stream, but with too much angular momentum for its ballistic trajectory to fall directly onto the surface of the primary. We consider two different cases, depending on the presence of magnetic field in the white dwarf. On the first hand, in the case of a non-magnetic white dwarf, the accreted material spirals inwards, forming an accretion disk inside which the matter

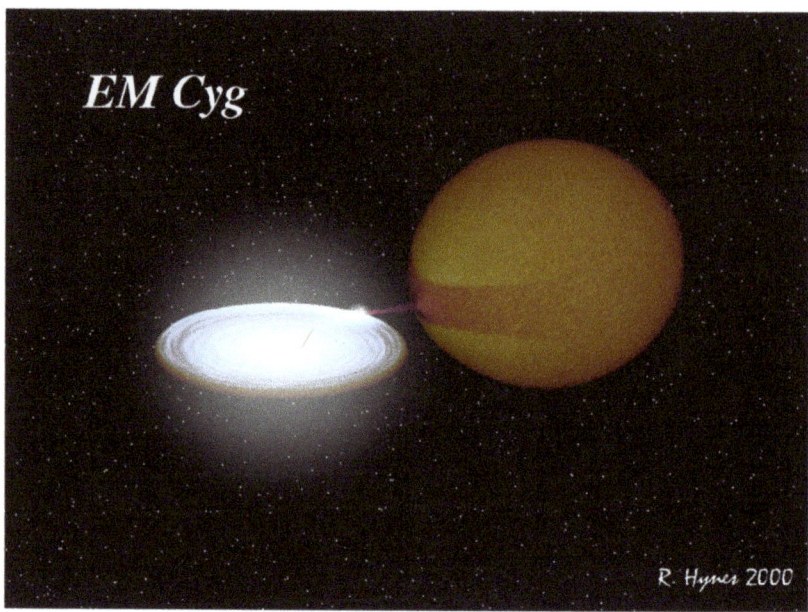

Figure 4.1. Schematic view of a cataclysmic variable, in this particular case is represented EM Cyg, a dwarf nova constituting of a low-mass star in orbit around a white dwarf. The accretion disk is much brighter that the secondary star. Reproduced with permission, Credit: Robert Hynes.

slowly spirals toward the primary, under effect of viscosity, and accumulates onto its surface. At the meeting point between the accretion flow and the disk, a *hot spot* forms. On the other hand, in the case of a magnetic white dwarf, the accreted material, which is constituted mainly of ionized hydrogen, follows the magnetic field lines, up to the magnetic poles, before reaching the surface of the white dwarf.

Cataclysmic variables are numerous: we know over $\sim$12, 000 known (including candidate) CV from various sky surveys (Jackim et al. 2020), and we estimate to a few millions the total number of such systems in our Galaxy. They are nearby objects, located at a typical distance of $\sim$100 pc, and bright, thus relatively easy to observe, serving as reference for more complex/distant objects. Some CV are also detected at high energy, for instance by the γ-ray INTEGRAL satellite (see, e.g., Fortin et al. 2018; Lutovinov et al. 2020; Tomsick et al. 2016, 2020, 2021). The Galactic distribution of CV detected by the INTEGRAL satellite is shown in Figure 4.2.

CV exhibit short orbital periods ranging from $P_{\rm orb} \sim 1$ to 15 hr, corresponding to an orbital separation of less than a few $R_\odot$ (a few CV exhibit longer orbital periods of a few days, in which case the donor is an evolved sub-giant). Due to their small orbital separation, large tidal effect lead to rotational deformation of the secondary, and to synchrone rotation of both components constituting the binary system (see Zorotovic & Schreiber 2020; and references therein). The characteristic time of mass transfer for CV, taking into account a mass for the secondary of $M_2 \leqslant 1 M_\odot$ and a mass transfer rate of $\dot{M} \geqslant 10^{-9} M_\odot \ {\rm yr}^{-1}$:

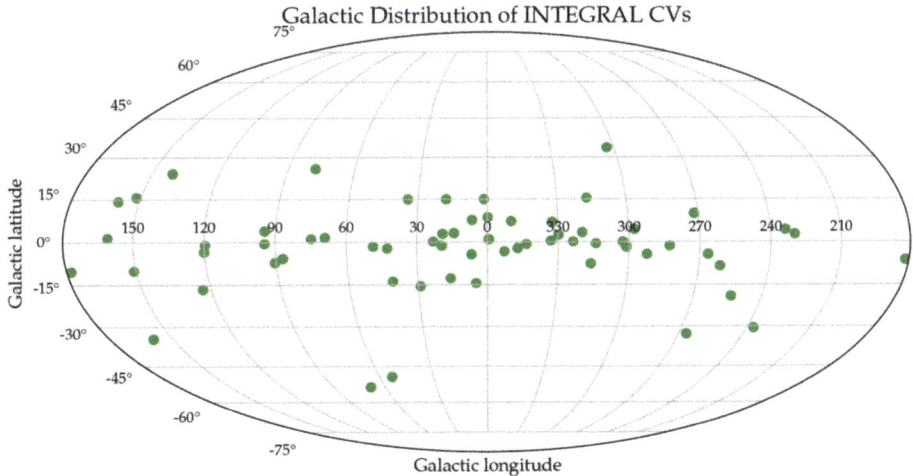

Figure 4.2. Galactic distribution, in Galactic coordinates, of CV detected by the INTEGRAL satellite. CV are distributed all over the Galaxy, with an overabundance within the Galactic bulge and plane, due to the old secondary stars. Only the 64 CV detected by the INTEGRAL satellite (most of them of polar or intermediate polar type) are shown here. Reproduced with permission, Credit: Francis Fortin.

$$t_M = \frac{M_2}{\dot{M}} \lesssim 10^9 \text{ years} \tag{4.1}$$

is largely inferior to Hubble time, at least while the mass transfer rate is not too small. The largest observed rates of mass transfer, of the order $\dot{M} \sim 10^{-8} M_\odot$/year, are attained during time intervals ranging from 6 to 40×10^6 yr.

Most of the luminosity emitted by a CV comes from accretion process onto the primary, the accreting luminosity being usually greater by an order of magnitude than the luminosity of the secondary ($\alpha \sim 0.1$):

$$L_{\text{acc}} = \frac{GM_1\dot{M}}{R_1} = \alpha\dot{M}c^2 => L_X \simeq 10^{33} - 10^{34} \text{ erg s}^{-1} \tag{4.2}$$

where M_1 and R_1 are respectively the mass and radius of the primary, and $\dot{M}$ the accretion rate. This accretion phenomenon can lead to thermonuclear explosions at the surface of the white dwarf, making these binaries bright enough to appear visible with the naked eye. The brightness of these systems granted them the term of *cataclysmic variables*. It is in the 1960s that they were understood as binary systems, mainly from spectroscopic studies showing periodically varying radial velocities (see e.g., Kraft 1964).

While initial studies were mainly focused onto the optical domain, due to instrumental development, each component of the system—companion star, white dwarf, accretion disk and hot spot—is detected via multi-wavelength observations, covering a large part of the electromagnetic domain. Emission in the optical and infrared domain mainly emanates from the companion star, usually a late spectral type red dwarf star, from G0 (exhibiting many Fe absorption lines) to M8 spectral type (with NaI, TiO in the optical and H_2O, CO in the infrared), or even very cool L or T types (with strong molecular absorption bands, see Smith 2006). The accretion disk radiates from the visible to ultraviolet, or even X-ray domain, which renders more difficult the determination of their bolometric luminosity, and of their mass transfer rate. Finally, the white dwarf, at a temperature between 10 to 40×10^3 K, is detected from UV to X-rays, dominating in the (extreme)-ultraviolet, with spectra well studied by IUE, HST and FUSE satellites.

CV exhibit photometric variability on various timescales, from low-amplitude (0.01–0.2 mag) flickering on short timescale (from tens of seconds to a few minutes), slow variations due to tidal deformation of the secondary, to long (orbital) timescales, via eclipses of disk and/or hot spot passing behind the secondary. The flickering, mainly due to the bright spot where the accretion stream hits the accretion disk, exhibits a maximum amplitude before the eclipse, before disappearing during the eclipse.

4.2 Various Types of Cataclysmic Variables

We now describe the various types of CV (for more details see the reviews by Warner 1995; Ritter & Kolb 2003; Smith 2006; Zorotovic & Schreiber 2020, and references therein).

4.2.1 Novae

4.2.1.1 Classical Novae

Classical novae have been known since antiquity, because of their extreme variability amplitude, typically of 9 magnitudes,[1] but ranging from 6 up to 19 magnitudes in a few days, before gradually fading over month-to-year timescale. This explains why it is only at the beginning of the 20th century that novae were distinguished from *supernovae*. In average, a nova brighter than visual magnitude $m_V = 3$ is detected once every 10 years. Fast novae reach their maximum brightness, and then decrease by a few magnitudes in a few days, while for slow novae this process can take months. Novae are detected on the whole electromagnetic spectrum, as shown for instance by the detection with the Fermi γ-ray space telescope of the nova V407 Cyg (Abdo et al. 2010).

A classical nova outburst is basically a giant instability of hydrogen thermonuclear fusion, occurring in the hot shell of matter slowly accumulating at the surface of the white dwarf, with a mass transfer rate from the companion of $\dot{M} \sim 10^{-8} M_\odot$ yr^{-1}. At this stage, because of its previous evolution, the white dwarf has already lost the majority of the hydrogen it possessed at its formation, its final composition depending on its mass: mostly composed of He if $M < 0.5 M_\odot$, a mixture of C and O if $0.5 < M < 1.2 M_\odot$, with an O/Ne core for $M > 1 M_\odot$. The material accreted from the companion is essentially made up of hydrogen, accumulating in a layer of increasing density on the surface of the white dwarf, where mixing occurs, and enriching this hydrogen layer in carbon, nitrogen, and oxygen (CNO).

---

[1] The stellar magnitude is defined as $\frac{m}{m_0} = -2.5 \log_{10} \frac{F}{F_0}$, thus an increase of 2.5 magnitudes corresponds to a factor of 10 in flux.

When density and pressure inside this layer are such that electrons in the ionized hydrogen become degenerate, the temperature at the basis of this layer reaches a point when explosive thermonuclear fusion of hydrogen is triggered. When such nuclear reactions occur in a stellar core (like hydrogen fusioning in helium during main sequence for instance), where thermal gas pressure dominates, the rise in temperature is accompanied by a rise in pressure, which regulates the nuclear fusion. This is however not the case in a degenerate gas, in which the pressure is independent of temperature, thus the rise in temperature further increases the hydrogen fusion rate (highly depending on the temperature, by a factor T^{40}), which leads to a thermal runaway fusion of hydrogen. A nova explosion thus occurs, accompanied by the release of energy in the range $10^{38} - 10^{40}$erg s^{-1}. This explosion is so violent that it ejects into the interstellar medium, at a speed of ~ 1000 km s^{-1}, the whole envelope of hydrogen, as a nova shell of 10^{-5} to $10^{-3}M_\odot$, rich in metals (metallicity $Z \sim 0.4$). This violent phenomenon, rapidly consuming all hydrogen present at the surface of the white dwarf, leads to a nova, which will remain bright in the sky for a few months.

Classical novae exhibit composite spectra with several components of various temperature, the dominant contribution being from the accretion disk, characterized either by optically thin (thick) continuum emission and strong emission (absorption) lines of hydrogen and helium, either coming from upper layers of the disk, or from irradiation from the white dwarf. The continuum emission is modeled as a multi-color blackbody emanating from the accretion disk, by adding rings at different distance r and temperature $T(r)$, the temperature being at its maximum close to the white dwarf, and decreasing when r increases. A more accurate alternative is to model each ring by a stellar atmosphere, characterized by its own surface gravity and effective temperature.

As mentioned above, a nova outburst leads to the ejection of the accreted matter, and disruption of the accretion disk, however it does not affect the binary system. This phenomenon is therefore recurrent, hydrogen accumulating again as a nova shell, with a recurrence time depending both on the mass transfer rate and on the mass of the white dwarf. A nova detected only once is usually called a *classical nova*, while a system that has been seen to repeat is called a *recurrent nova*. According to the standard theory of nova outbursts, the mass ejected during an eruption is similar to the mass accreted from the last eruption, which allows for repeated eruptions. There are of the order of a dozen known recurrent novae in the Galaxy, which exhibit an increase in amplitude of typically 9 magnitudes, with a recurrence timescale ranging from a few years such as U Sco, to a few decades such as IM Nor or T Cor Bor which host a red giant star as the secondary, the maximum being of nearly one century for V2487 Oph. Theoretical models predict a recurrent timescale that can go up to $10^4 - 10^5$ years. Systems such as T Pyx, with a white dwarf mass close to the Chandrasekhar limit of $M_1 = 1.4M_\odot$, are likely progenitors of Type Ia supernova, in which the white dwarf undergoes a thermonuclear explosion (Anupama & Kamath 2012).

We finally mention the *novae-like* objects, including all other accreting binaries hosting white dwarfs, sharing many nova characteristics described above, in their lightcurve and/or spectra, but for which no outburst has been detected. However, a detailed study of these sources usually allows to reclassify them in better defined groups.

4.2.1.2 Dwarf Novae

Dwarf novae, such as U Gem, first detected in outburst in 1855, were not recognized as a distinct class from classical novae, until the early 20th century. They exhibit quasi-periodic outbursts of ~2 to 5 magnitudes, rise in a timescale shorter than a day, typically last a few days before fading in a week timescale, and repeat more frequently than classical novae, on a week-to-year timescale. However, these parameters highly vary from one source to another. The lightcurves of these sources also show variability on orbital timescale. Contrarily to classical novae, their outbursts, of various shapes and durations (even for the same source), are explained by a thermal and viscous instability of the accretion disk, triggered when the mass transfer rate within the disk reaches a critical threshold. This instability is likely due to material of the accretion disk quickly falling toward the white dwarf, the energy released thus coming from the gravitational potential well of the white dwarf. Modeling of these outbursts allows astronomers to constrain the viscosity within the accretion disk. Some dwarf novae are detected at high energy, for instance in X-rays with the INTEGRAL satellite (Tomsick et al. 2016, 2020, 2021). Dwarf novae are divided in many sub-classes, according to the shape of their outbursts. Archetypical systems from the main sub-types are listed below:

- SS Cyg systems exhibit brightness increase of the order 2–6 mag in optical, in only a few days, before returning to original luminosity in several days. The archetypical system SS Cyg has been observed for more than a century (see Figure 4.3);
- SU UMa systems, like ER UMa and WZ Sag, mostly exhibit normal outbursts, and sometimes a brighter and longer *super-outburst*, characterized by *super-humps*, which are periodic brightening likely due to the presence of an elliptical accretion disk, its long rotational axis precessing in resonance with the orbital period of the system;
- Z Cam systems (Z Cam was first detected in 1904) exhibit a high transfer rate, close to stability, exhibiting bright and long states of persistent outbursts, separated by quiescent phases lasting many months, characterized by a constant luminosity, and frequent low-level flares.

What happens during an outburst within the accretion disk is described by the thermal-viscous accretion *disk instability model* (DIM, Cannizzo et al. 1982; Hameury et al. 1998; Dubus et al. 2001), with the disk at the interface between the secondary star and the white dwarf. According to this model, transferred matter piles up in an optically thin torus, which eventually becomes optically thick, causing the resulting convective structure to become thermally unstable. At this point the matter flows inward, and the sudden conversion of gravitational potential energy to radiation triggers the dwarf nova outburst. We now describe the behavior according to the DIM, in general agreement with most observations of dwarf nova outbursts, following the general discussion in Smith (2006).

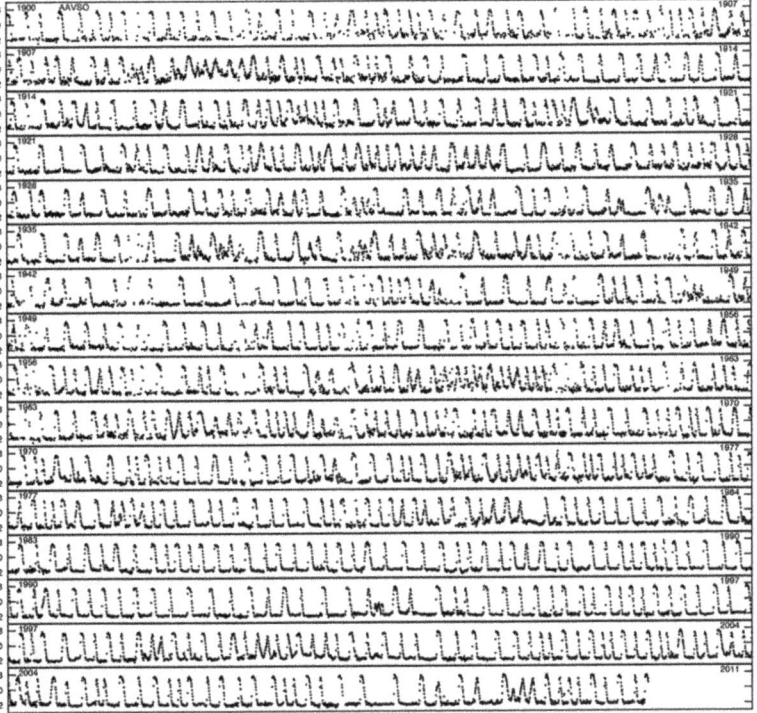

Figure 4.3. Historical lightcurve of the cataclysmic variable SS Cygni, from 1900 to 2010. Reprinted with permission from AAVSO (https://www.aavso.org/vsots_sscyg).

Let us call F_M the rate of mass flowing through the disk, set by the viscosity. There are two possibilities:

1. The first one, when $F_M > \dot{M}$, the mass flows within the disk at a rate greater than the mass transfer, leading to a decrease of the total mass within the disk, without triggering any outburst;

2. In the second one, $F_M < \dot{M}$, the mass flows within the disk at a rate smaller than the mass transfer, leading to accumulation of mass within the disk. Here there are two alternatives:

   (a) On the first hand, if viscosity ν is anti-correlated with surface density Σ, then an increase in Σ decreases ν and F_M, with the effect of increasing even more the mass within the disk, leading to an unstable mass transfer;

   (b) On the second hand, if ν is correlated with density, then an increase in Σ increases ν and F_M, with the effect of reducing the total mass within the disk, until reaching a stable mass transfer where $F_M = \dot{M}$, characterized by the absence of outburst, like in case 1.

This paradoxal situation has been solved by introducing two values in the ν–Σ relationship at the temperature of $\sim 10^4$ K (the opacity increases above this temperature, due to hydrogen ionization): a convective solution at low F_M and a radiative solution at high F_M. These two—minimum and maximum—values alternate between quiescent periods and outbursts, as schematically shown by the so-called *s-curve* (see Figure 4.4), reporting ν (or F_M) versus Σ. This curve, corresponding to thermal equilibrium, by equalizing heating and cooling rate, shows that there exists two stable solutions AB and CD, ν increasing with Σ. The AB segment, characterized by low ν, low F_M and low temperature (where hydrogen is neutral), the energy being carried away by convection, corresponds to *quiescence*. The BC segment corresponds to the prompt rise to maximum. The CD segment, characterized by high ν, high F_M, and high temperature (where hydrogen is ionized), the energy being carried away by radiation, corresponds to a slow decline, after the outburst maximum. Finally, the DA segment corresponds to the true return to quiescence. Dwarf nova outbursts thus occur when the mass transfer rate $\dot{M}$ corresponds to the unstable branch DB.

The accretion disk is quasi-stationary during flares. However, inbetween two outbursts, the local rate of mass transfer is much higher in external regions, implying that the disk mass increases. Another fact that can be tested observationally is the starting point of the DIM, the place at which the disk becomes brighter: either close to the inner part of the disk, or toward the external part.

Dwarf nova oscillations (DNO), less stable than the flickering, with a period stability parameter $Q = |\dot{P}|^{-1} \sim 10^4 - 10^6$, are detected only during outbursts, at low amplitude, via power spectrum analysis: they are likely due to magnetic accretion. *Quasi-periodic oscillations* (QPO), with larger amplitude and longer periods, but lower $Q \sim 1 - 10$ parameter, are likely due to oscillations or traveling waves within the accretion disk. Oscillations in CV follow a relationship $\frac{P_{QPO}}{P_{DNO}} \simeq 15$ (Warner et al. 2003), a value similar to the relationship $\frac{P_{HFQPO}}{P_{LFQPO}}$ seen in low-mass X-ray binaries,

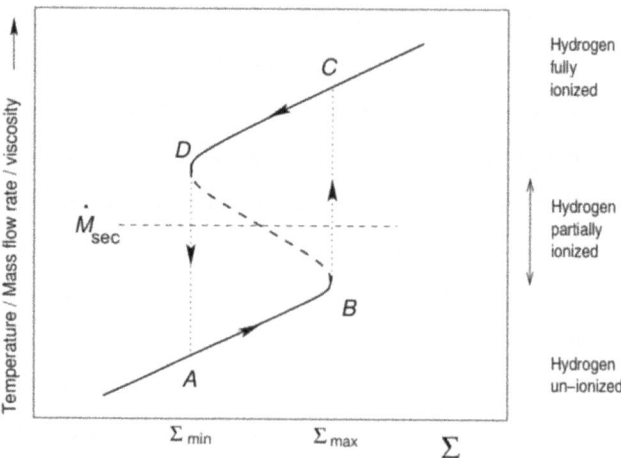

Figure 4.4. *S-curve* of CV, where is plotted the temperature, mass flow rate and viscosity ν, versus the surface density Σ. Reprinted with permission from Smith (2006).

hosting neutron stars (see Section 5.2.8) and black holes (see Section 5.3.5), suggesting a mechanism of similar origin in systems of different nature (see Figure 4.5 showing this correlation over a range of nearly six orders of magnitude in frequency).

4.2.1.3 Symbiotic Systems

Symbiotic systems, also called *symbiotic stars*, host a secondary star that is an evolved red giant (typically M5 III spectral type) instead of a main sequence star, with a long period of luminosity increase up to about 4 magnitudes, remaining close to near maximum brightness for a period that can reach up to 10 years (Lutovinov et al. 2020). To be able to host a red giant, the orbital separation in a symbiotic is of the order of 1 au, instead of $\sim 1 R_\odot$ like for most CV. Orbital periods range from 1 to 4 years, and even from 2 to 3 years for most systems. The mass transfer from the red giant to the white dwarf occurs via an accretion disk, when the companion fills its Roche lobe, but in many cases the star underfills it, and in this case accretion occurs through inhomogeneous stellar wind (Bollimpalli et al. 2018), like for symbiotics hosting a pulsating Mira variable. Archetypical symbiotic systems include CI Cygni (see Figure 4.6), AX Per, R Aqu, Z And, AG Dra, and also V407 Cygni, a symbiotic seen in γ-rays by the Fermi space telescope (Abdo et al. 2010). The number of known symbiotic systems amounts to ~ 400, with many new candidates coming from surveys of the Galactic bulge and plane (such as GALAH, Munari et al. 2021). The life of symbiotic systems is roughly divided into two major phases:

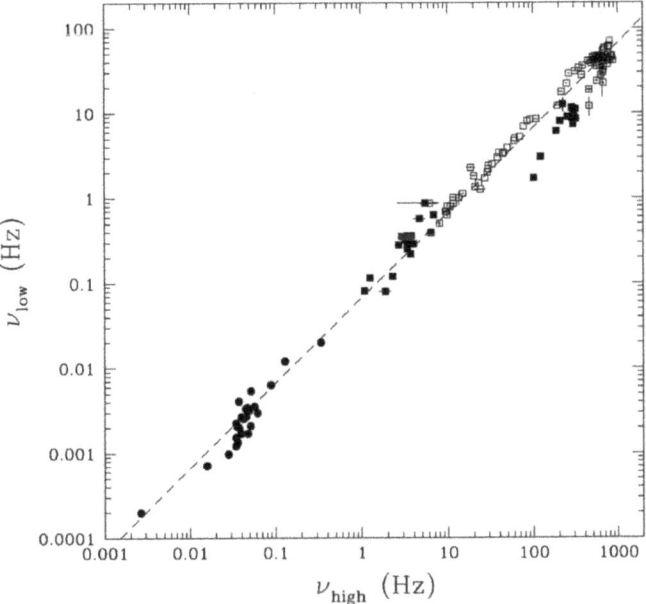

Figure 4.5. QPO diagram for 26 CV (filled circles) and low-mass X-ray binaries (open squares for neutron star binaries and filled squares for black hole binaries). X-ray binary data are from Belloni & Psaltis (2002). The dashed line marks $\frac{P_{\mathrm{QPO}}}{P_{\mathrm{DNO}}} \simeq 15$. Reprinted with permission from Warner et al. (2003).

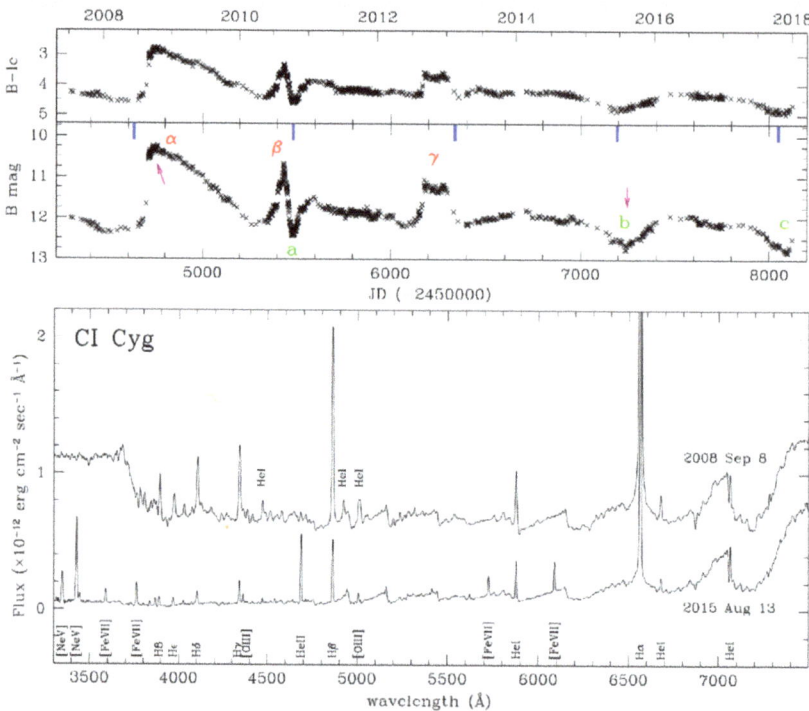

Figure 4.6. Photometric variability (*top*) and emission-line rich spectrum (*bottom*) of the prototype symbiotic system CI Cygni, hosting a cool red M giant companion in orbit with $P_{orb} = 855$ days around a white dwarf. Reprinted with permission from Munari (2019).

- Symbiotic systems spend most of their time in a quiescent *accreting-only phase*, during which they exhibit optical spectra dominated by the red giant, with no or weak emission lines. In this phase, accreted material accumulates onto the surface of the white dwarf, with mass-accretion rates of $0.1 - 1 \times 10^{-9} M_\odot$/year for a $1 M_\odot$-white dwarf.
- When eventually nuclear burning of accreted material ignites on the surface of the white dwarf, the symbiotic enters an active *burning phase*. On the first hand, if the accreted matter is constituted of electron-degenerate gas, the symbiotic ejects most of the accreted envelope at high velocity (ranging from ~1000 to ~6000 km s^{-1}), in the form of expanding optically-thick bipolar collimated outflows, which can be resolved at radio wavelengths. The system then simultaneously displays a rich emission-line spectrum and a strong nebular continuum. The emission lines originate from the red giant stellar wind, largely photo-ionized by the hot and luminous white dwarf. On the other hand, if the accreted matter is not made of electron-degenerate gas, nuclear burning proceeds in thermal equilibrium, and no matter is ejected. After a few years the symbiotic reaches its peak brightness, exhibiting low X-ray luminosities in the range $0.1 - 10 L_\odot$, burning the accreted envelope during many decades, or even centuries, before returning to the quiescent phase.

This outburst cycle can repeat on timescales as short as years/decades if the white dwarf mass remain inferior to $M_{\text{Chandrasekhar}}$. Since the white dwarf in symbiotics can eventually grow in mass and reach this upper mass limit, symbiotics are also good candidate progenitors to Type Ia supernova (see the review by Munari 2019 for more details).

4.2.2 Magnetic CV

Magnetic CV host a magnetized white dwarf, with magnetic field strength obtained either from polarization, from broad cyclotron lines, or from Zeeman splitting of absorption lines, in optical and UV domains. Many of these CV have been detected by the X-ray ROSAT satellite in the 1990s (Voges et al. 1999b, 1999a).

4.2.2.1 Polars

This class is also called AM Her-type, because of its archetypical system: AM Herculis. In this class, the primary is highly magnetized, with magnetic fields in the range $10^7 \leqslant B \leqslant 10^8$ G, and volume magnetization $M = \frac{\mu}{V}$ (with μ the magnetic moment) reaching or even exceeding $M \geqslant 10^{34}$ G cm^{-3}, i.e., 10^3 times more than for pulsars (Fabian et al. 1977). The magnetic field of the white dwarf is so high that it creates a magnetosphere surrounding the white dwarf, controlling the whole binary system, with the rotation of the primary synchronized with the secondary and the orbital period of the system. In these systems there is no accretion disk, and the ionized accretion stream coming from the companion star via the Lagrange point L_1, follows the magnetic field lines down to the surface of the white dwarf. When the accretion flow encounters the magnetosphere, close to the surface of the white dwarf, it becomes tied to and accelerated along the magnetic field lines, forming a shock at the top of an *accretion column* near the poles of the white dwarf, of small section ($\sim 1/10^4$ of the total surface of the white dwarf). At this point, the temperature of the shocked plasma reaches the virial temperature:

$$kT = \frac{GM_1 m_p}{R_1} \simeq 10 \text{keV} \tag{4.3}$$

which is hot enough to radiate in hard X-rays by free–free and cyclotron radiation, created by electrons spiraling along magnetic field lines, polarized in the optical domain (thus the name of *polar* to this class). Most of the radiation is emitted in soft X-rays, a substantial fraction of accretion energy being thermalized below the photosphere of the white dwarf, which explains why most cataclysmic variables detected by the high-energy INTEGRAL satellite are polars (Lutovinov et al. 2020).

4.2.2.2 Intermediate Polars

In this class, in which the members are also called DQ Her (see Figure 4.7), the white dwarf is moderately magnetized, and accretes from a cool main-sequence companion star. Because the magnetic field is weaker than in polars ($B \geqslant 10^6$ G), the rotation period of the white dwarf is not synchronized with the orbital period, the

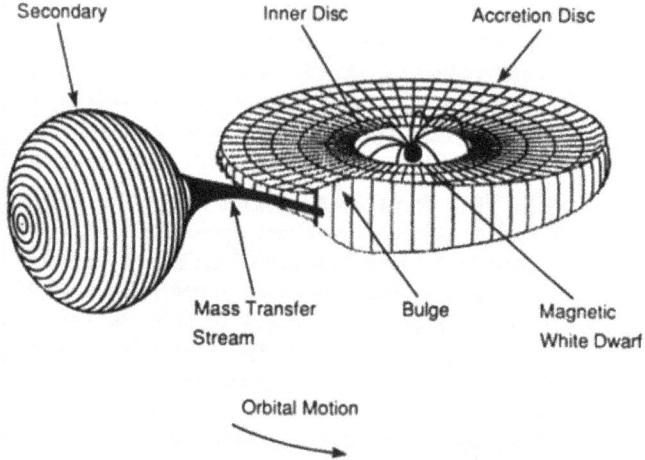

Figure 4.7. Schematic view of an intermediate polar, in this particular case is represented DQ Her, a low-mass star in orbit around a magnetized white dwarf. Credit: NASA.

white dwarf having minute-timescale spin period (71.1s for DQ Her, with a period stability parameter $Q > 10^{12}$), i.e., 10 times faster than the orbital period, due to an increase of angular momentum during the accretion phase from the secondary to the white dwarf. Because the magnetic field is weaker, the magnetosphere is smaller in size, and an accretion disk usually forms, however it is truncated at the point where the magnetic field of the white dwarf is intense enough to control the accretion flow. Close to the white dwarf, accretion stream meets the magnetosphere at the inner edge of the disk, and an *accretion curtain* forms, larger than the accretion stream of polars, and emitting in hard X-rays. Cyclotron lines are detected in infrared, accompanied in some cases by a low level of infrared polarization. Because of the presence of the accretion disk, these systems show dwarf nova-type outbursts. In many cases, accretion gives rise to a detectable X-ray emission, sometimes extended as in DQ Her (see Figure 4.8), interpreted in this case as non-thermal X-rays, related to a magnetized bipolar jet structure extending beyond the shell ejected during a nova event detected in optical back in 1934 (Toalá 2020). Finally, some intermediate Polars are also detected at high energy, as shown by numerous detections by the INTEGRAL satellite (Lutovinov et al. 2020; Tomsick et al. 2016, 2020, 2021).

4.2.3 Supersoft X-Ray Sources

Supersoft X-ray sources (SSS) were detected by the ROSAT satellite (Voges et al. 1999b, 1999a). They are luminous X-ray sources ($L \sim 10^{37}$erg s^{-1}), emitting a thermal radiation with a characteristic temperature of $\sim 10^5$K. They are white dwarfs accreting matter from a companion at a high rate ($\dot{M} \sim 10^{-7}M_\odot$/year), 100–1000 times higher than in CV, leading to continuous thermonuclear fusion of hydrogen occurring in a non-degenerate—stable—regime in a shell at the surface of the white dwarf. Both their high luminosity and low effective temperature are

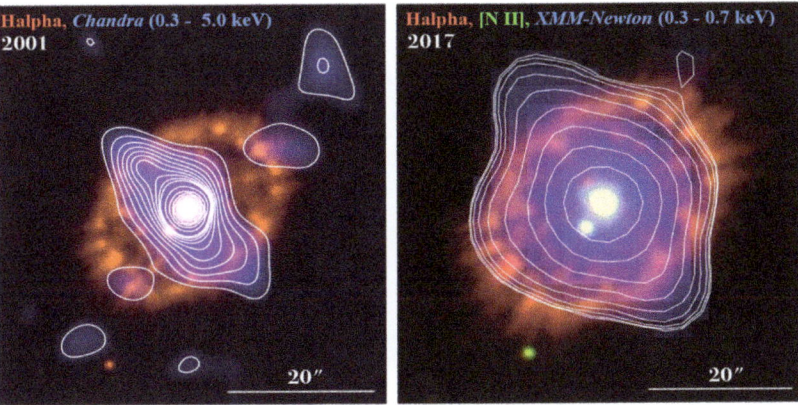

Figure 4.8. Magnetized jet from intermediate polar DQ Her, with Chandra (*left*) and XMM (*right*) extended X-ray emission contours superimposed on an Hα image. Reprinted with permission from Toalá (2020).

explained by the energy released by nuclear fusion (~ 0.007 mc^2), i.e., ~ 30 times higher than energy released by accretion onto the white dwarf (see table 2.2), thus emitting close to the Eddington limit. Since matter accumulates onto its surface, the mass of the white dwarf increases substantially, potentially exceeding the stability limit $M_{\text{Chandrasekhar}} = 1.4 M_\odot$, eventually exploding into a Type Ia supernova (see Figure 4.9). The secondary is usually a sub-giant star. SSS can be associated with radio emission, likely emanating from synchrotron emission arising from the outflow (Ogley et al. 2002).

4.2.4 AM CVn

AM Canum Venaticorum (AM CVn) are binaries in which a white dwarf accretes matter from an evolved, low-mass, H-depleted and He-rich secondary, instead of a main sequence star like in the case of CV. This secondary can be either non-degenerate, or even another white dwarf. Their mass–radius relationship allows them to reach short orbital periods $P_{\text{orb}} \leqslant 65$ mn, which can be as short as 3 mn (see for instance Cannizzo 1984; Warner 1995; Kalomeni et al. 2016). The particularity of these systems, is that the mass transfer is driven by gravitational-wave radiation (Paczyński 1967). Since the radius of the accretor represents a large fraction of the binary orbital separation, it is likely that the mass-transfer stream directly impacts the accretor, even before an accretion disk can form. This process is known as *direct-impact accretion*, which, when stable, characterizes AM CVn. They are observed at optical wavelengths as faint blue and variable objects. Thanks to intensive observing campaigns in UV and X-rays, we know ~ 30 of these systems in our Galaxy, and potentially several thousand of them will emit low-frequency gravitational waves detectable by the LISA space-based interferometer, well before their merging (Nelemans et al. 2004), the whole population contributing to the Galactic unresolved gravitational-wave background. AM CVn systems are thus ideal multi-messenger candidate sources, detected both in GW and electromagnetic facilities.

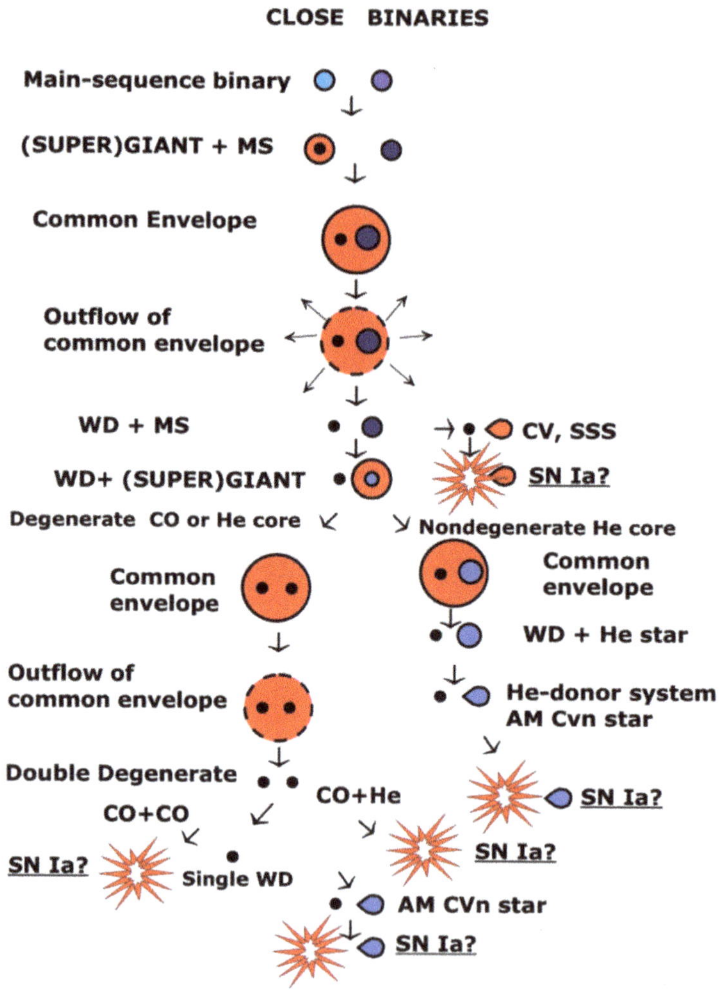

Figure 4.9. Evolution scenario of cataclysmic variables and related binaries (SSS, AM CVn), eventually leading to SN Ia. Reprinted with permission from Postnov & Yungelson (2014).

There are two main proposed formation channels for these systems, likely resulting from fine-tuned binary evolutionary path. In the first one, a highly evolved cataclysmic variable, with H-depleted and He-rich non-degenerate secondary, eventually collapses into a helium white dwarf. In the second one, a binary made of two white dwarfs, shrinks down its orbital separation due to GW radiation, eventually triggering mass transfer. These binaries subsequently evolve through one or two common envelope phases, can produce helium novae and sub-luminous explosions during their evolution, and eventually lead to Type Ia supernova, at the end of their overall evolution (see Figure 4.9). A review of these systems, including a modeling of the internal structure of both the donor and accretor, disk structure, disk atmosphere, and also their overall evolution, is described in Solheim (2010).

4.2.5 Binary White Dwarfs

Even if they are not properly speaking CV, we point out binary white dwarf systems, which are binary systems composed of two detached white dwarfs, emitting gravitational waves (Paczyński 1967), that will be within the LISA sensitivity range, when having orbital periods $\leqslant 1$ hr, emitting at frequencies $\geqslant 3$ mHz (Nelemans et al. 2004). Using a galactic population synthesis, Kremer et al. (2017) predict ~2700 of these systems to be observable by LISA during a significant fraction of their lifetimes, with a prominent negative orbital frequency evolution (chirp). He–He and He–CO double white dwarfs within the Galactic disk and bulge will be the most numerous sources for LISA, amounting to thousands of individually resolved systems, according to Lamberts et al. (2019); or even up to 10^4 (Lau et al. 2020). The number of detections linearly scales with the mass of the observed galaxy, as shown by Korol et al. (2020).

4.3 Formation, Evolution and Final Fate

4.3.1 Isolated Binary Evolution of a CV

As represented in Figure 4.9, the classical evolution of a CV is more complicated than just the juxtaposition of two isolated stars evolving together, mainly driven by loss of angular momentum, as a result of two mechanisms: *gravitational radiation* due to gravitational energy loss with a degenerate secondary filling its Roche lobe (Paczynski & Sienkiewicz 1981) and *magnetic braking* of the rotation, due to stellar wind magnetically coupled to the secondary, via coronal magnetic field lines rooted in the convection zone (Verbunt & Zwaan 1981). In the latter mechanism, the loss of angular momentum per unit mass in the stellar wind is very large, the outflowing matter being forced by the magnetic field to corotate with the star, out to large distances. This is justified by observations of isolated main sequence stars with convective envelopes, slowing down with age.

Originally, the CV consists of two main sequence stars of mass $M_1 > M_2$, orbiting with an orbital period P_{orb} of a few years. Star #1 evolves first and expands in a red giant, filling its Roche lobe, and transferring mass to star #2, leading to shorter orbital separation, and thus higher mass transfer rate, becoming dynamically unstable. Star #2 thus receives mass at a rate faster than thermal timescale, and eventually becomes literally engulfed within the expanding envelope of star #1, triggering the onset of the common envelope phase (see Section 3.4). Friction of star #2 orbiting within this envelope leads to decrease of orbital separation, with star #2 spiraling inwards, and transfer of angular momentum to the expanding envelope. This spiral-in process stops when the envelope is ejected, which also has the effect of stopping the decrease of orbital separation. At the end of this phase, the binary system is characterized by a tight orbital separation. This evolutionary phase is justified by the observation of many planetary nebulae surrounding a close binary (Jones & Boffin 2017), consisting of star #2 orbiting the naked core of star #1. The binary system is now formed of star #2—a red dwarf—orbiting star #1, which will collapse into a cooling white dwarf (if its mass is below the critical threshold mass of

~2.8$M_\odot$), with an orbital period of a few days. Such a system, called a *post-common envelope binary* (PCEB), is a good progenitor to become a CV. Low-mass WD are found in large number in samples of detached post-common envelope binaries (Schreiber et al. 2016). The orbital period still needs to be reduced, either by mechanism such as magnetic braking, or by tidal forces between the two components of the binary, leading to circularization of the orbit and synchronization of the stellar rotation period with the orbital period of the system. The magnetic braking mechanism then allows to get rid of angular momentum of the whole binary system, bringing the two components into close contact, leading to enhanced mass-transfer rates and high X-ray luminosities. In the course of the whole binary evolution, the white dwarf thus descends from a red giant star, characterized by a radius several hundred times larger than the typical orbital separation between the two components of a CV.

4.3.2 Distribution of Orbital Period

In Figures 4.10 and 4.11 we report the observed orbital period distribution of cataclysmic variables, showing the following striking properties:
1. The rarity of systems with an orbital period $P_{orb} \geqslant 40$ hr (called *period maximum*);
2. the lack of (or more precisely the presence of fewer) systems in the orbital period range $2 \leqslant P_{orb} \leqslant 3$ hr (*period gap*);
3. and the absence of systems below an orbital period of $P_{orb} \leqslant 50$ mn (*period minimum*).

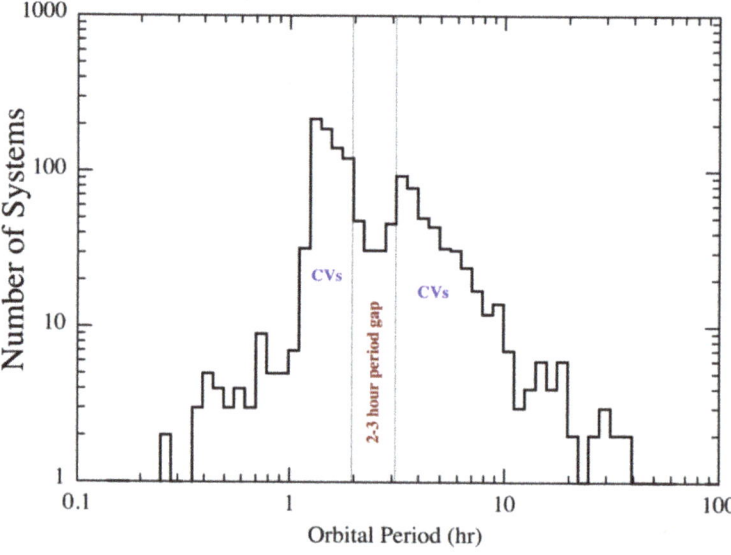

Figure 4.10. Distribution of orbital periods of CV from the Ritter & Kolb (2003) catalog. Ultracompact binaries, such as AM CVn, are located on the left of the distribution. Reprinted with permission of the AAS from Kalomeni et al. (2016).

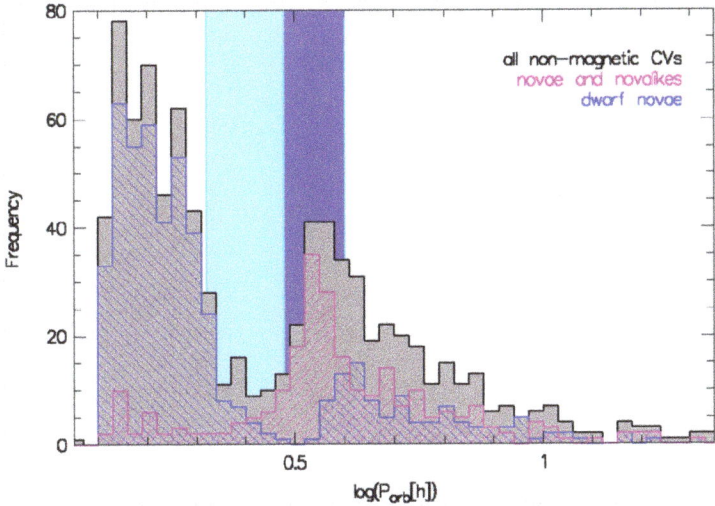

Figure 4.11. Orbital period distribution of various sub-types of non-magnetic CV: number of systems versus orbital period (hour), built using data from Ritter & Kolb (2003); update RKcat7.18, 2012. The position of the period gap is indicated by the light blue bar, the darker bar indicates the 3–4 h range dominated by systems with high mass transfer. Reprinted with permission from Schmidtobreick (2015).

4.3.2.1 Period Maximum

The period maximum is explained by the fact that the secondary star fills its Roche lobe. In addition, as every main sequence star, its radius is constrained by its mass, following the mass–radius relation. The stellar radius, equalizing the radius of the Roche lobe, is thus constrained by the orbital separation, with $M(M_\odot) \propto 0.1P_{orb}$ (hr). Since the mass of the secondary $M_2 < M_{WD} \sim 1.2M_\odot < 1.44M_\odot$ we obtain that $P_{orb} < 12$ hr. A few systems, like the extreme GK Per with $P_{orb} = 2$ days, can be explained by an evolved secondary star.

Observations of CV above the period gap, up to the period maximum, are consistent with this scenario, with higher accretion rates and larger radii of secondaries above the gap, compared to main sequence stars. On the opposite, at short orbital period ($P_{orb} \lesssim 3$ hr), the spectral type of the secondary is closer to the main sequence, likely due to a nuclear evolution, and to mass transfer (see Beuermann et al. 1998; Zorotovic et al. 2016; and Figure 4.12).

4.3.2.2 Period Gap

The period gap in the distribution of orbital periods begins to be populated when magnetic CV are included, thus the magnetic field must play an important role in the depletion of CV in this orbital period range. Since magnetic braking causes a loss of angular momentum, with a mass transfer rate $\dot{M} \sim 10^{-9} M_\odot$ yr^{-1}, the orbital separation decreases on a timescale of $10^8 - 10^9$ years, bringing all systems to pass through this period gap in the course of their evolution. The only viable explanation to see fewer systems must then be that systems reaching an orbital period of

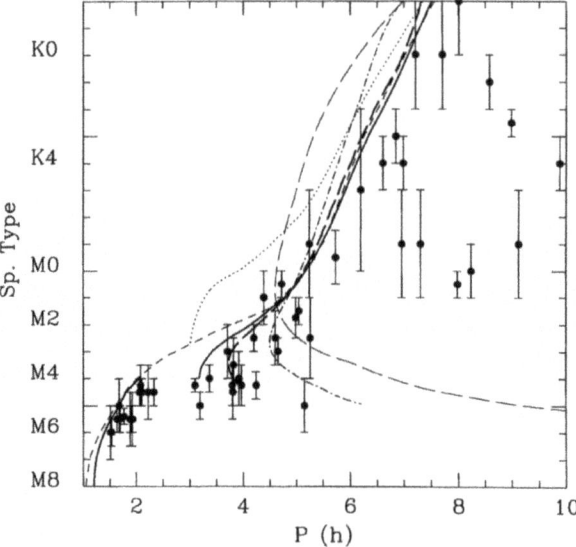

Figure 4.12. Spectral type of CV secondary versus orbital period P of the system, with evolutionary tracks representing unevolved donor stars and different mass transfer rates. The observed data are taken from Beuermann et al. (1998), while the short-dashed curve corresponds to solar metallicity stellar models on the ZAMS. This figure shows that at short orbital period ($P_{orb} \leqslant 3$ hr), the spectral type of the secondary is closer to isolated main sequence stars than at longer orbital period. Reprinted with permission from Baraffe & Kolb (2000).

$P_{orb} \leqslant 3$ hr become undetected, likely because mass transfer, and thus accretion and also X-ray emission, stops at this stage. But then the question becomes: what would cause the mass transfer to stop?

To answer this question, we have to consider the internal structure of stars. A main sequence star, such as the Sun, consists of a radiative core and a convective envelope, with the magnetic field created at the level of the *tachocline* (Spiegel & Zahn 1992), a transition zone between the radiative and convective zones. Let us consider a binary hosting a WD orbiting a main sequence star of mass $M \sim 0.3 M_\odot$, filling its Roche lobe, when reaching $P_{orb} \sim 3$ hr. The red dwarf has, at this stage of its evolution, lost so much mass, that its radiative core vanishes, and thus also its transition zone. The secondary becomes fully convective, unable to generate any magnetic flux, due to a sharp decline in the magnetic dynamo. Observations of single main sequence stars show a discontinuity in the magnetic braking, and a change in the magnetic field topology around the fully convective boundary. The magnetic braking mechanism then becomes inefficient, which implies a temporary loss of contact within the binary, strongly reducing, or even preventing, the loss of angular momentum and mass transfer. The red dwarf, previously over-filling its Roche lobe as a result of mass transfer, eventually relaxes, reducing its radius toward thermal equilibrium, and temporarily ceasing the mass transfer. The binary becomes a detached white dwarf in orbit around a main sequence star. This phenomenon, explaining the period gap, is the core of the standard model, so-called *disrupted magnetic braking* (DMB), a model initially proposed by Rappaport et al. (1983) and Spruit & Ritter (1983).

In the meantime, in the absence of magnetic field, orbital separation still continues to decrease, with loss of angular momentum due to gravitational radiation (Paczyński 1967), which becomes more efficient because the orbital separation is smaller, until it reaches an orbital period of $P_{orb} = 2$ hr. At this point, the Roche lobe has shrunk enough that the red dwarf now fills it again, resuming mass transfer, the system becoming again detectable as a CV.

In this model, the period gap would then just result from the secular evolution of CV, stopping any X-ray emission due to absence of mass transfer at $P_{orb} = 3$ hr, and resuming it when mass transfer reoccurs, when P_{orb} reaches 2 hr. In magnetic CV, since the two components are coupled by magnetic field lines, accretion can continue, even when crossing this period gap, this fact explaining their presence in the period gap.

4.3.2.3 Period Minimum

The distribution of observed orbital periods in CV exhibits a minimum orbital period of ~80 mn (McAllister et al. 2019). This period minimum is again due to the internal structure of the secondary star, as predicted by Rappaport et al. (1982) and Paczynski & Sienkiewicz (1983). Indeed, after reaching $P_{orb} \sim 2$ hr, the binary continues to evolve toward shorter orbital separation, losing angular momentum due to gravitation radiation, until it reaches $P_{orb} \sim 60-80$ min. At this point, the red dwarf has been almost completely consumed, reaching a threshold mass of $0.08 M_{\odot}$. Since it corresponds to a temperature too low to allow for hydrogen thermonuclear burning, the secondary becomes degenerate. Below this mass is the domain of the *brown dwarfs*, for which stellar contraction is stopped by electron degeneracy in the stellar core, similarly to white dwarfs. Because of the change in the mass–radius relation when a main sequence star evolves to a degenerate core (see Chapter 2), the response of the degenerate secondary becomes opposite to mass loss: instead of shrinking, it now expands, the binary now evolving back toward increasing orbital separation (*period bouncers*). Thus, the structural change of the secondary star, evolving from main sequence to degenerate core, results in the binary reaching the minimum of the orbital period of CV.

To summarize, the whole orbital period distribution reflects the secular evolution of the companion star in a CV, starting with a main sequence star, and finishing up with a degenerate core, which contains more helium than a *normal* brown dwarf, due to thermonuclear burning in the previous evolutionary phase.

4.3.2.4 The White Dwarf Mass Problem

An open question concerning the evolution of CV is related to the mass of the white dwarf: does it grow, as a consequence of mass transfer? Based on binary evolution arguments of envelope ejection, we would expect the contrary: the mass ejected during a nova outburst is similar to the mass accreted between two outbursts.

The mass of the WD should thus remain constant, at least during the CV phase, and only the more massive WD ($1.0 \leqslant M \leqslant 1.38 M_\odot$), accreting at very high rates, are expected to grow slowly but steadily in mass, eventually leading to SNIa progenitors (Hillman et al. 2015).

By comparing the average WD mass in CV above and below the period gap, McAllister et al. (2019) confirm the absence of trend in white dwarf mass with orbital period, showing that there is no evidence of mass growth for WD within CV. They show however a systematic and significant higher average mass of white dwarfs in CV, compared to isolated WD, or to detached progenitors of CV.

Early binary population models of CV predicted a mean WD mass $M \sim 0.6 M_\odot$ (Kolb 1993) and large numbers of CV containing He core. However, measurements gave masses in the range $M \sim 0.8 - 1.2 M_\odot$ (Ritter & Kolb 2003), now refined to an average WD mass of $\sim 0.83 M_\odot$ (McAllister et al. 2019), and no system hosting a He-core WD has been identified (Zorotovic & Schreiber 2011, 2020). The reason of this discrepancy, first interpreted as a selection effect (brighter systems host larger mass WD), is likely related to a high loss of frictional angular momentum, generated by mass transfer, and anti-correlated with the WD mass, as shown by binary population synthesis models such as *empirical consequential angular momentum loss* (eCAML), which reproduces the observed WD mass distribution (Schreiber et al. 2016). This angular momentum loss, which increases when the WD mass decreases, has the effect to drive CV hosting low-mass WD toward short orbital periods, into dynamically unstable mass transfer, thus disappearing as CV, favoring instead CV hosting higher mass WD (see Figure 4.13; McAllister et al. 2019).

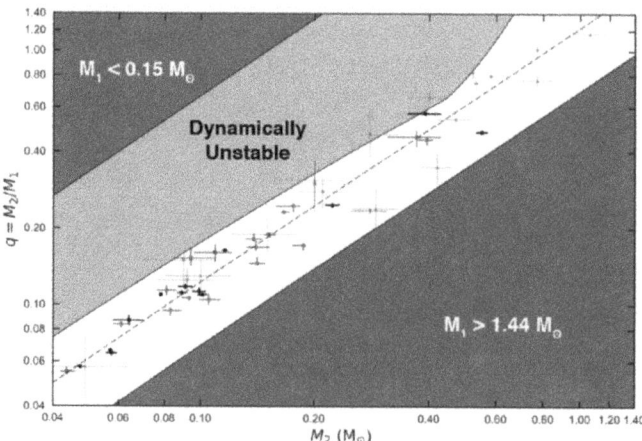

Figure 4.13. Figure of q versus white dwarf mass M_2 in CV. The dark gray regions are theoretically prohibited, excluding unrealistically low WD mass ($M \leqslant 0.15 M_\odot$) and mass greater than the Chandrasekhar mass limit ($M \geqslant 1.44 M_\odot$), while the light gray region is forbidden by the empirical consequential angular momentum loss (eCAML) model from Schreiber et al. (2016). The dashed gray line represents the average WD mass value ($M_1 = 0.83 M_\odot$; McAllister et al. 2019). The various points represent WD masses obtained from eclipse modeling, contact phase timing, or radial velocity. Reprinted with permission from McAllister et al. (2019).

4.4 Reviews, Catalogs and References

4.4.1 Reviews

- Reviews on the nature and evolution of CV: Warner (1995); Ritter & Kolb (2003); Smith (2006); Solheim (2010); Munari (2019); Lutovinov et al. (2020); Zorotovic & Schreiber (2020);
- Review on CV seen in X-rays: Kuulkers et al. (2010).

4.4.2 Catalogs

- *Open cataclysmic variable catalog*: all publicly available data from several catalogs and ongoing sky surveys (Jackim et al. 2020), including CV, nova, and dwarf nova (candidate and confirmed): https://depts.washington.edu/catvar
- *Catalog of cataclysmic binaries, low-mass X-ray binaries and related objects* (seventh edition): Ritter & Kolb (2003);
- *Catalog and Atlas of Cataclysmic Variables* (frozen on 2006 February 1): Downes & Shara (1993); Downes et al. (1997, 2001): https://archive.stsci.edu/prepds/cvcat/index.html
- *Catalog of symbiotic stars*: http://www.iphas.org/symbiotics

References

Abdo, A. A., Ackermann, M., Ajello, M., et al. 2010, Sci, 329, 817

Anupama, G. C., & Kamath, U. S. 2012, BASI, 40, 161

Baraffe, I., & Kolb, U. 2000, MNRAS, 318, 354

Belloni, T., Psaltis, D., & van der Klis, M. 2002, ApJ, 572, 392

Beuermann, K., Baraffe, I., Kolb, U., & Weichhold, M. 1998, A&A, 339, 518

Bollimpalli, D. A., Hameury, J. M., & Lasota, J. P. 2018, MNRAS, 481, 5422

Cannizzo, J. K. 1984, Natur, 311, 443

Cannizzo, J. K., Ghosh, P., & Wheeler, J. C. 1982, ApJL, 260, L83

Downes, R., Webbink, R. F., & Shara, M. M. 1997, PASP, 109, 345

Downes, R. A., & Shara, M. M. 1993, PASP, 105, 127

Downes, R. A., Webbink, R. F., Shara, M. M., et al. 2001, PASP, 113, 764

Dubus, G., Hameury, J. M., & Lasota, J. P. 2001, A&A, 373, 251

Fabian, A. C., Pringle, J. E., Rees, M. J., & Whelan, J. A. J. 1977, MNRAS, 179, 9P

Fortin, F., Chaty, S., Coleiro, A., Tomsick, J. A., & Nitschelm, C. H. R. 2018, A&A, 618, A150

Hameury, J-M., Menou, K., Dubus, G., Lasota, J-P., & Hure, J-M. 1998, MNRAS, 298, 1048

Hillman, Y., Prialnik, D., Kovetz, A., & Shara, M. M. 2015, MNRAS, 446, 1924

Jackim, R., Szkody, P., Hazelton, B., & Benson, N. C. 2020, RNAAS, 4, 219

Jones, D., & Boffin, H. M. J. 2017, NatAs, 1, 0117

Kalomeni, B., Nelson, L., Rappaport, S., et al. 2016, ApJ, 833, 83

Kolb, U. 1993, A&A, 271, 149

Korol, V., Toonen, S., Klein, A., et al. 2020, A&A, 638, A153

Kraft, R. P. 1964, ASPL, 9, 137

Kremer, K., Breivik, K., Larson, S. L., & Kalogera, V. 2017, ApJ, 846, 95

Kuulkers, E., Norton, A., Schwope, A., & Warner, B. 2010, in Cambridge Astrophysics Ser. 39, Compact Stellar X-ray Sources, ed. W. Lewin, & M. van der Klis, (Cambridge: Cambridge Univ. Press), 421

Lamberts, A., Blunt, S., Littenberg, T. B., et al. 2019, MNRAS, 490, 5888

Lau, M. Y. M., Mandel, I., Vigna-Gómez, A., et al. 2020, MNRAS, 492, 3061

Lutovinov, A., Suleimanov, V., & Manuel Luna, G. J. 2020, NewAR, 91, 101547

McAllister, M., Littlefair, S. P., Parsons, S. G., et al. 2019, MNRAS, 486, 5535

Munari, U. 2019, in The Impact of Binary Stars on Stellar Evolution, ed. G. Beccari, & M. Boffin (Cambridge: Cambridge Univ. Press)

Munari, U., Traven, G., Masetti, N., et al. 2021, MNRAS, 505, 6121

Nelemans, G., Yungelson, L. R., & Portegies Zwart, S. F. 2004, MNRAS, 349, 181

Ogley, R. N., Chaty, S., Crocker, M., et al. 2002, MNRAS, 330, 772

Paczyński, B. 1967, AcA, 17, 287

Paczynski, B., & Sienkiewicz, R. 1981, ApJL, 248, L27

Paczynski, B., & Sienkiewicz, R. 1983, ApJ, 268, 825

Postnov, K. A., & Yungelson, L. R. 2014, LRR, 17, 3

Rappaport, S., Joss, P. C., & Webbink, R. F. 1982, ApJ, 254, 616

Rappaport, S., Verbunt, F., & Joss, P. C. 1983, ApJ, 275, 713

Ritter, H., & Kolb, U. 2003, A&A, 404, 301

Schmidtobreick, L. 2015, in The Golden Age of Cataclysmic Variables and Related Objects—III (Golden2015), 34

Schreiber, M. R., Zorotovic, M., & Wijnen, T. P. G. 2016, MNRAS, 455, L16

Smith, R. C. 2006, ConPh, 47, 363

Solheim, J. E. 2010, PASP, 122, 1133

Spiegel, E. A., & Zahn, J. P. 1992, A&A, 265, 106

Spruit, H. C., & Ritter, H. 1983, A&A, 124, 267

Toalá, J. A., Guerrero, M. A., Santamaría, E., Ramos-Larios, G., & Sabin, L. 2020, MNRAS, 495, 4372

Tomsick, J. A., Bodaghee, A., Chaty, S., et al. 2020, ApJ, 889, 53

Tomsick, J. A., Coughenour, B. M., Hare, J., et al. 2021, ApJ, 914, 48

Tomsick, J. A., Krivonos, R., Wang, Q., et al. 2016, ApJ, 816, 38

Verbunt, F., & Zwaan, C. 1981, A&A, 100, L7

Voges, W., Aschenbach, B., Boller, T., et al. 1999, VizieR On-line Data Catalog: IX/10A

Voges, W., Aschenbach, B., Boller, T., et al. 1999, A&A, 349, 389

Warner, B. 1995, in Cataclysmic Variable Stars (Cambridge: Cambridge Univ. Press)

Warner, B., Woudt, P. A., & Pretorius, M. L. 2003, MNRAS, 344, 1193

Zorotovic, M., & Schreiber, M. R. 2020, AdSpR, 66, 1080

Zorotovic, M., Schreiber, M. R., & Gänsicke, B. T. 2011, A&A, 536, A42

Zorotovic, M., Schreiber, M. R., Parsons, S. G., et al. 2016, MNRAS, 457, 3867

Accreting Binaries
Nature, formation, and evolution
Sylvain Chaty

Chapter 5

Low-mass X-Ray Binaries (LMXB)

*"Two celestial bodies intertwined
gravity keep them connected
such a beauty to watch in motion
two distinct bodies move as one"*
—R. S. Mallari, *Binary Stars (Celestial Lovers), 2011*

We describe in this chapter the nature, formation, and evolution of low-mass X-ray binaries (LMXB), which are binary systems hosting a low-mass star orbiting a compact object—either accreting neutron star or black hole. We first review the nature of this population, distinguishing systems hosting neutron stars from those hosting black holes, with their different characteristics: accretion luminosity, temporal variability, and ellipsoidal variations. For the first type—neutron star systems—we describe Atoll sources, Z sources, X-ray bursters—exhibiting Type I and Type II X-ray bursts—dippers, accreting milli-second pulsars, ultra-compact X-ray binaries, and symbiotic X-ray binaries. We then mention their variability properties and frequency correlations, and describe the main quasi-periodic oscillation (QPO) models. For the second type—black hole systems—we describe the soft X-ray transients, the modeling of their emission, along with their evolution during an outburst, their spectral states, and short-timescale variability properties. We then discuss the observational evidence of an event horizon, before describing the fundamental plane of black hole activity. We finally describe the whole formation, evolution and final fate of LMXB, before closing up this chapter by giving useful links to reviews, catalogs, database, and references.

5.1 Nature of Low-Mass X-ray Binaries

A large majority (>90%) of accreting binaries, hosting a compact object that can be either a neutron star (NS) or a black hole (BH), are split in two different classes, according to the nature of their companion star. In the first class, the companion star (secondary) is a low-mass ($\leqslant 1 - 2M_\odot$) star, they are called *low-mass X-ray binary* (LMXB), and emit most of their emission in the X-ray domain (see Figure 5.1). The second class, called *high-mass X-ray binary* (HMXB), in which the companion is a massive star (O or B spectral type star), and which emits mostly in the optical domain, will be described in Chapter 6. We focus in this chapter on LMXB, where we distinguish systems hosting neutron stars from those hosting black holes, although, as we will see in this chapter, both share a wealth of common properties.

5.1.1 Population of LMXB

Since LMXB emit in an energy range in which the interstellar medium is nearly transparent, it is possible to detect them through the whole Galaxy, and sensitive soft X-ray surveys have been conducted, using for instance Uhuru, ROSAT and ASCA satellites, providing soft-X-ray images of the Galactic bulge, where are concentrated most LMXB.

We currently know ~218 LMXB in our Galaxy (we only count here confirmed LMXB, meaning that they are detected in X-rays, possess an optical and/or infrared counterpart, and that the spectral type of the secondary has been determined spectroscopically, following the method described in Fortin et al. 2018), to which we add LMXB within Magellanic Clouds and close-by galaxies, with X-ray luminosities ranging from 10^{34} to 10^{38} erg s^{-1}. Their spatial distribution (see Figure 5.2), concentrated in the Galactic center, bulge, and disk, within 0.6 kpc of the Galactic plane, clearly outlines the Galactic structure. This distribution points toward a relatively old stellar population, older than ~10^9 years. Globular clusters, although they contain a very small fraction of the Galactic mass, also host a substantial fraction of LMXB, revealing a particular way of LMXB formation in these dense stellar environments. There are fewer LMXB than cataclysmic variables (CV), first because stars of higher mass are less numerous (see the initial mass function in Figure 2.2), and second due to unbound systems during the supernova event involved in the formation of the compact object (see Section 6.5.3 on natal kicks).

5.1.2 Accretion Luminosity

LMXB share similar properties with CV (see Chapter 4). The companion star is a low-mass late-type main sequence star, as shown in the generic image of Figure 5.1, representing GRO J1655-40. They are semi-detached systems, the secondary filling its Roche lobe, with matter going through the L_1 Lagrange point, accumulating in an accretion disk surrounding the compact object (see Figure 3.4). The main difference between LMXB and CV is the nature of the compact object, replacing the white dwarf by a neutron star, or even a black hole, increasing the potential well by a factor 10^3 times deeper in LMXB, compared to CV (see Table 2.1). LMXB are thus highly luminous objects, with a typical X-ray (0.5 − 100 keV) luminosity (a good tracer of the total accretion rate) naturally higher than CV, spanning a wide range of accretion luminosity ($\alpha \sim 0.1$):

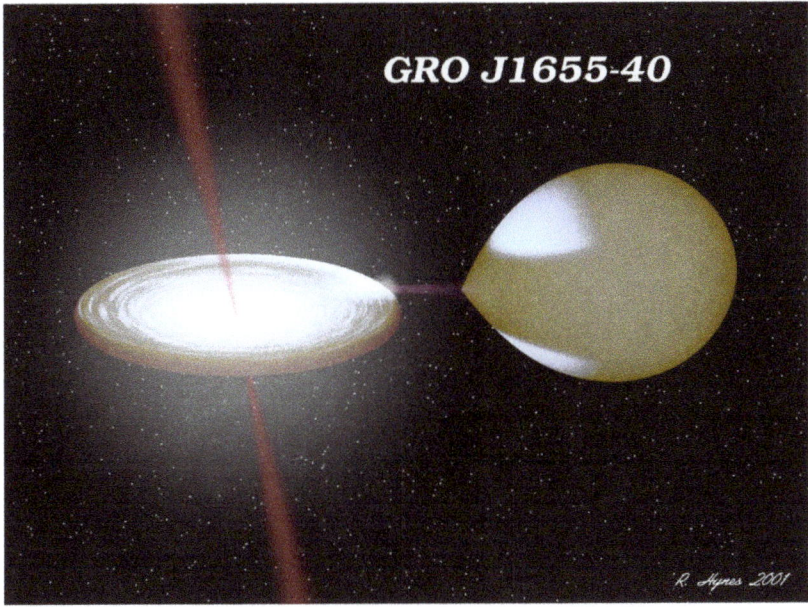

Figure 5.1. Representation of a low-mass X-ray binary, in this case GRO J1655-40, a low-mass star (F6 IV spectral type) orbiting a $6M_\odot$ black hole, this system being also a microquasar with matter ejected at relativistic velocity $v = 0.92$ c. Reproduced with permission, Credit: Robert Hynes.

$$L_{acc} = \frac{GM_1\dot{M}}{R_1} = \alpha\dot{M}c^2 => L_X \sim 10^{35} - 10^{38} \text{ erg s}^{-1} \qquad (5.1)$$

sometimes even reaching a maximum close to $\sim 10^{39}$ erg s^{-1}. An accretion disk surrounding a neutron star or a black hole emits a thermal blackbody radiation of temperature of the order of a few keV (typically between 5 and 10 keV), thus emitting in the soft X-ray domain, at energies higher than for most CV, which granted them the name of *X-ray binaries*. In practice, their radiation is not exactly the one of a blackbody, but of a multi-color blackbody, as described in Section 3.6.3.

The emission from an accretion disk dominates the broad-band spectrum, at least from X-ray to optical domain, which explains why the secondary can be detected in transient systems only when accretion onto the compact object occurs at a very low rate. In most systems, the rate between X-ray and optical luminosities is confined within the range $10^2 \leqslant \frac{L_X}{L_{opt}} \leqslant 10^4$. In addition, effects of accretion disk and irradiation of the companion by X-ray emission are substantial, the companion receiving an X-ray flux orders of magnitude greater than its intrinsic flux due to its own nuclear reactions.

5.1.3 Temporal Variability

LMXB can be divided in *persistent* sources, usually active and showing typical X-ray luminosities exceeding $L_X \sim 10^{36}$ erg s^{-1}, and *transient* sources, spending most of

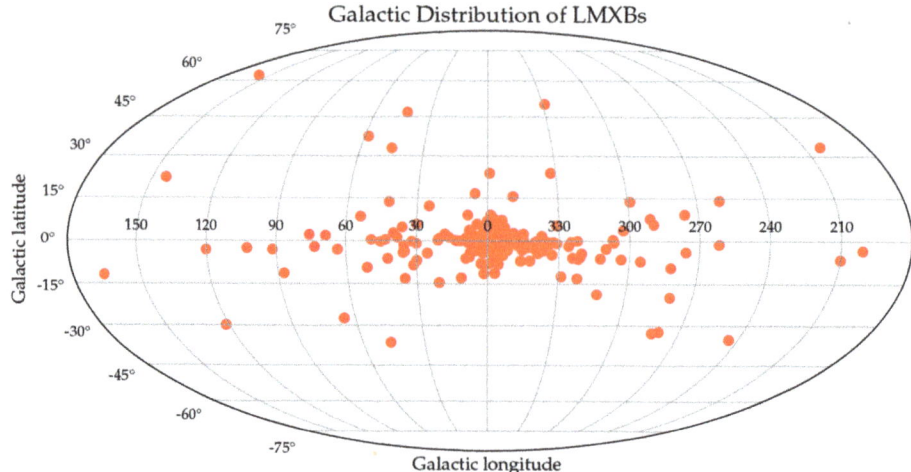

Figure 5.2. Galactic distribution, in Galactic coordinates, of LMXB, which happen to be all located within less than 2 degrees from the Galactic center. All LMXB known to date are included, based on a catalog of 218 LMXB. Reproduced with permission, Credit: Francis Fortin.

their life in quiescence ($L_X \sim 10^{30} - 10^{34}$ erg s^{-1}), with occasional bright outbursts ($L_X \sim 10^{37} - 10^{38}$ erg s^{-1}). In general, LMXB hosting neutron stars tend to be persistent sources, while black hole systems are more likely transient sources.

Temporal variability of LMXB is extremely rich, with a timescale ranging from milli-second to tens of years. While the accretion disk in these systems extends up to $10^5 - 10^7$ km, most of their X-ray emission L_X emanates from internal regions of the accretion disk, within the close environment of the compact object, inside a radius of a few tens of km from the compact object, corresponding to a period of a Keplerian orbit of only 0.5–10 ms. Characteristic velocities near the compact object $v \sim \sqrt{\frac{GM}{R}} \sim 0.5\,c$ lead to orbital motion of short dynamical timescale $\sqrt{\frac{r^3}{GM}} \sim 0.1$ ms at 15 km and 2 ms at 10^2 km from a $1.4M_\odot$ neutron star, and a timescale of $\sim$1 ms at $3R_s \sim 10^2$ km from a $10M_\odot$ black hole (van der Klis 2006). Orbital periods range from 10 minutes for tight binaries (such as a white dwarf orbiting a neutron star), up to $\sim$100 days for wide binaries. This distribution of orbital periods differs from CV, due to the presence of more evolved companions in LMXB (Figure 5.2).

5.1.4 Ellipsoidal Variations

In LMXB, the presence of a companion star is detectable when the K or M spectral type star dominates the emission in the visible domain, thus during quiescence periods. It is even possible to detect the deformation of this rotating companion star, by analyzing visible lightcurves of quiescent binary systems. This method allows observers to constrain the mass of both components (see Section 3.2.2), leading for instance to the non-ambiguous presence of a black hole as the compact object within

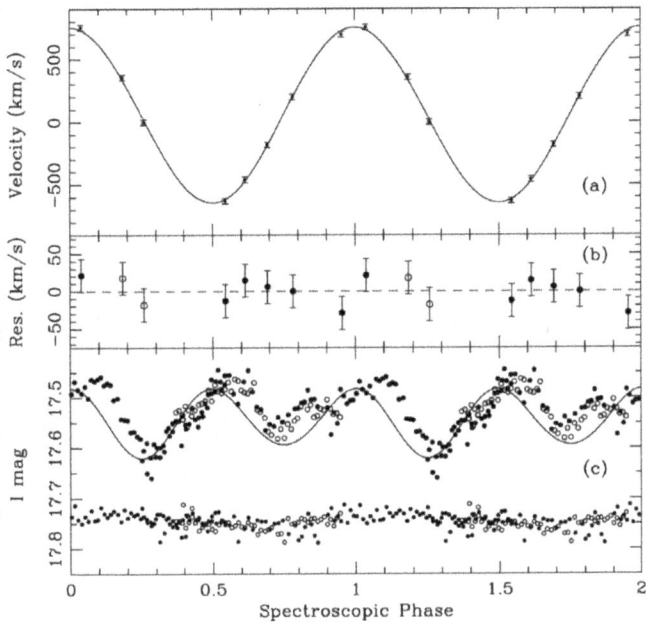

Figure 5.3. Ellipsoidal variations of the LMXB XTE J1118+480. *Top:* Radial velocity fit of photometric and spectroscopic data, folded on $P_{\rm orb}$, assuming a circular orbit of the companion star around the BH. *Middle:* Residuals between data and fit. *Bottom:* I-band lightcurve of XTE J1118+480 (top trace), with a superposed ellipsoidal model (different depths of minima due to star luminosity depending on surface gravity), and intensity of a comparison star (bottom trace). Reprinted with permission of the AAS from McClintock et al. (2001a).

binary systems, when the mass of the accretor mass is greater than the mass limit of a neutron star ($M_{\rm NS}^{\rm max} < 2.6 M_\odot$), such as A 0620-00 (McClintock & Remillard 1986), GRO J1655-40 (Orosz & Bailyn 1997), XTE J1118+480 (see Figure 5.3; and McClintock et al. 2001a), and GS 1354-64 (see Figure 5.4; and Casares et al. 2009). This is the main method used to derive the parameters reported in Table 5.1 (see Casares & Jonker 2014 for a complete description of the method).

To model the visible lightcurve of a binary system, one has to take into account the Roche geometry (the solid angle of the star varying with the orbital phase). The main parameters of such ellipsoidal variation model include: effective temperature, semimajor axis, inclination of orbit, mass ratio, radii, luminosities, etc. These models also have to take into account the gravity darkening, due to the fact that when a star is not spherically symmetric, the surface gravity is not uniform. For instance, an isolated star, flattened at its poles by a rapid rotation, will have an effective gravity (sum of gravitational and centrifugal accelerations) lower at its equator than at its poles. The same effect occurs for a star within a binary system, elongated by tidal effect: its effective surface gravity is weaker on the axis joining the two centers of mass, than on the axis perpendicular to the orbit. The direct

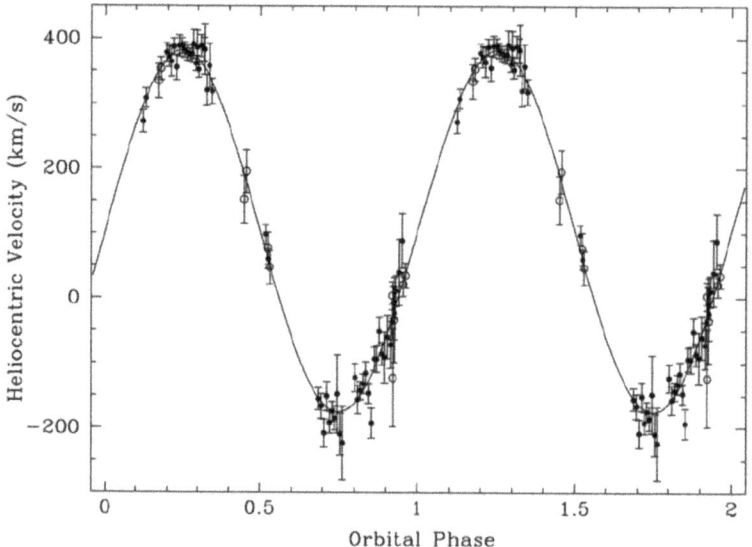

Figure 5.4. Ellipsoidal variations of the LMXB GS1354-64 (= BW Cir): Radial velocity fit of the companion star in BW Cir orbiting the black hole, folded on P_{orb}. Reprinted with permission of the AAS from Casares et al. (2009).

consequence is that the effective temperature depends on the effective gravity, with the following relationship:

$$T_{eff} \propto g^{\beta} \Rightarrow \frac{T_{equator}}{T_{pole}} = \left(\frac{g_{equator}}{g_{pole}} \right)^{\beta} \qquad (5.2)$$

where T_{eff} is the effective temperature, g the gravity, $\beta = 0.25$ for a radiative envelope, and $\beta = 0.08$ for a convective one. The gradient of gravity darkening is important when the star fills its Roche lobe, since $g = 0$ at L_1 (Lagrange point on an equipotential gravity, see Section 3.3). Atmosphere models are used to correct for center-edge darkening.

The maximum luminosity of the lightcurve occurs when the star is seen with the greatest solid angle, and minimum at conjonction or opposition. Both minima have different depth, the side of the star facing the primary being less luminous than the opposite, because of the lower surface gravity, as shown by equation (5.2). The amplitude of luminosity variation (magnitude Δm), initially computed for the black hole system A 0620-00 (McClintock et al. 1983) is:

$$\Delta m = \frac{3}{2} \frac{M_2}{M_1} f^3 \left(\frac{R_2}{a} \right)^3 \sin^2 i (1 + \tau_0) \frac{15 + u}{15 - 5u} \qquad (5.3)$$

where f is the Roche lobe filling fraction, τ_0 corresponds to gravity darkening ($\tau_0 = 0.57$ for a K5 star), and u to center-edge darkening, depending on effective temperature T_{eff} and gravity g ($u = 0.85$ for a K5 star).

Table 5.1. Dynamically-determined Binary Parameters and Effective Spin of Black Holes within LMXB Systems.

Name	Spectral type	Porb (h)	K2 (km s^{-1})	f(M1) $M_\odot$	M1 $M_\odot$	q	i (°)	$v_{rot}\sin i$ (km s^{-1})	Spin	Ref.
LMXB										
GRO J0422+32 (V518 Per)	M4-5V	$5.091\,85 \pm 510^{-6}$	378 ± 16	1.19 ± 0.02	8.5 ± 6.5	$0.11^{+0.05}_{0.02}$	$10 - 50$	90^{+22}_{-27}	—	0422
A0620-00 (V616 Mon)	K2-7V	$7.752\,337\,2 \pm 210^{-7}$	437 ± 2	2.79 ± 0.04	6.6 ± 0.3	0.074 ± 0.006	51.0 ± 0.9	—	0.12 ± 0.19	0620
GRS 1009-45 (N. Vel 93)	K7-M0V	$6.844\,9 \pm 310^{-4}$	475 ± 6	3.2 ± 0.1	$\geqslant 4.4$	0.055 ± 0.010	$37 - 80$	87 ± 5	—	1009
XTE J1118+480 (KV UMa)	K7-M1V	$4.078\,414 \pm 510^{-6}$	709 ± 1	5.27 ± 0.04	7.55 ± 0.65	0.024 ± 0.009	$68 - 79$	96^{+}_{-11}	—	1118
GRS 1124-684 (Nova Mus 91)	K3-5V	$10.382\,54 \pm 710^{-5}$	407 ± 3	3.02 ± 0.06	5.65 ± 1.85	0.079 ± 0.007	$39 - 65$	85 ± 3	$0.63^{+0.16}_{-0.19}$	1124
Swift J1357.2-0933	M2-4V	2.8 ± 0.3	967 ± 49	11 ± 2	$\geqslant 8.3$	~ 0.04	~ 90	—	—	1357
XTE J1550-564 (V381 Nor)	K2-4IV	$37.008\,8 \pm 10^{-4}$	363 ± 6	7.7 ± 0.4	11.70 ± 3.89	0.03	75 ± 4	55 ± 5	$0.34^{+0.37}_{-0.45}$	1550
XTE J1650-500	K4V	7.69 ± 210^{-2}	435 ± 30	2.7 ± 0.6	$\leqslant 7.3$		$\geqslant 47$		0.79 ± 0.01	1650
1H J1659-487 (GX 339-4)	>GIV	42.14 ± 10^{-2}	$>317 \pm 10$	5.8 ± 0.5	$\geqslant 6.0$	$\leqslant 0.125$			$\geqslant 0.95$	1659
MAXI J1659-152 (GRB 100925A)	M2-M5V	2.414 ± 510^{-3}			$4.7 - 7.8$	$0.019 - 0.053$			—	m1659
H 1705-250	K3-M0V	12.51 ± 310^{-2}	448 ± 4	4.9 ± 0.1	6.45 ± 1.55	$\leqslant 0.053$	$48 - 80$	$\leqslant 79$	$\geqslant 0.92$	1705

(*Continued*)

Table 5.1. (*Continued*)

Name	Spectral type	Porb (h)	K2 (km s^{-1})	f(M1) $M_\odot$	M1 $M_\odot$	q	i (°)	$v_{\rm rot} \sin i$ (km s^{-1})	Spin	Ref.
(N. Oph 77)										
XTE J1859+226 (V406 Vul)	K5V	$6.58 \pm 5\,10^{-2}$	541 ± 70	4.5 ± 0.6	$\geqslant 5.42$	$\leqslant 70$			—	1859
GRS 1915+105 (V1487 Apl)	K1-5III	812 ± 4	126 ± 1	7.0 ± 0.2	12.0 ± 2.0	0.042 ± 0.024	60 ± 5	21 ± 4	$0.88^{+0.06}_{-0.13}$	1915
GS 2000+251 (QZ Vul)	K3-7V	$8.258\,095 \pm 5\,10^{-6}$	520 ± 5	5.0 ± 0.1	7.5 ± 1.65	0.04 ± 0.01	$54 - 60$	86 ± 8	—	2000
GS 2023+338 (V404 Cyg)	K3III	$155.311 \pm 2\,10^{-3}$	208.5 ± 0.7	6.08 ± 0.06	$9.0^{+0.2}_{-0.6}$	0.067 ± 0.005	67^{+3}_{-1}	39 ± 1	$\geqslant 0.92$	2023

Notes. This table is an updated and completed version from Corral-Santana et al. (2016). Spin values are taken from Reynolds (2021). References: **0422:** Kuulkers et al. (1996); Gelino & Harrison (2003); **0620:** McClintock & Remillard (1986); Kuulkers et al. (1999); González Hernández & Casares (2010); **1009:** Goldoni et al. (1998); Shahbaz (1996); **1118:** Chaty et al. (2003); Hynes et al. (2000), (2003); McClintock et al. (2001b); **1124:** Orosz et al. (1996); Shahbaz et al. (1997); Chaty et al. (2002b); **1357:** Corral-Santana et al. (2013); **1550:** Dubus & Chaty (2006); Chaty et al. (2011); Orosz et al. (2002), (2011); Curran & Chaty (2013); **1650:** Orosz et al. (2004); Curran & Chaty (2012); **1659:** Grindlay (1979); Callanan et al. (1992); Chaty et al. (2002b); **m1659:** Corral-Santana et al. (2018); **1705:** Harlaftis et al. (1997); **1859:** Hynes et al. (2002); Corral-Santana et al. (2011); **1915:** Chaty et al. (1996); Mirabel et al. (1996); Martí et al. (2000); Harlaftis et al. (2001); Harlaftis & Greiner (2004); **2000:** Ioannou et al. (2004); **2023:** Khargharia et al. (2010).

5.2 Low-mass X-ray Binaries Hosting Neutron Stars

Soon after the discovery of the first X-ray source outside the solar system—Sco X-1 (Giacconi et al. 1962), which happens to be the brightest persistent accreting binary hosting a neutron star—it was realized that accretion of matter onto the surface of a neutron star was a powerful source of X-ray emission, emanating from the boundary layer between its surface and the accretion disk. These observations, related to fundamental parameters such as radius, mass, and spin of neutron stars, are a necessary tool to constrain their equation of state. There are two main ways to study the X-ray emission of neutron star systems.

- On the first hand, neutron star systems exhibit a rich phenomenology of spectral states, related to the solid surface and magnetic field anchored to this surface. Spectral variation studies allow astronomers to better understand the underlying physical mechanisms. The X-ray *color–color diagram* (CCD), splitting low and high-energy bands, is a useful tool to classify neutron star spectral states, from *hard state*—generally characterized by low variability frequencies and luminosities—toward *soft state*, with higher variability frequencies and luminosities (see Figure 5.5 and van der Klis 1998). In this section we describe the zoology of weakly magnetized accreting neutron stars, such as Atoll sources and Z sources, historically distinguished by their timing and spectral properties. The shape of the track traced in a CCD allows X-ray observers to distinguish the main differences between distinct types of sources, mainly depending on their accretion rate, significantly higher in Z sources than in Atoll sources.
- On the second hand, neutron star systems also exhibit a rich phenomenology of temporal variability and pulsations, useful to study, in order to constrain the accretion flow in the inner part of the accretion disk, closely linked to the neutron star spin. However, the first observations of LMXB hosting neutron stars did not allow to detect any pulsation nor periodic signal, because

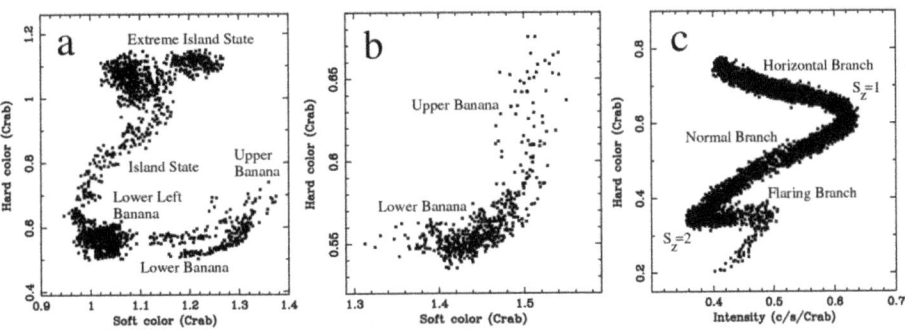

Figure 5.5. Color–color diagram of NS–LMXB from RXTE/PCA data; *left:* Atoll source 4U 1608-52; *middle:* GX Atoll source GX 9+1; *right:* Z source GX 340+0. Soft color $\frac{3.5-6}{2-3.5}$ keV; hard color $\frac{9.7-16}{6-9.7}$ keV; intensity 2–16 keV, all normalized to Crab. Reprinted with permission from van der Klis (2006).

accretion of matter onto the surface of the neutron star had spun it up so much, that it reached frequencies higher than the sensitivity limit attained by X-ray satellites such as EXOSAT and Ginga, allowing to observe timing signatures up to a maximum frequency of 200 Hz. The typical frequency sampled by more recent X-ray satellites ranges from a few mHz (ks timescale) to a few kHz (ms timescale). In particular, the Rossi X-ray Timing Explorer (RXTE) satellite, orbiting the Earth from 1995 to 2012, changed the face of the world—at least in what concerns neutron stars—discovering hundreds of aperiodic pulsations in the X-ray domain, in the $\sim$10 Hz to $\sim$2 kHz frequency range, in many accreting neutron star systems located within the Galactic bulge.

In the following of this section, we further classify LMXB hosting neutron stars, according to the characteristics of their rich spectral and temporal variability.

5.2.1 Atoll Sources

Atoll sources exhibit a weak magnetic field ($B \sim 0.1 - 1 \times 10^8$ Gauss), and accrete at low luminosity from $L \sim 10^{-3}$ to $\sim 0.5 \times L_{Edd} = 10^{38}$ erg s^{-1}. They draw a "U" or "C"-shape in the color–color diagram (see Figure 5.5, left panel), taking their name from the similarity of this curve to a coral atoll, exhibiting two main states: the curved *banana* state, characterized by a high flux and soft spectrum, and the *island* state, characterized by a low flux and hard spectrum. The banana state is further divided into *lower* and *upper* branches, at lower and higher luminosities respectively, sources moving back and forth between these two extremes on timescales of hours to a day. Accretion rate increases from island state to upper banana. At the extreme of the island state is located a hard *extreme island* state. Atoll sources exhibit X-ray bursts at $L < 2 \times 10^{37}$ erg s^{-1}. Their spectra are dominated by non-thermal Comptonized emission from a hot region close to the neutron star, accompanied by a weak thermal component from the inner accretion disk, modeled using multi-color blackbody. Comptonization come either from a transition layer between the inner disk and the neutron star (Sunyaev & Titarchuk 1985) or from an extended corona above the accretion disk (White & Stella 1988). In this context, the hard spectrum in the island state is due to a higher electron temperature of the central Comptonizing region. In addition, many Atoll sources exhibit Type I X-ray bursts (see Section 5.2.3.1).

Atoll sources exhibit coherent oscillations, their Fourier spectra showing an horizontal plateau, from low frequencies up to a break frequency $\nu_b \sim 10^{-2} - 10$ Hz. In the upper banana the power law noise of frequency $\leqslant 1$ Hz dominates, in the lower banana there are several 10 Hz band-limited noise, and in the lower left banana twin kHz *quasi-periodic oscillations* (QPO) are observed (see Section 5.2.8). At higher frequency, the spectrum decreases as a power law of index $\alpha \sim 2$, this large-band component being called *shot noise*.

In term of luminosity, in between Atoll and Z sources (see Section 5.2.2), lie the *bright* Atoll sources, also referred to as *GX* sources, characterized by higher luminosity in the range $L \sim 0.2 - 0.5 \times L_{Edd}$. These sources only exhibit lower and upper branches of the banana state (see Figure 5.5, middle panel).

5.2.2 Z Sources

Z sources exhibit higher magnetic field ($B \sim 0.1 - 1 \times 10^9$ Gauss) and luminosity—accreting near or above the Eddington luminosity $L \sim 0.5 - 1 \times L_{Edd} \geqslant 10^{38}$ erg s^{-1}—than Atoll sources. They draw, on hour to day timescale, a Z-shape in the color–color diagram, exhibiting three different spectral shapes in hardness–intensity diagram: *horizontal branch* (HB), *normal branch* (NB) and *flaring branch* (FB), reflecting physical changes within the sources, accretion rate increasing from HB to FB (see Figure 5.5, right panel). Their spectra consist of both a strong thermal component, and a non-thermal Comptonized emission. Z sources do not usually exhibit X-ray bursts, but instead display strong flares lasting a few 1000 s on the flaring branch. Hard X-ray tails (power law with a photon index $\Gamma \sim 2 - 3$) have been detected in the spectra of Z sources, while located in the HB of their CCD, with the intensity dramatically decreasing in NB and FB.

On the normal branch, the X-ray intensity and luminosity regularly increase, following the mass accretion rate $\dot{M}$. At the same time, the neutron star blackbody temperature increases by a factor 2, accompanied by radiation pressure, disruption of the inner accretion disk, and ejection of relativistic jets, detected in radio domain (see Chapter 8). In Z sources such as Sco X-1, high radiation pressure, caused by continuous flaring, explains the continuous presence of relativistic jets (Hjellming et al. 1990). On the flaring branch, the blackbody luminosity increases, suggesting an additional energy source arising from the neutron star. At such high luminosities, the mass accretion rate per surface $\dot{m}$ reaches a threshold when nuclear burning onto the surface of the neutron star becomes unstable.

Similarly to Atoll sources, Fourier spectra of Z sources exhibit an horizontal plateau, with three main types of low-frequency oscillations, detected in the different spectral states HB, NB, and FB: kHz QPO and a 15 – 60 Hz QPO (called *horizontal branch oscillation*–HBO) on the HB and upper NB, a ~6 Hz QPO (called *normal branch oscillation*–NBO) on the lower NB, and mostly power-law noise up to a break frequency $\nu_b \sim 1$ Hz. At higher frequency, the spectrum exhibits a large-band *red noise* component, decreasing as a power law of index ranging from −1 to −2.

To explain the various states between Atoll and Z sources, many parameters need to be invoked, such as differences in magnetic field, variations in the accretion rate $\dot{M}$, and/or in the inner disk radius r_{in}. We refer the reader to reviews (van der Klis 2006; Motta et al. 2017; Sazonov et al. 2020) for more details on the links between accretion states and oscillations in neutron star LMXB. In the catalog of Liu & van Paradijs (2007), there are ~25 Atoll sources, thus more numerous than the ~8 Z sources. Atoll sources tend to be fainter in radio and X-rays, with harder X-ray spectra and a more regular bursting activity (van der Klis 2006), completing the whole Z pattern along the CCD on longer timescales (years versus hours/days) than Z sources.

5.2.3 X-ray Bursters

5.2.3.1 Type I X-ray Bursts

Type I X-ray bursts are the most frequent X-ray emission from LMXB (usually called *X-ray bursters*, or even *bursters*; Martí et al. 1998; Mignani et al. 2002), constituting an unambiguous proof of the presence of a neutron star, since the matter can only accumulate on a solid surface. Several thousands of such bursts have been observed so far by various X-ray missions, for instance 1187 Type I X-ray bursts, emanating from 48 accreting neutron stars, have been observed by the RXTE satellite, during a time interval spanning more than 10 years (Galloway et al. 2008). They are due to unstable sudden thermonuclear fusion of accreted material (hydrogen and/or helium), accumulating within a shell onto the surface of the neutron star, with a relatively weak magnetic field ($B \sim 10^8$ G). The mechanism responsible for these bursts is the thermal and viscous instability of the accretion disk, similar to dwarf nova outbursts (see Section 4.2.1 and Figure 5.6), except for the gravity ($g \sim 10^6$ higher at the surface of the neutron star than in white dwarfs), and its size (a neutron star radius being $\sim 10^2$ smaller than a white dwarf radius, see Table 2.1). The mass of non-degenerate gas accumulating on the surface of the neutron star is of the order of $10^{-12} M_\odot$, i.e., 7–9 orders of magnitude smaller than in the case of white dwarfs. While temperature is high enough to allow a stable hydrogen burning in regions where the gas is not degenerate, this is not the case for helium thermonuclear reactions occurring at high temperature and density. As soon as the explosion is triggered, first localized at the hot spot of the burst, a deflagration wave quickly propagates over the whole surface of the neutron star, filled by degenerate gas, toward the NS equator and poles.

This phenomenon of Type I X-ray burst, characterized by an unambiguous fast (~ 1 s) rise, followed by a slow (from a few seconds to several minutes) quasi-exponential decay, is also less spectacular than for dwarf novae (luminosity L increasing by $10 - 10^3$ in a short ~ 10 s timescale), and a typical luminosity of $10^{39} - 10^{40}$ erg s^{-1}, reaching, or even overtaking, the Eddington limit. Of course, these bursts exhibit different profiles and energetics, depending on the composition of the accreted material, and on the local accretion rate (Strohmayer & Bildsten 2006). During the fast (~ 1 s) rising phase of the burst, when the accreted material is burned in a few seconds, atmospheric shells of the neutron stars are pushed away by the radiation pressure, to an altitude of ~ 100 km, before gradually falling back on the surface, the burst then experiencing an exponential decrease, as seen in Figure 5.7. The X-ray spectrum is thermal, the spectral evolution indicating a temperature decay with time, with a recurrence time of a few hours to day-timescale between outbursts. Such a short recurrence time indicates that all accreted material is not burned during a burst, with a partial deflagration on the NS surface, producing periodic pulsations. In contrast, for some sources, called *hidden bursters*

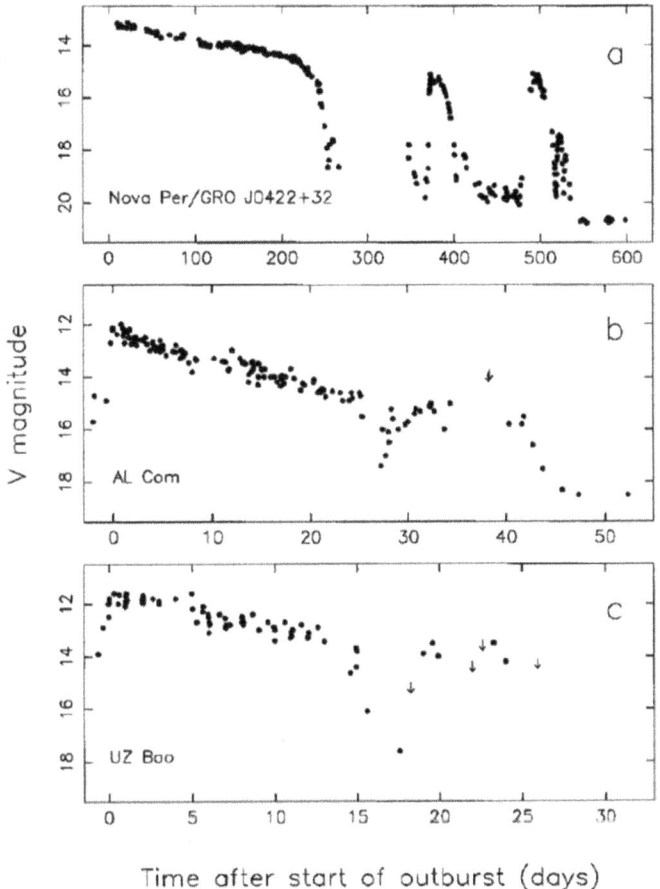

Figure 5.6. Outburst lightcurves of soft X-ray transient GRO J0422+32 (Nova Per 1992) (*top*), and dwarf novae AL Com (*middle*) and UZ Boo (*bottom*). Note the similarity in the lightcurves, with both the SXT and dwarf novae exhibiting, after the general decay of the main outburst, a temporary minimum, followed by subsequent increase to a high flux level. Reprinted with permission of the AAS from Kuulkers et al. (1996).

or *burst-only* sources, only one burst was detected, with no detection of any appreciable accretion flux, sometimes even within a few hours of the burst. The likely explanation is that in this case, thermonuclear explosions occur with a long characteristic recurrence time of several months to tens of years, due to slow accumulation of a critical mass of material, required for ignition on the NS surface (Strohmayer & Bildsten 2006).

More than 100 Type I X-ray bursters are currently known,[1] and as pointed out by Sazonov et al. (2020), with the development of wide-field X-ray monitor instruments on board satellites over the last thirty years, capable of detecting thermonuclear bursts from distant regions of our Galaxy, it is likely that most of the bright,

---

[1] See updated list in https://personal.sron.nl/~jeanz/bursterlist.html

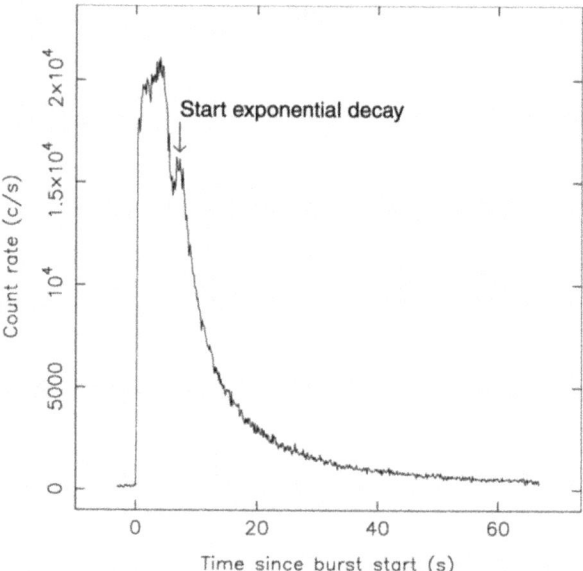

Figure 5.7. Typical lightcurve showing the profile of a Type I X-ray burst from the neutron star LMXB 2S 0918-549, from the start of the burst. Note the complex profile at the beginning of the burst, followed by an exponential decay during the rest of the burst, explained by the cooling of the neutron star surface, heated by the explosion of accumulated gas. Reprinted with permission of the AAS from Jonker et al. (2001).

persistent, and regularly flaring, bursters have already been discovered within the Milky Way.

In these X-ray bursters, we can distinguish four distinct cases of ignition (Galloway et al. 2017):

1. He-ignition in mixed H/He accreted material;
2. He-ignition in pure He accreted material, following exhaustion of accreted H by steady burning;
3. ignition in (almost) pure He accumulated from an evolved donor in an ultra-compact X-ray binary (see Section 5.2.6);
4. ignition of carbon fuel produced as a by-product of H/He bursts (*superburst*).

Transient coherent milli-second pulsations, at a frequency between ~270 to 620 Hz, are usually detected in the decrease phase of these sources. This frequency increases, in a few seconds, by 1–2 Hz, toward an asymptotic value, which likely corresponds to the spin period $P_{\rm spin}$ of the neutron star. These pulsations are due to anisotropic emission of matter, projected from the hot spot at an altitude of a few tens of meter above the surface of the neutron star, with a velocity lower than neutron star spin because of angular momentum conservation, before falling back onto the surface of the neutron star, and recovering their initial velocity corresponding to the neutron star spin. For a 10 km radius neutron star, spinning at 400 Hz, the rotation velocity at the equator is $v_{\rm spin} \sim 0.1$ c. Measurements of the spin have shown that the variation rate of the frequency of an accreting neutron star is of the order of $\sim 2 \times 10^{-6}$ Hz yr^{-1}.

We point out the interesting system **IGR J17062-6143**, exhibiting *intermediate duration* Type I X-ray bursts, and on which Neutron Star Interior Composition Explorer (NICER) observed the tail end of the X-ray burst cooling phase, with a very low flux—relatively to its average value—during a 3-day period right after the burst. Bult et al. (2021) interpret this intensity dip as the inner accretion disk gradually restoring itself, after being perturbed by the burst irradiation. For other examples of intermediate duration Type I X-ray bursters, see Alizai et al. (2020). X-ray observations of NS–LMXB in quiescence also allowed observers to measure temperature and luminosity of neutron stars, such as in the binary system SAX J1750.8-2900. The spectrum, fit with a classical blackbody, and a non-magnetized, pure hydrogen neutron star atmosphere model, led to a bolometric luminosity of $\sim 10^{34}$ erg s^{-1}, making it the most luminous NS–LMXB in quiescence, likely due to an unusually high core temperature of the NS (Lowell et al. 2012).

5.2.3.2 Type II X-ray Bursts

Type II X-ray bursts are much rarer, observed in sources such as the *rapid burster* MXB 1730-33, and the *bursting pulsar* GRO J1744-28, an LMXB constituted of a neutron star orbiting an evolved giant star (Court et al. 2018). They are characterized by a different lightcurve (or burst profile) from Type I X-ray bursts: lasting for tens of seconds, they start and stop abruptly with no gradual decay from peak, contrary to Type I bursts which exhibit a rapid rise followed by a slow decline. They can also show rapid successions of bursts, with a recurrence time of a few minutes, amounting up to ~ 20 bursts per hour, followed by a long quiescence period. These bursts, with a harder X and γ-ray spectrum, do not show any temporal spectral evolution. They are likely associated to viscous instabilities within the accretion flow, leading to an increase in the accretion rate from the companion star toward the compact object (Lewin et al. 1988, 1993; Strohmayer & Bildsten 2006).

5.2.4 Dippers

Several LMXB, called *dippers* or *dipping sources* (such as Cyg X-2, Casares et al. 2010; Bałucińska-Church et al. 2011), present occultations in the form of periodic temporal decrease at orbital modulations in the X-ray lightcurve, which are called *dips* (22 such sources in Liu & van Paradijs 2007), as well as *partial/total eclipses* (13 sources in Liu & van Paradijs 2007). Since these dips in X-ray intensity take place at every orbit, and are usually associated with high inclination systems ($i \geqslant 60°$), they are interpreted as photo-electric absorption of the X-ray source, likely by cold and/or partially ionized matter of the accretion flow, concentrated within the outer part of the accretion disk.[2] During the dip phenomenon, the dominant Comptonized emission decreases, suggesting an extended emitting region of variable thickness. Dip ingress timing, representing the beginning of the occultation, show that the radial size of the accretion disk corona ranges from 10^4 to 7×10^5 km, based on

---

[2] Dips are thus different from eclipses, which occur when the X-ray emission emanating from the central source is blocked by the companion star.

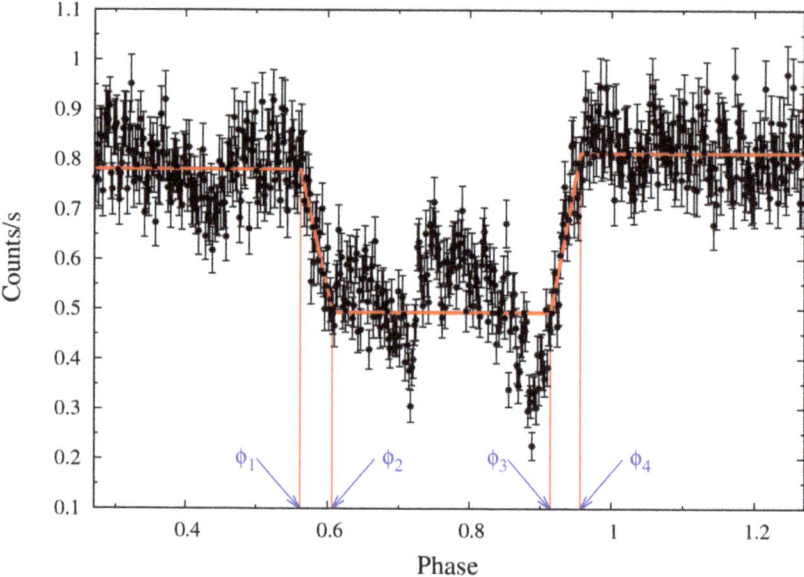

Figure 5.8. Chandra X-ray lightcurve, folded on the orbital period, of the dipping neutron star LMXB 4U 1323-619. The red line is the best-fit function of the dip. The blue arrows highlight the phases of ingress (ϕ_1 and ϕ_2) and egress (ϕ_3 and ϕ_4) of the dip. Reproduced with permission from Astronomy & Astrophysics, © ESO (Gambino et al. 2016).

X-ray luminosity of the dipping source during X-ray bursts (Church et al. 2014). While the shape of a dip is generally irregular and varies from one cycle to another (see Figure 5.8), their periodic occurrence in the lightcurve is strictly connected to the orbital motion, both the X-ray source and the disk being fixed in the co-rotating frame of the binary system (Gambino et al. 2016, 2017).

5.2.5 Accreting Milli-second Pulsars

Coherent quasi-periodic oscillations (QPO) were detected by RXTE during X-ray bursts, at a frequency of a few hundreds of Hz, revealing the high spin period P_{spin} of neutron stars within LMXB, of the order $\sim 1-5$ ms. These *milli-second pulsars* usually exhibit weak magnetic field in the range $10^8 - 10^{11}$ G, two orders of magnitude lower than in normal pulsars, characteristic of a relatively old stellar population.

An isolated pulsar continuously radiates its rotational energy, and as a consequence naturally slows down. With a characteristic decay time of magnetic field of the order of 10^6 years, milli-second pulsars simultaneously appear as the oldest pulsars when considering their magnetic field, and the youngest ones when taking into account their rotation. They were then proposed to be the outcome of the evolution of an old neutron star accreting from a companion star within an LMXB, spinning up toward $1.6 \leqslant P_{spin} \leqslant 10$ milli-seconds, due to transfer of mass and angular momentum during long ($\sim 0.1-1$ Gyr) X-ray bright accretion phases (Papitto et al. 2020). LMXB are thus considered as the natural progenitors to recycle old neutron stars into rapidly spinning, weakly magnetized, neutron stars, such as SAX J1808.4-3658, the first *accreting milli-second X-ray pulsar* (AMSP) detected within an LMXB (Wijnands & van der Klis 1998). As long as $P_{spin} > P_{eq}$ (the equilibrium period), further accretion of matter onto the surface of the neutron star remains possible: while ordinary single pulsars have a typical lifetime of 10 to 100 Myr, the recycled milli-second pulsars present a lifetime of several Gyr, due to their low values of spin down rate $\dot{P}$, because of weak magnetic braking. In the course of its long life with a fast spinning pulsar, the companion star can be either totally accreted, or evaporated by the intense X-ray emission of the pulsar, giving birth to systems called *black widows*.

Other AMSP have been studied since, such as IGR J17591-2342, detected by the INTEGRAL satellite, which exhibits an unusual outburst profile with multiple peaks in the X-rays. A combined study, using Chandra and NICER satellites, allowed to reveal an outflowing wind at a velocity of 2800km s^{-1}, with a neutron star which likely formed from the collapse of a white dwarf, producing a rare calcium-rich Type Ib supernova explosion (Nowak et al. 2019).

Multi-wavelength observations are necessary to reveal the nature of these sources, first in X-rays, such as for instance Chandra observations of IGR J17511-3057, a NS–LMXB among the few AMSP exhibiting X-ray bursts (Paizis et al. 2012), and then in optical/infrared domain, to pinpoint the nature of the system by unambiguously identifying the spectral type of the companion star (Curran & Chaty 2011).

5.2.6 Ultra-compact X-ray Binaries

Ultra-compact X-ray binaries (UCXB) constitute a sub-class of LMXB characterized by short orbital periods $P_{orb} \leqslant 1$ hr, implying a hydrogen-poor donor (which can be either an O/Ne/Mg, C/O or He white dwarf, or even a non-degenerate He star), of low mass $M_2 \sim 0.01 M_\odot$, accreting onto a NS or a BH, with luminosities $L_X \leqslant 7 \times 10^{36}$ erg s^{-1}. Most of known UCXB exhibit thermonuclear X-ray bursts, implying accretion onto a neutron star. In addition, some UCXB are accreting milli-second X-ray pulsars. Since they combine a short orbital period with a peculiar chemical composition, UCXB are unique astrophysical laboratories, especially concerning binary evolution. Indeed, these sources are expected to be strong emitters of gravitational waves, within the frequency range $(10^{-4} - 1$ Hz) accessible to the ESA Laser Interferometer Space Antenna (LISA) mission (Tauris 2018). While their predicted number in our Galaxy amounts up to 10^6 (Nelemans & van Haaften 2013), we only know 38 confirmed or candidate UCXB (Sazonov et al. 2020).

5.2.7 Symbiotic X-ray Binaries

We already mentioned symbiotic systems in Section 4.2.1 (in the chapter on CV), evolved binaries in which an evolved red giant orbits a white dwarf (Munari 2019). The equivalent, in the realm of LMXB, is again an evolved late-type (K1—M8) red giant, this time orbiting a magnetized neutron star, or even a black hole. Such symbiotic systems are unambiguously identified by their peculiar optical spectra inherent to the nature of the evolved red giant. A complete list of symbiotic X-ray binaries, along with their parameters—P_{spin}, P_{orb}, L_X and distance—is given in table 1 in Yungelson et al. (2019). In this book is also presented a model of population synthesis via their binary evolution, which involves surviving a common envelope (CE) phase, and the kick of the supernova event. From this model, only a few dozen such systems are expected in our Galaxy. Indeed, unambiguously identified systems are rare, with only a few percent of symbiotic systems hosting a neutron star, instead of a white dwarf. Even rarer, a symbiotic X-ray binary composed of a K-type giant companion orbiting a black hole, has been identified in the INTEGRAL source IGR J17497-2821 (Paizis et al. 2007).

5.2.8 Variability Properties

5.2.8.1 Different Kinds of Variability

Accreting binaries hosting neutron stars exhibit very rich X-ray variability, on a wide range of timescales from minutes down to milli-seconds, corresponding to frequencies from mHz to kHz. This variability is constituted of broad structures called *broad-band noise* and narrow features, detected in the Fourier domain *power density spectra* (PDS) $P_\nu(\nu)$ (squared modulus of the Fourier transform of the lightcurve), with power spectra subtracted from Poisson noise and fit with empirical models, such as multiple superposed Lorentzian functions (Belloni et al. 2002). We can distinguish:

- *Power-law noise* follows a power law $P_\nu \propto \nu^{-\alpha}$, with a power-law index, or *slope* α, typically between 0 and 2. $\frac{1}{f}$ noise is characterized by $\alpha = 1$, white noise is constant with $\alpha = 0$, and red noise corresponds to $\alpha = 2$, or any kind of noise where P_ν decreases with ν.
- *Band-limited noise* corresponds to noise steepening toward higher frequency, either abruptly with a *break* at frequency ν_{break}, or gradually. *Flat-topped noise* describes (approximately constant) white noise, and *peaked noise* describes noise for which P_ν has a local maximum at $\nu \geqslant 0$.
- *Quasi-periodic oscillations* (QPO) are finite-width peaks in the PDS, described by a Lorentzian, characterized by the centroid frequency of the fundamental peak (ν_0, corresponding both to the Lorentzian peak frequency and the first harmonic), full width at half maximum (w in Hz, inverse of coherence length), amplitude (r), and a quality factor $Q = \frac{\nu_0}{w}$, describing both how narrow a peak is, and the number of successive coherent oscillations. By convention, an oscillation is called QPO when the quality factor $Q \geqslant 2$, and peaked noise when $Q \leqslant 2$. A *sharp* QPO is characterized by a high Q. QPO simultaneously detected at different energy bands sometimes exhibit phase lags, due to a time lag at their emission. By convention, lags are called *hard* when low-energy photons precede high-energy ones, and *soft* on the opposite.

Since QPO are produced in the vicinity of relativistic objects such as neutron stars (and also black holes, see Section 5.3.5), they are tightly correlated with spectral states, reflecting oscillations of the turbulent accretion flow in the strong field regime, in the General Relativity frame. They bring us information on the internal structure of the accretion disk, such as over-density within the disk, rotating at Keplerian velocity. Because they yield accurate centroid frequencies linked to traveling waves, inhomogeneous motions, and/or accretion-related timescales, they allow to explore the accretion flow, which would instead remain inaccessible via spectral analysis alone. PDS of accreting neutron stars, such as the ones revealed by the RXTE satellite, show numerous QPO components, with a few dominant features in more than 25 binary systems: band-limited white noise, low-frequency (Hz) oscillations, and high frequency (kHz) oscillations. The Z source Sco X-1 exhibits for instance these three types of oscillations in most of the states: a break at ~6 Hz, low-frequency QPO at ~45 and ~90 Hz, and twin high-frequency QPO in the range ~800 – 1100 Hz (see Figure 5.9). We now describe QPO properties at various frequencies.

- *Low-frequency (Hz) QPO* (LFQPO), first detected in the neutron star system GX 5-1 (van der Klis et al. 1985), are observed in the range 5–60 Hz, with quality factor $Q > 2$, and amplitudes between 1 and 10%. They are observed in horizontal branch (HB)—horizontal branch oscillations (HBO)—normal branch (NB)—normal branch oscillations (NBO)—and flaring branch (FB) —flaring branch oscillations (FBO)—of Z sources (see right panel of Figure 5.10, and van der Klis 2006 for a complete review). In the HB branch, there exists a strong correlation between frequency and X-ray intensity. This correlation has been explained in the context of the magnetospheric beat-frequency model (see Section 5.2.9). QPO of frequency ~1 Hz are sometimes detected in dipping systems.

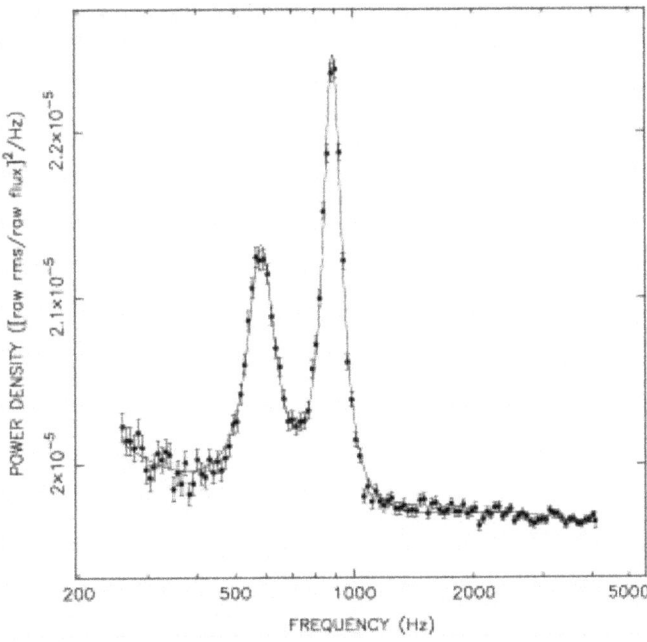

Figure 5.9. PDS of Sco X-1 showing double kilohertz QPO peaks, with best fit superimposed. Reprinted with permission of the AAS from van der Klis et al. (1997).

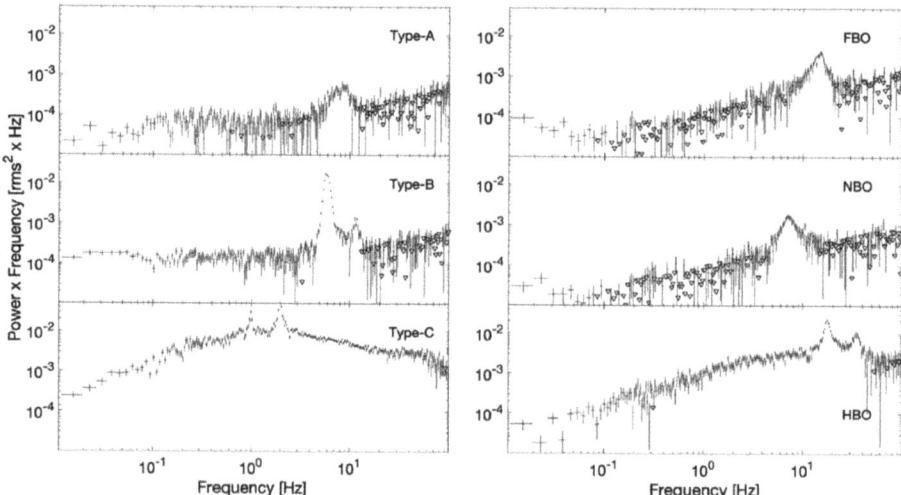

Figure 5.10. *Left:* PDS showing LFQPO detected in black hole systems, *top:* XTE J1859+226, *middle and bottom:* GX 339-4. *Right:* PDS showing QPO detected in neutron star systems, *top/middle:* GX17+2, *bottom:* Cyg X-2. FBO: flaring branch oscillations; NBO: normal branch oscillations; HBO: horizontal branch oscillations. Reprinted with permission from Elsevier, Ingram & Motta (2019).

- *Middle-frequency (hHz) QPO* (MFQPO), observed as a coherent band-limited white noise with an approximately constant frequency in the range between 100 and 200 Hz, are detected with similar characteristics in most states of Atoll sources, and in accreting milli-second pulsars, such as SAX J1808.4-3658. They are characterized by a low quality factor, sometimes reaching $Q = 2$, thus coherent enough to be classified as QPO. The similarity of these QPO across various sources points toward a link with neutron star fundamental properties.

- *High-frequency (kHz) QPO* (HFQPO), the fastest variability components detected in accreting binaries with frequencies ranging from 200 to 1300 Hz, cannot be related to the neutron star spin, since their frequency varies for a same source. They are however of similar order than the dynamical timescale of the innermost regions of the thin accretion disk, allowing astronomers to investigate the motion of clumps of matter orbiting in the strong gravitational field of the neutron star, and of hotspots co-rotating on the neutron star surface. Such QPO were first detected in the sources 4U 1728-34 and Sco X-1, with peaks at ∼700 Hz and ∼1100 Hz, thus apparently not connected (at least directly) with the NS spin period. Many kHz QPO have been detected, and although only one peak is sometimes detected, they often come in a pair, separated by several hundreds of Hz. These *twin peak* QPO (respectively *upper* and *lower* kHz QPO) move together up and down in the 200 – 1300 Hz frequency range in the power spectrum, in correlation with the source state. Such QPO are detected in all Z sources—in HB and upper NB—in Atoll sources—in island state and lower left banana state—and in a few milli-second pulsars, down to the lowest accretion rate $\dot{M}$, becoming undetected at the highest $\dot{M}$. The quality factor of these coherent QPO is very high for Atoll sources, reaching up to $Q \sim 200$, and lower in Z sources: $Q \sim 2-10$, increasing with the QPO frequency.

5.2.8.2 Frequency Correlations

Frequency correlations between various QPO are seen both in Atoll and Z sources, in some cases over nearly three orders of magnitude in frequency, suggesting that they are originating from similar physical phenomena (Wijnands & van der Klis 1999; van der Klis 2006).

- A correlation exists between twin (upper and lower) high frequency kHz QPO, with low-frequency ($\leqslant$80 Hz) QPO (HBO and NBO) and/or broad noise features in neutron star LMXB, especially in Atoll and Z sources. In particular, the frequencies between low-frequency QPO and lower kHz QPO follow a tight correlation (see Figure 5.11), which might find its origin in an empirical model based on physical dependence of various frequencies, due to a combination of the sonic point and magnetospheric beat-frequency models (see Section 5.2.9). A similar relationship for dwarf nova ($P_{QPO}/P_{DNO} = 15$; Warner et al. 2003) suggests a similar mechanism underlying such oscillations in systems of distinct nature (see Figure 4.5, showing this correlation between dwarf nova, NS–LMXB and BH–LMXB oscillations). Frequencies of kHz QPO are strongly correlated with other timing and spectral features (Psaltis et al. 1999a).
- A correlation also exists between the upper (ν_u) and lower (ν_l) twin high frequency kHz QPO (see Figure 5.12), fitted either by a power-law function

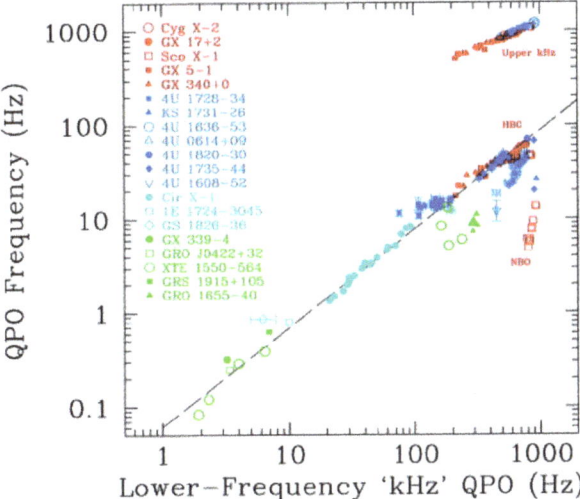

Figure 5.11. Frequency correlation between low-frequency and high-frequency (lower kHz) QPO in LMXB, including Z sources (in red), Atoll sources (in blue), other NS–LMXB (in cyan) and BH–LMXB (in green). The dashed line represents the frequency correlation between HBO and lower kHz QPO in Z sources in the 200–550 Hz range, extrapolated by $\geqslant$2 orders of magnitude toward lower frequencies. The error bars are typically similar to the size of the symbols. Reprinted with permission of the AAS from Psaltis et al. (1999a).

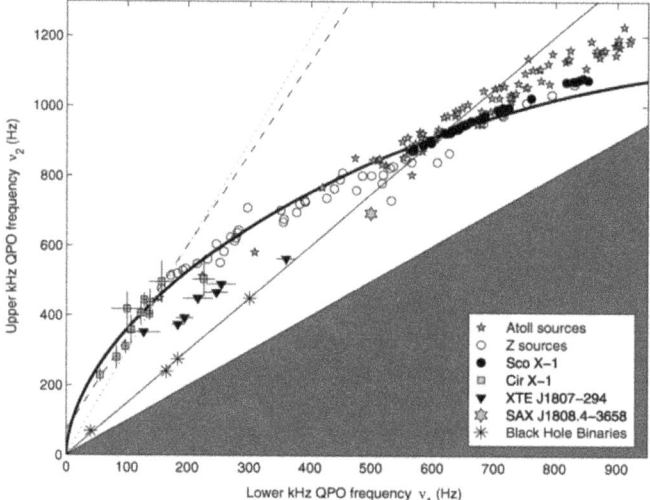

Figure 5.12. Frequency correlation of the twin (upper versus lower) kHz QPO in NS and BH systems. The thin solid (dotted) line represents a fixed 3:2 (3:1) ratio. The dashed line is a fit to the Cir X-1 points (excluding the highest frequency). The thick solid line represents the relation between the periastron-precession and Keplerian frequencies for $M = 2M_\odot$. In the relativistic precession model, the two frequencies respectively correspond to the lower and upper kHz QPO. Asterisks indicate simultaneous high-frequency QPO in BH systems. Reprinted with permission from Belloni et al. (2007).

$\nu_l \sim \nu_u^b$ with different values of b index for systems of distinct nature, such as Atoll and Z sources (Psaltis et al. 1998; Zhang et al. 2006), or by a linear relationship $\nu_u = a\nu_l + b$ (Belloni et al. 2007). While the frequency of these twin HFQPO increases with luminosity on a hour-timescale, the frequency difference between them $\Delta\nu = \nu_u - \nu_l$ decreases in some sources—like for instance in Sco X-1—and increases in other—like Circinus X-1.

5.2.9 QPO Models

Various theoretical models have been imagined to explain both the presence of these QPO and their correlations. Before describing the main models, we first give the timescale and lengthscale corresponding to the orbital motion of a test particle within the innermost region of the accretion disk surrounding a compact object such as a neutron star, occurring at the Keplerian frequency:

$$\nu_{orb} = \sqrt{\frac{GM}{4\pi^2 r_{orb}^3}} \sim 1200\text{Hz}\left(\frac{r_{orb}}{15\text{ km}}\right)^{-3/2}\sqrt{\frac{M}{1.4M_\odot}} \tag{5.4}$$

and at the Keplerian radius:

$$r_{orb} = \left(\frac{GM}{4\pi^2\nu_{orb}^2}\right)^{1/3} \sim 15\text{km}\left(\frac{\nu_{orb}}{1200\text{ Hz}}\right)^{-2/3}\left(\frac{M}{1.4M_\odot}\right)^{1/3}. \tag{5.5}$$

These timescale and lengthscale are of the same order than those implied by the kHz QPO detected in LMXB hosting neutron stars, in the range 200–1300 Hz, naturally suggesting that these QPO might arise from the orbital motion of accreting matter very close to the compact object. Two main explanations arise: first, the interaction of the innermost accretion flow with the magnetosphere of the neutron star, with frequencies of the twin peaked QPO directly related to orbital epicyclic frequencies in a 3:2 (or 3:1) parametric resonance (Abramowicz & Kluźniak 2003; Wijnands et al. 2003), and second, frame-dragging relativistic effects, and different oscillation modes of surrounding plasma, in the frame of the relativistic precession model (Stella & Vietri 1999).

5.2.9.1 Magnetospheric Beat-frequency Models

This model invokes the interaction of clumps with the magnetosphere of the neutron star, a thick corona consisting of cool and dense plasma, to explain the commensurability of the twin lower (ν_l) and upper (ν_u) frequency kHz QPO. On the first hand, ν_u is identified as the frequency of dense clumps of matter orbiting at Keplerian velocity, in the innermost boundary of the accretion flow close to the surface of the neutron star, at the magnetic radius r_{mag} at which the magnetic field drains the accretion flow onto the poles (assuming the same rotation direction of neutron star and disk). On the second hand, ν_l would correspond to a frequency due to a resonance—a rotational beat interaction—between the neutron star spin ν_{spin} and general relativistic orbital and epicyclic frequencies (Abramowicz & Kluźniak 2003; Wijnands et al. 2003). There cannot exist any higher-frequency QPO, because no stable orbit exists at smaller radius, the plasma being directly accreted onto the surface of the neutron star. This model is based on a nearly constant frequency difference between the twin kHz QPO $\Delta\nu = \nu_u - \nu_l$, close to the coherent frequency observed during X-ray bursts, and interpreted as the spin frequency of the neutron star ν_{spin} (or $\nu_{spin}/2$, depending on the source). Since the magnetic axis is not aligned with the rotation axis of the neutron star, X-ray emission is modulated when the magnetic axis crosses the orbiting clump, at the beat frequency $\nu_{beat} = \nu_u - \nu_{spin}$, identified as ν_l. This model was further developed, to be able to interpret the 15–60 Hz HBO in Z sources.

An alternative *sonic-point* beat-frequency model invokes sound speed and drifting velocity of accreting material toward the neutron star. Here the frequency of the lower kHz QPO is the difference between the Keplerian frequency at a radius close to the sonic point and the neutron star spin frequency. This model naturally explains the difference between twin kHz QPO frequencies, close, but not equal, to ν_{spin} (Psaltis et al. 1999b; Bozzo et al. 2009).

5.2.9.2 Relativistic Precession Model

The *relativistic precession model* (RPM) invokes general relativity in the strong field regime, within the accretion flow in the vicinity of the spinning neutron star: a test particle orbits in a plane which is mis-aligned with respect to the equatorial plane of neutron star, due to relativistic Lense–Thirring effect, dragging the surrounding spacetime, and inducing nodal precession and disk oscillations in inclined orbits.

This model, developed by Stella & Vietri (1999) to explain high-frequency QPO in black hole systems such as GRO J1655-40 and XTE J1550-564 (see Figure 5.18, and Section 5.3.5 on QPO detected in black hole systems), can also explain kHz QPO in neutron stars. In this model, ν_u is identified as the orbital frequency, and we find either $\nu_{\text{burst}} = \nu_{\text{spin}}$ or $\nu_{\text{burst}} = 2 \times \nu_{\text{spin}}$, depending on the presence of one or two hotspots on the neutron star surface, during the burst. In addition to the spin, other parameters such as width and amplitude of the QPO, coherence, energy spectrum, and time lags, have to be taken into account, to constrain underlying physical mechanisms (see, e.g., Ingram & Motta 2019). As already noted by Wijnands & van der Klis (1999), the fact that some correlations occur in neutron stars as well as black holes likely means that they arise within the accretion disk. This model has been widely tested, see for instance Psaltis et al. (1999b), Barret et al. (2005), Belloni et al. (2007), and van Doesburgh & van der Klis (2019), for a comparison between magnetospheric beat-frequency model and Lense–Thirring effect.

5.3 Low-mass X-ray Binaries Hosting Black Holes

LMXB hosting black holes are also called *black hole binaries*. There are ~60 BH systems known within our Galaxy, ~20 hosting dynamically confirmed black holes via radial velocity obtained with spectro-photometric observations (see Section 5.1.4, and a list of dynamically confirmed black holes within LMXB systems in Table 5.1, updated from Corral-Santana et al. 2016, and completed), and ~40 hosting candidate black holes, sharing many observational properties with confirmed ones. The total number of LMXB hosting black holes in our Galaxy has been estimated, through a population synthesis study, to be greater than 10^4 (Yungelson et al. 2006).

After the discovery of Cyg X-1 in 1964—a rare persistent X-ray source ($L_X \geqslant 10^{37} \text{ erg s}^{-1}$) hosting a black hole, see Chapter 1—additional and more frequent transient systems were discovered during their outbursts. Among them were A 0620-00, the first transient source hosting a dynamically confirmed black hole, LMC X-1 and LMC X-3 in the Large Magellanic Cloud, and then GX 339-4, V 404 Cygni, Nova Muscae 1991, GRO J1655-40, XTE J1859+226, H 1743-322, XTE J1752-223, 4U 1630-47, IGR J17091-3624, etc. These systems (see Table 5.1) are discovered in two ways: either serendipitously at early stages of their bright X-ray outburst (in the hard spectral state) at a rate of ~2 outbursts per year on average, and then studied thanks to simultaneous multi-wavelength observing campaigns, or during surveys and monitoring observations of the Galactic plane and center. When a system enters an outburst, its X-ray luminosity significantly increases, peaks up to luminosity $L \sim 10^{38} - 10^{39} \text{ erg s}^{-1}$, which can be higher than the Eddington luminosity, and then decays, following the varying accretion rate onto the black hole, which increases by a few orders of magnitude. In between outbursts, lasting from weeks to months, it enters a much longer quiescent phase, characterized by a low X-ray luminosity $L_X \sim 10^{30} - 10^{31} \text{ erg s}^{-1}$. These systems have a recurrence period ranging from several years to decades, with a low duty cycle of a few percent.

High-energy (first in X-rays up to 10 keV, then to a few MeV) observations, with instruments such as Ginga, CGRO, Granat, RXTE, XMM-Newton, Chandra and

INTEGRAL, began to reveal common properties of such systems: their variability on short timescales, and significant photometric and spectroscopic changes. These observations initially led to the identification of black hole high(-soft) and low (-hard) spectral states, based on observations of Cyg X-1, a peculiar source exhibiting a bimodal variability, remaining most of the time in low-hard and intermediate state (Malzac et al. 2006), until eventually more accurate observations required to add more states in order to fully describe these systems.

The first studies of X-ray spectral states, based on the modeling of the optically thick accretion disk emission via a superposition of multi-color blackbody spectra (see Section 3.6), allowed to measure the inner radius of the disk surrounding the black hole (Mitsuda et al. 1984), and later to reveal the connections between X-ray (arising mainly from the inner part of the accretion flow), infrared (outer part of the accretion flow), γ-ray (from the corona), radio (from the base of the outflow) emission (Mirabel et al. 1998). Over the decades of studying these systems via multi-wavelength facilities, it is now possible to connect the physical components at the origin of the X-ray energy spectra, and the short-timescale variability. However, such studies are rendered difficult by the fact that outbursts, spectral states, and temporal variability differ significantly from one source to another, or even from one outburst to another, for the same source.

5.3.1 Soft X-ray Transients

Black hole systems are usually transient sources, also called *soft X-ray transients*, or historically *X-ray novae* (for instance, the 1916 outburst of A 0620-00 was believed to be a classical nova—Nova Mon 1916—because of the similarity between optical and X-ray lightcurves). They remain undetected during periods going from a few months, to a few decades, with a luminosity less or of the order of $10^{32} - 10^{33}$ erg s^{-1}, and then reach the Eddington limit $L_X \sim 10^{38}$ erg s^{-1}, and simultaneously brighten in optical by up to 10 magnitudes during active phases, which last up to several weeks to months, similar to optical novae (Chaty et al. 2002a, 2006; Chaty & Bessolaz 2006; Paizis et al. 2011, 2015). Multi-wavelength observations of soft-X-ray transients during their outburst also allow to reveal irradiation of the accretion disk (Dubus et al. 1999, 2001), as clearly revealed by the systems XTE J1859+226 (Hynes et al. 2002) and XTE J1818-245 (Zurita Heras & Chaty 2011).

One outstanding case is the black hole candidate XTE J1118+480, for which a wealth of multi-wavelength observations were obtained, from radio to X-rays, including ultraviolet emission, thanks to a very low absorption, the source being located in the direction of Ursa Major, far from the Galactic plane (see the lightcurve in Figure 5.13 and the SED in Figure 5.14, both from Chaty et al. 2003) and see also Hynes et al. (2000, 2003), Esin et al. (2001), McClintock et al. (2001b), and Vieyro et al. (2016).

During quiescent phases, the secondary can be detected, allowing observers to measure the mass of the compact object (see Section 5.1.4 and Table 5.1). For more than 20 sources (such as A 0620-00, GRO J1655-40, and GRS 1915+105—which has shown decade-long outbursts above $0.5L_{\rm Edd}$), this mass is significantly higher

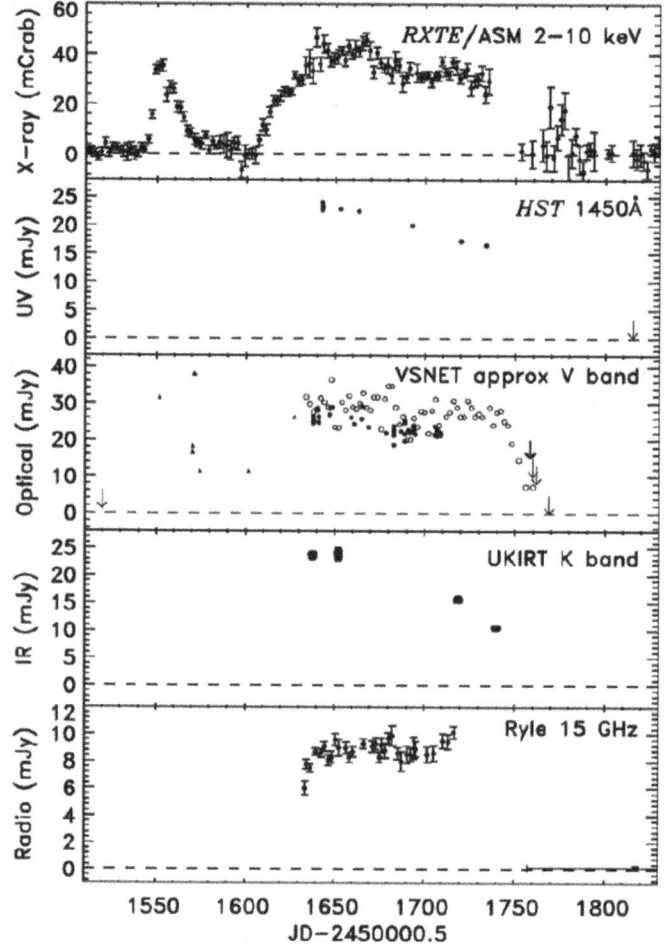

Figure 5.13. Multi-wavelength lightcurve of the black hole LMXB XTE J1118+480 during its outburst from March to November 2000. From top to bottom: X-ray RXTE/ASM 2–10 keV, far-UV HST/STIS 145 nm, optical VSNET ~V band, infrared UKIRT K band, and radio Ryle telescope 15 GHz data. Reprinted with permission from Chaty et al. (2003).

than $3M_\odot$, thus pointing toward a black hole. Orbital periods go from 8 hours to 6.5 days. Many points still have to be understood, such as the exact role of irradiation of accretion disk (Dubus et al. 1999), its geometrical form (such as warped disk), etc. LMXB hosting black holes have been largely compared to those hosting neutron stars, since both systems share many common properties, for instance in term of accretion and ejection processes, allowing to disentangle which property is characteristic to the fundamental nature of a black hole as an accretor, compared to a neutron star.

Only three BH–LMXB (1E 1740.7-2942, GRS 1758-258—both of them micro-quasars located in the Galactic bulge, see Chapter 8—and 4U 1957+115) are considered persistently bright X-ray sources (Tetarenko et al. 2016), since they did not switch on or off, during more than 30 years of continuous observation.

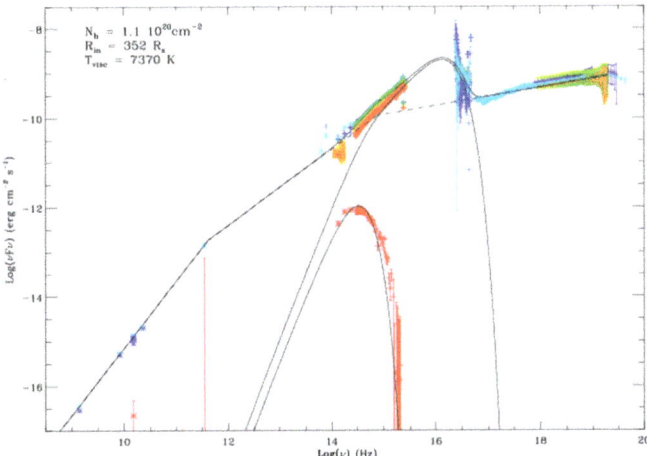

Figure 5.14. Spectral energy distribution of the BH–LMXB XTE J1118+480 during its outburst from March to November 2000, fit with a multi-color blackbody disk model: outer disk temperature T_{out} = 7370 K and inner disk radius of R_{in} = 352R_s, corresponding to a low-hard state. The absorption on the line of sight is N_H = 1.1 × 10^{20}cm^{-2}. The two solid lines in radio and X-ray domains correspond to non-thermal contributions, with power laws of spectral indices α = 0.5 and −0.8, respectively. The lower curve corresponds to the emission of the M1 V companion star. Multi-wavelength data come from VLA, RT, JCMT, UKIRT, HST, EUVE, Chandra and RXTE. Reprinted with permission from Chaty et al. (2003).

5.3.2 Modeling the Emission

Accreting binaries hosting black holes exhibit spectral variations in the X-ray domain (∼0.1 – 200 keV), which can be related to distinct accretion states, in the frame of a simple model constituted of three main components:

1. a *geometrically thin and optically thick accretion disk* (Shakura & Sunyaev 1973), truncated at a certain radius—the inner disk truncated radius—larger than the innermost stable circular orbit (ISCO, see Section 3.6.1), at the origin of persistent thermal multi-color blackbody emission at energy ⩽5 keV, peaking at a few keV, with the blackbody temperature increasing when approaching the black hole (see Section 3.6.1 for this type of accretion disks).

2. a *geometrically thick, optically thin and hot accretion flow*, made of a cloud of thermalized hot electrons surrounding the black hole, transferring energy by inverse Compton scattering, to the soft X-ray seed photons from the accretion disk, creating a highly variable, non-thermal, power-law hard X-ray emission at energy ⩾6 keV, with a high-energy cut-off linked to the characteristic Comptonizing electron temperature. This accretion flow, initially called *corona* in reference to the magnetic solar corona, has been modeled using ADAF prescription (see Section 3.6.2 for this type of accretion disks). The geometry and location of this large-scale accretion flow is not clear, either above the disk, or at the base of the outflow.

3. a *reflection component* can also be seen, superimposed to the spectral continuum, resulting from the irradiation of the hot flow emission on the

accretion disk. The irradiated flux is reprocessed in the disk upper atmosphere, leading to emission of characteristic features, including Fe K α fluorescence line at energy $\sim$6.4 keV, other Fe ionization states, and a broad *Compton hump* peaking at $\sim$20 – 30 keV. Both the relativistic motion of the accretion flow within the disk, and the gravitational pull of the black hole, cause distortion of the observed reflection spectrum, leading to an asymmetrically broadened iron line profile. In some sources, the reflection component includes radiation reprocessed within a local absorber (more details in Ingram & Motta 2019).

5.3.3 Evolution of the Outburst

We now describe the general evolution of an LMXB hosting a black hole during an outburst, including the characteristics of its spectral states and temporal properties. The long-term (daily to yearly) variability of black hole LMXB in outburst follow repeating cyclical patterns, based on the configuration of the accretion flow, according to the classification originally proposed by Homan & Belloni (2005).

5.3.3.1 Hardness–Intensity Diagram

We first have to introduce an useful tool, the X-ray *hardness–intensity diagram* (HID) where the X-ray intensity (luminosity, flux, or count rate) is plotted versus the hardness ratio (HR, ratio of intensity measured in two different energy bands, usually hard X-ray band $\sim$6 – 10 keV to soft X-ray band $\sim$4 – 6 keV) of the X-ray spectrum. HID are commonly used to describe the behavior of LMXB hosting black holes and neutron stars, and even AGN. As noted by Motta et al. (2021), HID can be viewed as the equivalent of Hertzsprung–Russell diagram, with the fundamental difference that while stars evolve in the HR diagram through different phases of their evolution in millions of years, LMXB evolve in the HID through different states in week-to-year timescale.

5.3.3.2 From LHS to HSS, and Back Again

Hardness–intensity diagrams reveal a hysteresis with a typical *Q-shaped* track, which is drawn counter-clockwise from the bottom right. We show for instance the lightcurve and hardness ratio of the 2002–2003 outburst of GX 339-4 in Figure 5.15, and its corresponding HID in Figure 5.16. At the beginning of the outburst, just awaking from the low-luminosity quiescence, the system is in the *low-hard state* (LHS, right vertical branch of HID in Figure 5.16), during which the X-ray spectrum has a hard power law shape up to a few tens of keV, signature of non-thermal processes in a hot, optically thin, plasma (i.e., the corona). The radio emission observed in this state, extending up to the infrared, is believed to arise from a steady, *compact* (unresolved) and mildly-relativistic jet of small size (a few tens of a.u.) and flat/inverted radio-to-infrared spectrum expected from a stratified self-absorbed synchrotron jet (Blandford & Königl 1979; see Section 3.5.2). Such a jet model based on synchrotron emission in the low-hard spectral state has been adjusted to fit the broad-band spectrum from radio to X-rays of a few black hole

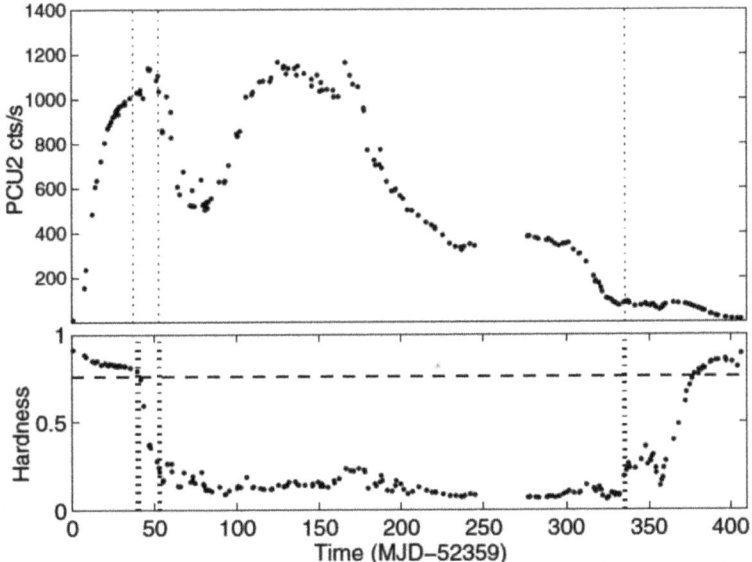

Figure 5.15. RXTE count rate lightcurve and hardness ratio of the black hole LMXB GX 339-4 during its 2002–2003 outburst. The dotted lines indicate major state transitions, and the horizontal dashed line shows the hardness level corresponding to the transition from the right branch to the left branch in the HID (see Figure 5.16). Reproduced with permission from Astronomy & Astrophysics, © ESO (Belloni et al. 2005).

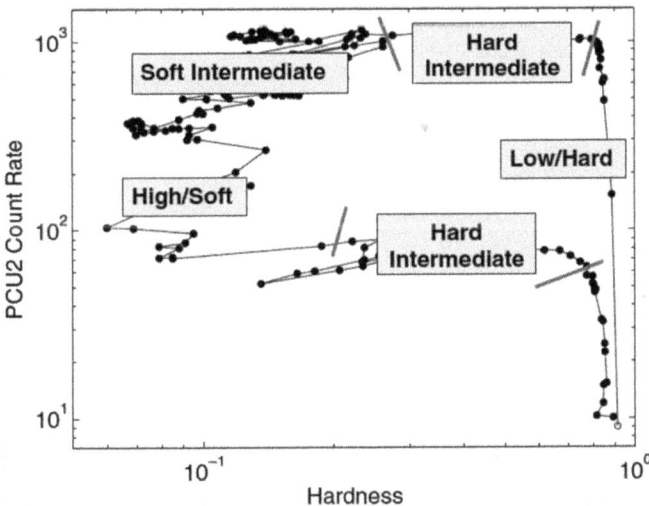

Figure 5.16. Hardness–intensity diagram of the black hole LMXB GX 339-4, with a typical hysteresis behavior, drawn counter-clockwise from bottom-right corner, with transitions marked by gray segments. The branches corresponding to the various basic states in the Q-shaped track are labeled. Reprinted with permission from Homan & Belloni (2005).

systems, in particular XTE J1118+480 (Markoff et al. 2001), GX 339-4 (Markoff et al. 2003), and also the systems Swift J1753.5-0127, GRO J1655-40, and XTE J1720-318 (Zhang et al. 2010).

Triggered by a rise in the accretion rate, due to disk instability, the source transits from the LHS to the HSS state, as explained by various models. For instance, the magnetic disk-outflow model assumes a large-scale magnetic field in the outer thin disk, advected by the inner advection dominated accretion flow (ADAF), and ejecting a fraction of the gas into the outflow, with a field strength of the outer disk proportional to the accretion rate of the disk (Cao et al. 2021). The outburst then goes on with the *high-soft state* (HSS, left branch), which usually closely follows, or even includes, the outburst peak. In this state, quasi-equatorial and highly ionized winds, detected through X-ray absorption lines, are launched at relativistic velocities. During the evolution of the outburst, the transitions between the two main (low and high) states take place at two different flux levels: first at high flux, when the source moves from LHS to HSS, and second at low flux, the outburst finishing off when the source comes back from HSS to LHS, after having completed a full hysteresis cycle. During the high flux transition (corresponding to the top branch), the source crosses first the *hard intermediate state* (HIS) and second the *soft intermediate state* (SIS), which are characterized by different timing properties. It is during these intermediate states that powerful transient relativistic radio ejections are observed, often resolved as synchrotron emitting blobs in high-resolution images obtained in radio (see Figure 8.6). We can even add to this list an additional *very high state* (VHS) or *ultra-luminous state* (ULS), characterized by a high luminosity, sometimes exceeding the Eddington limit. To summarize, the general evolution of an outburst follows the track from LHS, via HIS and SIS, up to HSS, and then back to SIS, HIS, to finish with LHS.

Some sources exhibit weaker secondary outbursts in the HSS, following the initial outburst by some weeks to a few months. These outbursts, sometimes called *re-flares* or *re-brightening*, are similar to those in CV or LMXB hosting neutron stars, thus suggesting a disk instability origin (Dubus et al. 2001).

5.3.3.3 *Failed Outbursts*

In addition, a few transient LMXB hosting black holes sometimes undergo unusual *failed (—transition)* outbursts, during which they do not go through the main accretion spectral states, basically remaining within the LHS during the whole outburst, without reaching the HSS. Most of these outbursts are under-luminous, which may suggest that they do not complete the cycle, because of early decrease of mass accretion rate, as seen during the 2001 and 2002 outbursts of XTE J1550-564 (Chaty et al. 2011; Curran & Chaty 2013), but also during the 2002 outburst of XTE J1650-500 (Curran & Chaty 2012), the 2008 outburst of H 1743-322 (Chaty et al. 2015), and during the 2012–2013 outburst of Swift J1745-26 (Chaty & Fortin 2020).

More generally, from studying a sample of 56 BH–LMXB undergoing 128 outbursts (using data from RXTE/PCA, Swift/BAT and MAXI), Alabarta et al. (2021) found that 36% of BH LMXB of their sample experienced at least one such failed outburst, which in total represent 33% of all the outbursts of the sample (see

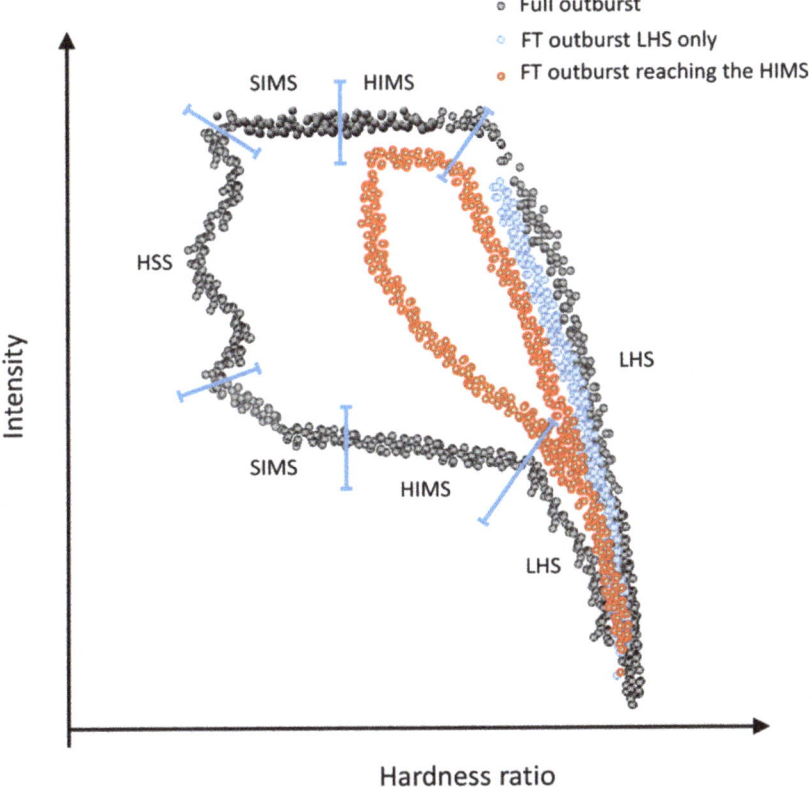

Figure 5.17. Hardness–Intensity diagram of full (gray) and failed (red and blue) BH–LMXB outbursts. The red track represents the sources that leave the LHS but do not reach the HSS, while the blue track represents the sources that stay in the LHS. Reprinted with permission from Alabarta et al. (2021).

Figure 5.17). This clearly suggests that failed outbursts are common events. They also show that failed and full outbursts cannot be distinguished from their X-ray lightcurves, HID or X-ray variability during the initial $10 - 60$ days after the outburst onset, suggesting that both types of outbursts are driven by the same physical process. There is however a difference between failed and full outbursts, revealed by the optical and infrared emission of BH–LMXB, as shown by the system GX 339-4, which exhibits higher optical and infrared flux before a failed outburst, compared to a full outburst. Alabarta et al. (2021) suggest that the optical/infrared flux is related to a physical process which can discriminate whether the outburst will eventually fail, or go through the whole cycle.

5.3.4 Spectral States

We now give details on the various spectral states encountered by LMXB hosting a black hole, following the outburst track (see for instance Alabarta et al. 2021):

1. *Quiescent state* (QS): for low accretion rate $\dot{m} < 10^{-2}$, the inverse Compton effect is weak, the X-ray flux smaller than the optical flux, and the radiative

efficiency is low, with most of the energy taken away out of black hole horizon.

2. *Low-hard state* (LHS): for accretion rate ranging in the interval $10^{-2} < \dot{m} < 10^{-1}$, this state is characterized by a combination of a weak (or even absent) X-ray thermal accretion disk and a dominant non-thermal spectrum with a hard power law (photon index $\Gamma \sim 1.4 - 2$) and a high-energy cut-off around $E \sim 100$ keV, produced by the inverse Compton process. This state corresponds to a simultaneous increase of both the inner radius of the truncated accretion disk and of the hot inner accretion flow. The luminosity and radiative efficiency rapidly increase with $\dot{m}$. Another signature of this state is its large amplitude of short-timescale variability, rms amounting up to $\sim 20 - 30\%$ rms in the $0.1 - 100$ Hz band, with a centroid frequency of a few mHz at low luminosity.

3. *Hard/soft intermediate states* (HIS/SIS): for accretion rate approaching the critical value $\dot{m}_{\text{crit}} \sim 10^{-1}$, when the source evolves from LHS to HSS, these states are characterized by a decrease of the inner disk truncation radius, the hot accretion flow being confined within central regions. The thermal emission from the accretion disk rises, and the emitted spectrum softens. In contrast, when the source evolves back from HSS to LHS, the inner disk truncation radius increases, the hot accretion flow extending from central to outer regions. The thermal emission from the accretion disk progressively decreases, and the emission from the hot flow increases, accompanied by a hardening of the spectrum. In these two states both thermal and non-thermal contributions are present, each component continuously decreasing or increasing, depending on the outburst evolution. The short-timescale variability decreases from the LHS to HSS track, with rms in the range $\sim 10 - 20\%$ in the HIS (centroid frequency ~ 10 Hz), and rms $\sim 5 - 10\%$ in the SIS (centroid frequency $\sim 5 - 6$ Hz), due to different amount of seed photons involved in the inverse Compton effect.

4. *High-soft state* (HSS): for high accretion rate $\dot{m} \geqslant \dot{m}_{\text{crit}}$, this state is characterized by the dominant emission of an accretion disk extending up to the ISCO, surrounding a reduced spherical and low density hot accretion flow. This creates a combination of a soft X-ray (peaking at $\sim 1 - 5$ keV) thermal spectrum emanating from the inner part of the accretion disk, and a weak non-thermal hard X-ray power law arising from a small population of electrons constituting the hot flow, the majority of which having cooled down after inverse Compton scattering by low-energy seed photons. In this state there is limited short-timescale variability amplitude rms ($\leqslant 5\%$).

5. *Very high state* (VHS): for accretion rate reaching the Eddington limit $\dot{m} \sim \dot{m}_{\text{Edd}}$, this state, also called the *ultra-luminous state* (ULS), is similar to the high-soft state, with hot corona surrounding the inner part of the disk. This state exhibits a low amplitude of short-timescale variability, with $\sim 5 - 10\%$ rms in the $0.1 - 100$ Hz band, and a centroid frequency amounting up to 30 Hz.

5.3.5 Variability Properties

The short—less than minute—timescale variability in LMXB hosting black holes is fully revealed in the Fourier domain, especially using the *power density spectrum* (PDS) already mentioned in Section 5.2.8 on variability of LMXB hosting neutron stars. PDS from black hole LMXB display various features, from broad-band noise spanning several decades in frequency ($\sim 1 - 100$ Hz), to narrow QPO features ($\sim 0.1 - 450$ Hz). QPO in black hole systems were first discovered—even before their detection in neutron star systems—in GX 339-4 (see Figure 5.11) with X-ray observations using the HEAO satellite (Samimi et al. 1979), and then during one of the first multi-wavelength campaign, involving Ariel 6 X-ray and optical ESO observations (Motch et al. 1983). Since QPO are related to specific spectral states and transitions, they allow astronomers to better understand the physical conditions at the basis of accretion states.

QPO in black hole systems are usually divided into two large groups, based on their frequency range: the *low-frequency QPO* (LFQPO with centroid frequency ⩽30 Hz, sub-divided into three types: A, B, and C) and the *high-frequency QPO* (HFQPO with centroid frequency ⩾60 Hz, see Figure 5.18).

5.3.5.1 LFQPO

LFQPO are the most commonly observed QPO within BH–LMXB, with high signal-to-noise ratio. While Type-A LFQPO are weak and broad, Type-B and C LFQPO are strong and narrow features. Type-A LFQPO are the least common type of QPO in black hole systems, with only 10 significant detections in the whole RXTE archive. Type-C LFQPO are the most common type, detected in all accretion states —especially HIS and LHS—apart from the SIS (which is closely associated to Type-B LFQPO), with their centroid frequency highly dependent on the spectral state, probably linked to propagating fluctuations in the mass accretion rate. Type-C LFQPO are also detected in most of the electromagnetic spectrum, at ultraviolet, optical, and infrared wavelengths. A similarity exists between Type-A, B, and C

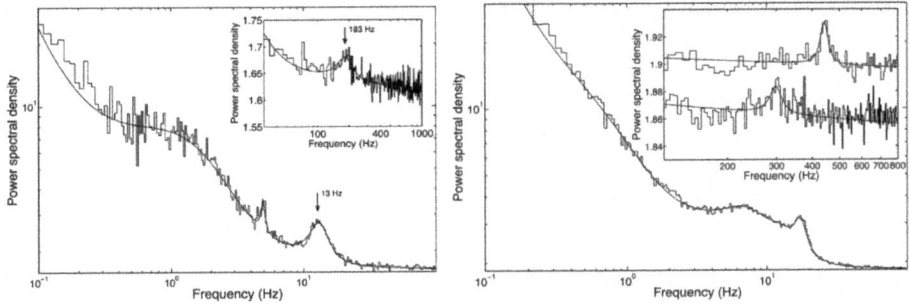

Figure 5.18. QPO detected in PDS of two black hole systems. *Left:* PDS of XTE J1550-564 shows three simultaneously detected narrow QPO: a Type-C QPO at ~ 13 Hz, a Type-B QPO at ~ 5 Hz in large panel, and an HFQPO at ~ 183 Hz in the inset. Figure reproduced from Motta et al. (2014b). *Right:* PDS of GRO J1655-40 shows three simultaneously detected narrow QPO: a Type-C QPO at ~ 17 Hz and the two (lower and upper) HFQPO at ~ 300 and ~ 440 Hz in the inset. Reprinted with permission from Motta et al. (2014a).

LFPQO detected in black hole systems, suggested to be the analogs of FBO, NBO, and HBO QPO, respectively, detected in neutron star systems (see Figure 5.10, and Ingram & Motta 2019).

5.3.5.2 HFQPO

HFQPO, representing the fastest black hole variability, are rarely detected in black hole systems, being weak and transient features, contrarily to neutron star systems. The most prolific source for HFQPO is the microquasar GRS 1915+105, with the first HFQPO detected in 1997 at 67 Hz (Belloni et al. 1997a, 1997b). Two additional sources exhibit HFQPO detected at significant level, and at even higher frequencies than GRS 1915+105: XTE J1550-564 (HFQPO at ∼180 and ∼280 Hz; Motta et al. 2014b) and GRO J1655-40 (HFQPO at ∼300 and ∼450 Hz; Motta et al. 2014a, see Figures 5.11 and 5.18). HFQPO, seen in the IS in black hole systems are likely explained in the frame of the *relativistic precession model* (already mentioned in Section 5.2; Stella & Vietri 1999), their frequency and broad-band noise components being commensurate (resonance 3:2 or 5:3) with the general relativistic epicyclic frequencies of accretion flow moving in close vicinity of the spinning black hole (from the ISCO to a few tens of gravitational radii R_S), with the spacetime dragged by the Lense–Thirring effect. This model allows to determine the spin of the black hole: $a = 0.34 \pm 0.04$ for XTE J1550-564 (Motta et al. 2014b) and $a = 0.290 \pm 0.003$ for GRO J1655-40 (Motta et al. 2014a), see also Reynolds (2021).

Using a modified RPM model, Rink et al. (2021) interpret the low-frequency QPO from XTE J1550-564 and GRO J1655-40 as nodal precession frequencies, correlated with the simultaneously observed high-frequency QPO in these black holes systems (see Figure 5.19). They also constrained deviations from general relativity, in the strong field regime, to less than one part per thousand. More details on black hole short-timescale variability can be found in Ingram & Motta (2019).

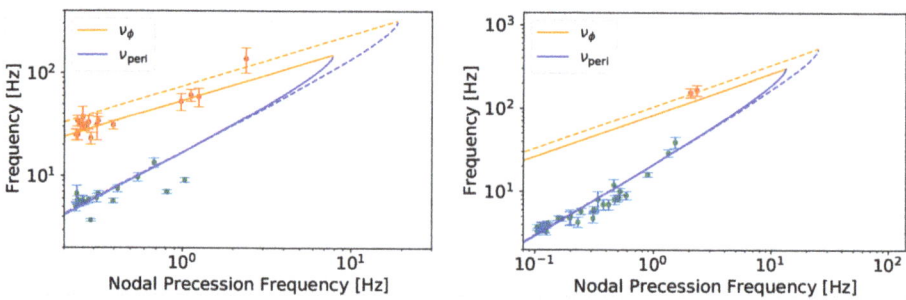

Figure 5.19. *Left:* high-frequency QPO of XTE J1550-564, versus simultaneously observed low-frequency QPO, interpreted as nodal precession frequencies. Dashed curves give the values predicted by the RPM model ($M = 9.1 M_\odot$ and $\frac{a}{M} = 0.34$; Motta et al. 2014b), and solid curves trace the best-fit model ($M = 19.0 M_\odot$ and $\frac{a}{M} = 0.31$). The *Kepler* frequency and the periastron precession frequency (ν_ϕ and $\nu_{\rm peri}$ respectively) are equal at the ISCO. *Right:* same figure, for GRO J1655-40, with dashed curves showing values predicted by the RPM model ($M = 5.3 M_\odot$ and $\frac{a}{M} = 0.29$; Motta et al. 2014a), and solid curves tracing the best-fit model ($M = 9.1 M_\odot$ and $\frac{a}{M} = 0.27$). Reprinted with permission from Rink et al. (2021).

5.3.5.3 *Mapping the Geometry of the Accretion Flow*

Combining both spectral fitting and fast X-ray variability observations allows observers to produce a more complete and accurate model, such as the *full spectral-timing model* developed by Kawamura et al. (2022) to map the geometry of the accretion flow of the black hole system MAXI J1820+070 during a bright low-hard state, as observed by NICER and NuSTAR. This model reproduces the variability by propagating mass accretion rate fluctuations in a spectrally inhomogeneous hot flow. By including reflection, it allows to reproduce the lag-frequency spectra, first by propagation through the variable flow (hard lagging soft at low frequencies), and second by reverberation from the hard X-ray continuum illuminating the disk (soft lagging hard at high frequencies). Kawamura et al. (2022) note that they obtain from this model a light travel time corresponding to a distance of $\sim 45 R_S$, in agreement with the truncated disk model geometry for the low-hard state.

5.3.6 Black Hole Horizon

There are a few observational signatures, allowing observers to distinguish between NS and BH systems, especially from their X-ray behavior. For instance, the detection of Type-I X-ray bursts unambiguously classifies the object as a NS–LMXB, while an X-ray nova-like behavior suggests a BH LMXB. There is another possibility to unambiguously distinguish NS from BH, which is to use the existence of the event horizon, a fundamental property of black holes.

In a hot accretion flow (i.e., of ADAF-type, see Section 3.6.2), the viscous energy is mainly stored in protons, within the flow. When accretion occurs onto a black hole, this energy is never radiated, and instead lost forever, as soon as it reaches the horizon event of the black hole. On the other hand, in case the flow encounters a solid surface, as in the case of accretion onto the surface of a neutron star, the energy is radiated, releasing a luminosity $\sim 0.1 \dot{m} c^2$. An absence of reprocessed radiation on a solid surface can thus be considered as the likely evidence of a black hole horizon. In this context, let us compare the luminosities in two cases, first during an outburst, and then during quiescence:

1. *During an outburst*, the accretion rate is high in both cases (neutron star and black hole systems), with $\dot{m} \sim 1 > \dot{m}_{crit}$. For a constant $\dot{m}$, observed luminosities L_{max} are proportional to the mass m of the compact object. Binaries hosting black holes, of higher mass than neutron stars, should thus exhibit L_{max} greater than for neutron star systems.

2. *During quiescence*, the accretion rate is low: $\dot{m} < \dot{m}_{crit}$ for both types of systems. However, there exists a difference: in the case of a neutron star, its surface reprocesses the radiation from the accretion flow, its luminosity remaining proportional to $\dot{m}$, while in the case of a black hole, the energy is lost with the flow crossing the event horizon. This results in a quiescent luminosity L_{min} lower for a black hole than for a neutron star system, with a luminosity proportional to $\dot{m}^2$, and an amplitude of luminosity variation ($L_{max} - L_{min}$) greater for a black hole than for a neutron star.

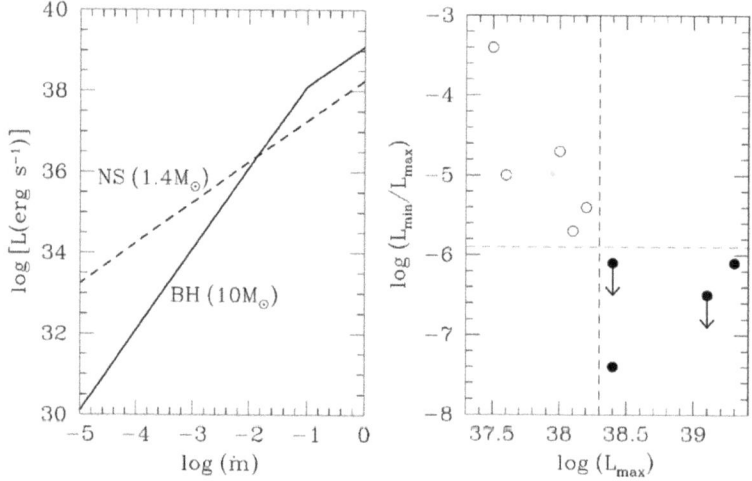

Figure 5.20. *Left:* expected luminosity variation of NS and BH systems versus the Eddington-scaled mass accretion rate $\dot{m}$ (the solid line corresponds to a $10M_\odot$ BH, and the dashed line to a $1.4M_\odot$ NS). *Right:* luminosity range L_{min}/L_{max} versus outburst luminosity L_{max} of NS (open circles) and BH (filled circles) systems. The vertical dashed line represents L_{Edd} for a $1.4M_\odot$ NS, and the horizontal dashed line shows that BH exhibit a lower luminosity range than NS systems, suggesting, in the frame of ADAF model, the evidence of an event horizon in BH systems. Reprinted with permission of the AAS from Narayan et al. (1997).

With the accretion rate $\dot{m}$ varying by a few orders of magnitude between outburst and quiescence, BH are expected to exhibit a substantially larger luminosity variation than NS systems, as it seems to be confirmed by comparing the observed luminosities of both types of systems, hosting black holes and neutron stars, as reported in Figure 5.20 (Narayan et al. 1997).

5.3.7 Fundamental Plane of Black Hole Activity

Multi-wavelength observing campaigns have revealed that a non-linear correlation exists between X-ray and optical, infrared and radio luminosity. This correlation, originally discovered in GX 339-4 (Corbel et al. 2000), and then detected in many other black hole systems, rapidly became known as the *fundamental plane* of black hole activity (see Figure 5.21), since it allows to quantify related contributions, on one side the thermal emission coming from the accretion disk, and on the other side the non-thermal emission arising from jets (Gallo et al. 2003; Russell et al. 2006). We show in Figure 5.22 the optical/infrared versus L_X version of this fundamental plane, including data from H1743-322, which experienced a failed outburst, remaining in the low-hard state during its whole outburst (Chaty et al. 2015), which explains why it follows an under-luminous track compared to other sources in the LHS. Another microquasar and black hole system, XTE J1550-564, exhibited a faint X-ray outburst in April 2003, remaining in the LHS. Optical and infrared observations reveal an abrupt change of slope from the infrared domain to the optical, attributed to a jet break, transiting from optically thick to thin regime, leading to an estimated

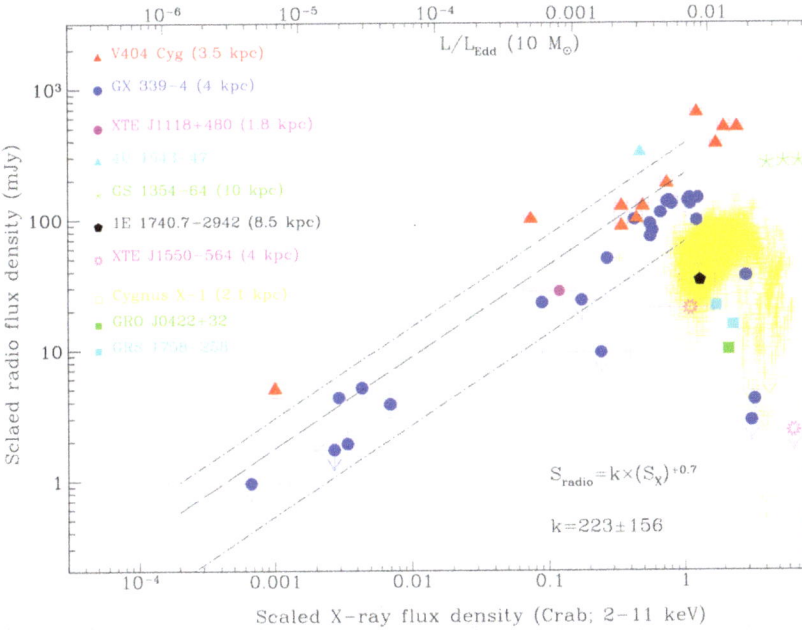

Figure 5.21. Fundamental plane of black hole activity, represented with radio versus X-ray (L_X) flux density diagram for LMXB hosting black holes in low-hard state. Reprinted with permission from Gallo et al. (2003).

characteristic size $\geqslant 2 \times 10^8$ cm and a magnetic field $B \leqslant 5 \times 10^4$ G for the jet base, assuming a homogeneous one-zone synchrotron model (Chaty et al. 2011).

5.3.7.1 What about Neutron Stars?

A tentative version of the fundamental plane of neutron star activity is shown is Figure 5.23; based on a sample of 36 accreting binaries hosting neutron star, observed in radio. This sample is constituted of 13 weakly magnetized ($B \leqslant 10^{10}$ G) and 23 strongly magnetized ($B \geqslant 10^{10}$ G) neutron stars, more than doubling the previous sample observed at current-day sensitivities (van den Eijnden et al. 2021). On the first hand, this study shows that strongly magnetized neutron stars are systematically fainter in radio than weakly magnetized neutron stars, not exceeding $L_R \sim 3 \times 10^{28}$ erg s^{-1}, and confirms that all magnetized neutron stars are fainter than accreting stellar-mass black holes (see Figure 5.23). On the second hand, it shows that strongly magnetized neutron star binaries do not reveal any global correlation between X-ray and radio luminosity, which van den Eijnden et al. (2021) caution may be a result either of sensitivity limits, or of effects of neutron star spin and magnetic field on radio luminosity and jet power.

5.3.7.2 Polarization

Finally, a useful tool to investigate properties of these black holes during low-hard states is through polarimetry at various wavelengths, in particular in the infrared

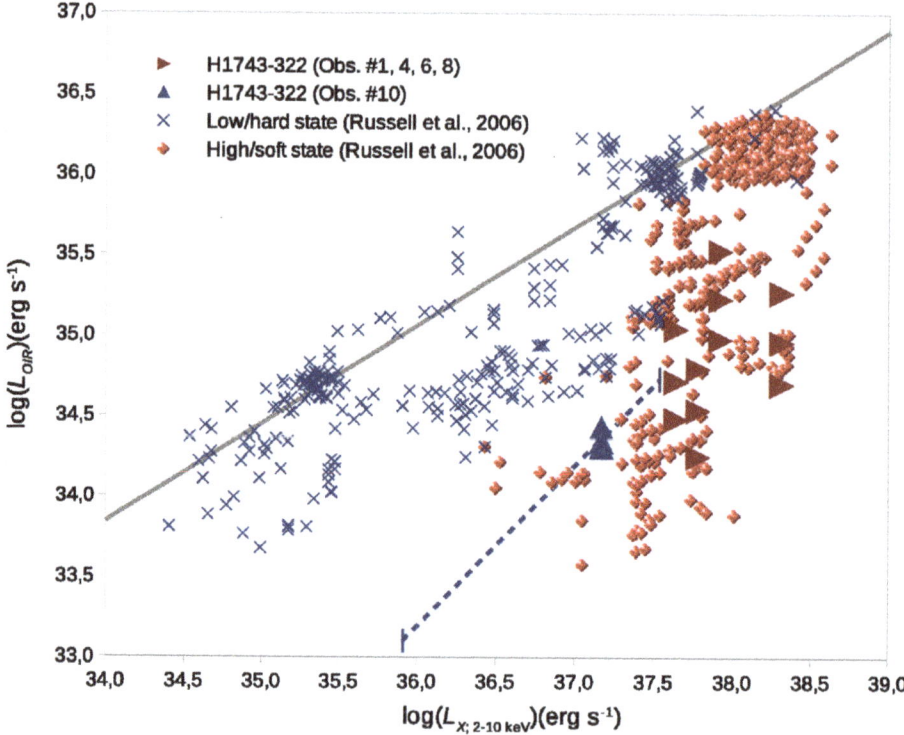

Figure 5.22. Fundamental plane of black hole activity, represented with optical/infrared (OIR) versus X-ray (L_X) luminosity diagram for LMXB hosting black holes in low-hard and high-soft state—data from Russell et al. (2006)—with data from the black hole system H1743-322, which experienced a failed outburst. Reproduced with permission from Astronomy & Astrophysics, © ESO (Chaty et al. 2015).

domain, located at the transition between accretion and ejection regimes (Dubus & Chaty 2006). High energy γ-rays are also fundamental in this domain, as shown by polarization measurements, made by INTEGRAL/IBIS in the energy range 400 keV–2 MeV, on the black hole system Cyg X-1 in the low-hard state. Spectral modeling of this polarization suggests two distinct emission mechanisms: while the first one, in the 250–400 keV energy range, is consistent with emission dominated by Compton scattering on weakly polarized thermal electrons, the second one, in the 400 keV–2 MeV energy range, is strongly polarized (Laurent et al. 2011). This strong polarization in the MeV domain could result either from the launching of the jet originally detected in the radio band (Götz et al. 2019), or from non-thermal processes in the hot and magnetized plasma of the corona surrounding the accreting black hole (Romero et al. 2014). The contribution of an ordered magnetic field associated to the accretion disk might be the clue to distinguish between these two origins, with gamma-ray polarization which should be measurable by future instruments, as suggested by the modeling of the black hole system XTE J1118 +480 (Vieyro et al. 2016).

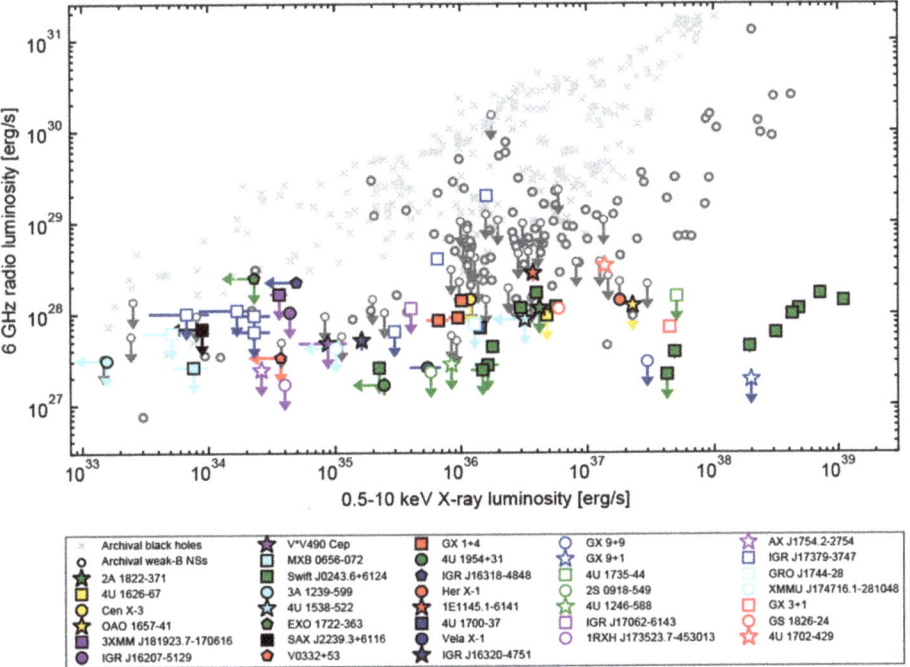

Figure 5.23. Fundamental plane of NS activity, represented with radio versus X-ray luminosity diagram for LMXB and HMXB hosting NS, superimposed on the fundamental plane of BH activity (shown as light gray crosses). While archival observations of weakly magnetized neutron stars are shown as dark gray circles, new radio observations of NS binaries are indicated with open markers for weakly magnetized NS, and filled markers for strongly magnetized NS. Reprinted with permission from van den Eijnden et al. (2021).

5.4 Formation, Evolution and Final Fate

A typical evolution scenario of LMXB is represented in the left panel of Figure 5.24. In low-mass systems, evolutionary processes are dominated by internal evolution, tidal friction, dynamo activity, stellar wind, and magnetic braking (Eggleton & Kisseleva-Eggleton 2005). The whole story begins with two zero-age main sequence (ZAMS) stars, star #1 evolves and fills its Roche lobe, triggering a first accretion episode. Then the system enters the common envelope phase (see Section 3.4), with spiral-in of star #2 within the envelope of star #1. If the envelope is successfully ejected, the binary survives to this crucial phase, initially invoked by Paczynski (1976) to tighten the orbits, by a reduction factor of ~100 compared to initial orbital separation (see Figure 3.5). The formation of LMXB with observed orbital periods between ~10 mn and ~10 hr can only be explained by invoking this phase. However, the fact that the envelope has been ejected, also means that star #1 has lost its entire H-envelope, and possibly also its He-envelope, thus becoming a *helium star*. Depending on its previous evolution, and initial mass, its envelope has been lost either during RLO episode, and/or during CE phase.

After this phase, star #1 collapses and explodes in a supernova event, if it is massive enough. The critical threshold mass, for a helium star to collapse into a

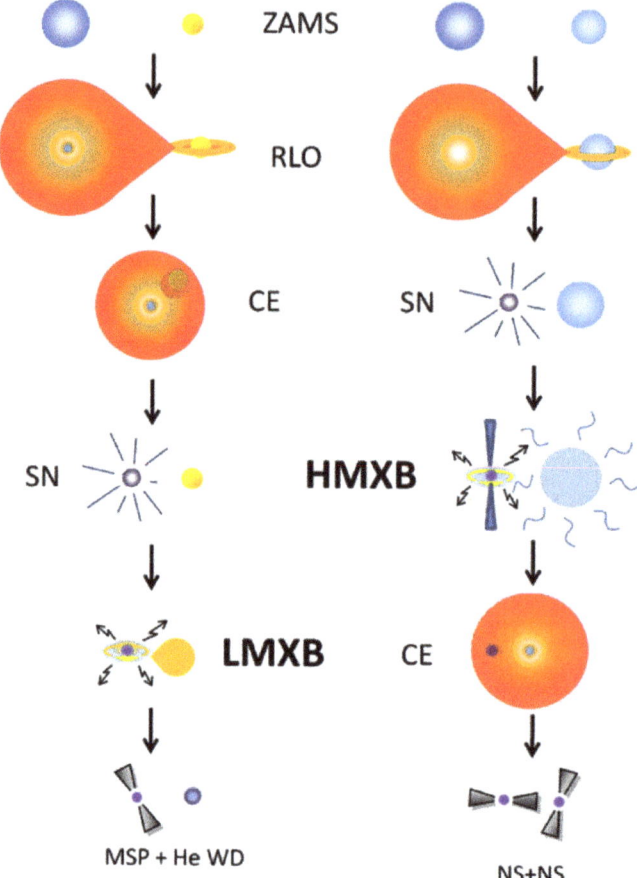

Figure 5.24. Schematic scenario of the isolated binary evolution channel for LMXB (left) and HMXB (right). Reprinted with permission from Springer, Ivanova et al. (2013).

neutron star, is $\sim 2.8 M_\odot$, corresponding to a ZAMS mass of $10-12 M_\odot$. The formation of some LMXB with short orbital periods can only be explained when invoking natal kicks—due to asymmetric supernova event—at the formation of the neutron star, with the net effect of reducing the orbital separation. Afterwards, it is the turn of star #2 to evolve, crossing the Hertzsprung gap, and to begin to fill its Roche lobe. At this stage, the low-mass X-ray binary phase begins, with a timescale of accretion ranging from 10^7 and 10^9 yr. Tauris & Savonije (1999) and Podsiadlowski et al. (2002) have shown that all evolution sequences with (sub)giant donors up to $2 M_\odot$ are stable against dynamical mass transfer. Then star #2 eventually collapses into a compact object (a white dwarf in Figure 5.24): we finish up with a binary system composed of a milli-second pulsar and a helium white dwarf.

We now know in our Galaxy ~ 300 binary systems hosting accreting milli-second pulsar (AMSP; Lorimer 2013; Papitto et al. 2020), divided into three classes: A, B, and C (Tauris 1996):

- The first group (class A) contains the wide-orbit ($P_{orb} > 20$ days) systems consisting of a milli-second pulsar orbiting a low-mass helium white dwarf companion ($M_{WD} < 0.45 M_{\odot}$).
- The second group (class B) contains the close-orbit ($P_{orb} \leqslant 15$ days) systems consisting of a milli-second pulsar orbiting a low-mass helium white dwarf companion ($M_{WD} < 0.45 M_{\odot}$). These systems can lead to single milli-second pulsar, in case the companion is destroyed or evaporates, likely during the accretion phase onto the neutron star (associated to X-ray emission).
- The third group (class C) contains the close-orbit ($P_{orb} \leqslant 15$ days) systems consisting of a milli-second pulsar in orbit around a high-mass CO/O-Ne-Mg white dwarf companion ($M_{WD} \geqslant 0.45 M_{\odot}$). These systems originate from intermediate mass X-ray binary (IMXB, with $2 \leqslant M_2 \leqslant 8 M_{\odot}$, see Chapter 7), after they lost a significant mass and angular momentum, likely during the common envelope phase.

The future of groups B and C depends on the evolution after an orbital bifurcation period at $P_{orb} \sim 2 - 3$ days. On one hand, converging systems evolve with decreasing P_{orb}, until the donor becomes degenerate, leading to the formation of an ultra-compact X-ray binary, due to magnetic braking and gravitational wave radiation. For a $1 M_{\odot}$ secondary, the initial period range leading to the formation of such ultra-compact systems (with minimum periods less than ~40 mn) lies within $P_{orb} \sim 13 - 18$ hr (Podsiadlowski et al. 2002). On the other hand, diverging systems evolve with increasing P_{orb} until the donor loses its envelope, leading to the formation of a wide and detached binary.

5.5 Reviews, Catalogs, Database and References

5.5.1 Reviews

- Reviews on the nature and evolution of LMXB systems, X-ray novae, X-ray bursts, soft X-ray transients, hard X-ray sources, etc.: Lewin et al. (1993, 1995); Tanaka & Shibazaki (1996); Tauris & van den Heuvel (2006); Krivonos et al. (2017, 2021); Sazonov et al. (2020);
- Review on accreting milli-second pulsars: Papitto et al. (2020);
- Reviews on QPO from neutron star systems: Lewin et al. (1988); Belloni et al. (2002); van der Klis (2006); Wang (2016b, 2016a);
- Review on QPO from black hole systems: Ingram & Motta (2019);
- Review on spins in black hole systems: Reynolds (2021);
- Review on black hole systems: Tauris & van den Heuvel (2006); Remillard & McClintock (2006); Casares & Jonker (2014); and particularly on those detected by the INTEGRAL satellite: Motta et al. (2021).

5.5.2 Catalogs

- Catalog on LMXB, including both neutron star and black hole systems: Liu & van Paradijs (2007);
- List of more than 100 Galactic X-ray bursters (along with P_{orb}, $\dot{m}$, ν, QPO): https://personal.sron.nl/~jeanz/bursterlist.html

- List of Type I X-ray bursts observed with the RXTE satellite: Galloway et al. (2008);
- Catalog of intermediate duration Type I X-ray bursts observed with the INTEGRAL satellite: Alizai et al. (2020);
- List of symbiotic X-ray binaries: table 1 in Yungelson et al. (2019);
- *WATCHDOG*, a comprehensive all-sky database of 77 Galactic black hole X-ray binaries: Tetarenko et al. (2016);
- *BlackCAT*, a catalog of 59 stellar-mass black holes in X-ray transients, with dynamical measurements, reported in Table 5.1: Corral-Santana et al. (2016).

5.5.3 Database

LMXB hosting neutron stars and black holes require by nature multi-wavelength observations, since the inner accretion disk emits in X-ray/gamma-ray, the outer accretion disk emits in optical/infrared, the X-ray heating (irradiation) on the star emits in optical/UV, the jets emit in radio and X-rays, and the companion star emits from ultraviolet to infrared. We refer the reader to Chapter 9, for more information on multi-wavelength facilities.

We give in Table 5.1 the list of dynamically-determined binary parameters (see Section 5.1.4) and effective spin of black holes within LMXB systems.

References

Abramowicz, M. A., & Kluźniak, W. 2003, GReGr, 35, 69

Alabarta, K., Altamirano, D., Méndez, M., et al. 2021, MNRAS, 507, 5507

Alizai, K., Chenevez, J., Brandt, S., & Lund, N. 2020, MNRAS, 494, 2509

Bałucińska-Church, M., Schulz, N. S., Wilms, J., et al. 2011, A&A, 530, A102

Barret, D., Kluźniak, W., Olive, J. F., Paltani, S., & Skinner, G. K. 2005, MNRAS, 357, 1288

Belloni, T., Homan, J., Casella, P., et al. 2005, A&A, 440, 207

Belloni, T., Méndez, M., & Homan, J. 2007, MNRAS, 376, 1133

Belloni, T., Méndez, M., King, A., van der Klis, M., & van Paradijs, M. 1997a, ApJ, 488, L109

Belloni, T., Méndez, M., King, A., van der Klis, M., & van Paradijs, M. 1997b, ApJ, 479, L145

Belloni, T., Psaltis, D., & van der Klis, M. 2002, ApJ, 572, 392

Blandford, R. D., & Königl, A. 1979, ApJ, 232, 34

Bozzo, E., Stella, L., Vietri, M., & Ghosh, P. 2009, A&A, 493, 809

Bult, P., Altamirano, D., Arzoumanian, Z., et al. 2021, ApJ, 920, 59

Callanan, P. J., Charles, P. A., Honey, W. B., & Thorstensen, J. R. 1992, MNRAS, 259, 395

Cao, X., You, B., & Yan, Z. 2021, A&A, 654, A81

Casares, J., González Hernández, J. I., Israelian, G., & Rebolo, R. 2010, MNRAS, 401, 2517

Casares, J., & Jonker, P. G. 2014, SSRv, 183, 223

Casares, J., Orosz, J. A., Zurita, C., et al. 2009, ApJS, 181, 238

Chaty, S., & Bessolaz, N. 2006, A&A, 455, 639

Chaty, S., Dubus, G., & Raichoor, A. 2011, A&A, 529, A3

Chaty, S., Fortin, F., & López-Oramas, A. 2020, A&A, 637, A2

Chaty, S., Haswell, C. A., Malzac, J., et al. 2003, MNRAS, 346, 689

Chaty, S., Mignani, R. P., & Israel, G. L. 2002a, MNRAS, 337, L23

Chaty, S., Mignani, R. P., & Israel, G. L. 2006, MNRAS, 365, 1387

Chaty, S., Mirabel, I. F., Duc, P. A., Wink, J. E., & Rodriguez, L. F. 1996, A&A, 310, 825

Chaty, S., Mirabel, I. F., Goldoni, P., et al. 2002b, MNRAS, 331, 1065

Chaty, S., Muñoz Arjonilla, A. J., & Dubus, G. 2015, A&A, 577, A101

Church, M. J., Gibiec, A., & Bałucińska-Church, M. 2014, MNRAS, 438, 2784

Corbel, S., Fender, R. P., Tzioumis, A. K., et al. 2000, A&A, 359, 251

Corral-Santana, J. M., Casares, J., Muñoz-Darias, T., et al. 2016, A&A, 587, A61

Corral-Santana, J. M., Casares, J., Muñoz-Darias, T., et al. 2013, Sci, 339, 1048

Corral-Santana, J. M., Casares, J., Shahbaz, T., et al. 2011, MNRAS, 413, L15

Corral-Santana, J. M., Torres, M. A. P., Shahbaz, T., et al. 2018, MNRAS, 475, 1036

Court, J. M. C., Altamirano, D., Albayati, A. C., et al. 2018, MNRAS, 481, 2273

Curran, P. A., & Chaty, S. 2013, A&A, 557, A45

Curran, P. A., Chaty, S., & Zurita Heras, J. A. 2011, A&A, 533, A3

Curran, P. A., Chaty, S., & Zurita Heras, J. A. 2012, A&A, 547, A41

Dubus, G., & Chaty, S. 2006, A&A, 458, 591

Dubus, G., Hameury, J. M., & Lasota, J. P. 2001, A&A, 373, 251

Dubus, G., Lasota, J., Hameury, J., & Charles, P. 1999, MNRAS, 303, 139

Eggleton, P. P., & Kisseleva-Eggleton, L. 2005, Ap&SS, 296, 327

Esin, A. A., McClintock, J. E., Drake, J. J., et al. 2001, ApJ, 555, 483

Fortin, F., Chaty, S., Coleiro, A., Tomsick, J. A., & Nitschelm, C. H. R. 2018, A&A, 618, A150

Gallo, E., Fender, R. P., & Pooley, G. G. 2003, MNRAS, 344, 60

Galloway, D. K., Goodwin, A. J., & Keek, L. 2017, PASA, 34, e019

Galloway, D. K., Muno, M. P., Hartman, J. M., Psaltis, D., & Chakrabarty, D. 2008, ApJS, 179, 360

Gambino, A. Г., Iaria, R., & Di Salvo, T. 2016, A&A, 589, A34

Gambino, A. F., Iaria, R., & Di Salvo, T. 2017, RAA, 17, 108

Gelino, D. M., & Harrison, T. E. 2003, ApJ, 599, 1254

Giacconi, R., Gursky, H., Paolini, F. R., & Rossi, B. B. 1962, PhRvL, 9, 439

Goldoni, P., Vargas, M., Goldwurm, A., et al. 1998, A&A, 329, 186

González Hernández, J. I., & Casares, J. 2010, A&A, 516, A58

Götz, D., Gouiffès, C., Rodriguez, J., et al. 2019, NewAR, 87, 101537

Grindlay, J. E. 1979, ApJ, 232, L33

Harlaftis, E. T., Dhillon, V. S., & Castro-Tirado, A. 2001, A&A, 369, 210

Harlaftis, E. T., & Greiner, J. 2004, A&A, 414, L13

Harlaftis, E. T., Steeghs, D., Horne, K., & Filippenko, A. V. 1997, AJ, 114, 1170

Hjellming, R. M., Stewart, R. T., White, G. L., et al. 1990, ApJ, 365, 681

Homan, J., & Belloni, T. 2005, Ap&SS, 300, 107

Hynes, R., Mauche, C., Haswell, C., et al. 2000, ApJ, 539, L37

Hynes, R. I., Haswell, C. A., Chaty, S., Shrader, C. R., & Cui, W. 2002, MNRAS, 331, 169

Hynes, R. I., Haswell, C. A., Cui, W., et al. 2003, MNRAS, 345, 292

Ingram, A. R., & Motta, S. E. 2019, NewAR, 85, 101524

Ioannou, Z., Robinson, E. L., Welsh, W. F., & Haswell, C. A. 2004, AJ, 127, 481

Ivanova, N., Justham, S., Chen, X., et al. 2013, A&ARv, 21, 59

Jonker, P. G., van der Klis, M., Homan, J., et al. 2001, ApJ, 553, 335

Kawamura, T., Axelsson, M., Done, C., & Takahashi, T. 2022, MNRAS, 511, 536

Khargharia, J., Froning, C. S., & Robinson, E. L. 2010, ApJ, 716, 1105

Krivonos, R. A., Bird, A. J., Churazov, E. M., et al. 2021, NewAR, 92, 101612

Krivonos, R. A., Tsygankov, S. S., Mereminskiy, I. A., et al. 2017, MNRAS, 470, 512

Kuulkers, E., Fender, R. P., Spencer, R. E., Davis, R. J., & Morison, I. 1999, MNRAS, 306, 919

Kuulkers, E., Howell, S. B., & van Paradijs, J. 1996, ApJL, 462, L87

Laurent, P., Rodriguez, J., Wilms, J., et al. 2011, Sci., 332, 438

Lewin, W. H. G., van Paradijs, J., & Taam, R. E. 1993, SSRv, 62, 223

Lewin, W. H. G., van Paradijs, J., & van den Heuvel, E. P. J. 1995, in X-Ray Binaries, ed. W. H. G. Lewin, J. Van Paradijs, & E. P. J. Van den Heuvel (Cambridge: Cambridge Univ. Press)

Lewin, W. H. G., van Paradijs, J., & van der Klis, M. 1988, SSRv, 46, 273

Liu, Q. Z., van Paradijs, J., & van den Heuvel, E. P. 2007, A&A, 469, 807

Lorimer, D. R. 2013, in IAU Proc. 291, Neutron Stars and Pulsars: Challenges and Opportunities after 80 years, ed. J. van Leeuwen (Cambridge: Cambridge Univ. Press), 237

Lowell, A. W., Tomsick, J. A., Heinke, C. O., et al. 2012, ApJ, 749, 111

Malzac, J., Petrucci, P. O., Jourdain, E., et al. 2006, A&A, 448, 1125

Markoff, S., Falcke, H., & Fender, R. 2001, A&A, 372, L25

Markoff, S., Nowak, M., Corbel, S., Fender, R., & Falcke, H. 2003, A&A, 397, 645

Martí, J., Mirabel, I. F., Chaty, S., & Rodríguez, L. F. 2000, A&A, 356, 943

Martí, J., Mirabel, I. F., Rodríguez, L. F., & Chaty, S. 1998, A&A, 332, L45

McClintock, J. E., Garcia, M. R., Caldwell, N., et al. 2001a, ApJ, 551, L147

McClintock, J. E., Haswell, C. A., Garcia, M. R., et al. 2001b, ApJ, 555, 477

McClintock, J. E., Petro, L. D., Remillard, R. A., & Ricker, G. R. 1983, ApJL, 266, L27

McClintock, J. E., & Remillard, R. A. 1986, ApJ, 308, 110

Mignani, R. P., Chaty, S., Mirabel, I. F., & Mereghetti, S. 2002, A&A, 389, L11

Mirabel, I. F., Dhawan, V., Chaty, S., et al. 1998, A&A, 330, L9

Mirabel, I. F., Rodriguez, L. F., Chaty, S., et al. 1996, ApJL, 472, L111

Mitsuda, K., Inoue, H., Koyama, K., et al. 1984, PASJ, 36, 741

Motch, C., Ricketts, M. J., Page, C. G., Ilovaisky, S. A., & Chevalier, C. 1983, A&A, 119, 171

Motta, S. E., Belloni, T. M., Stella, L., Muñoz-Darias, T., & Fender, R. 2014a, MNRAS, 437, 2554

Motta, S. E., Muñoz-Darias, T., Sanna, A., et al. 2014b, MNRAS, 439, L65

Motta, S. E., Rodriguez, J., Jourdain, E., et al. 2021, NewAR, 93, 101618

Motta, S. E., Rouco Escorial, A., Kuulkers, E., Muñoz-Darias, T., & Sanna, A. 2017, MNRAS, 468, 2311

Munari, U. 2019, in Cambridge Astrophysics Ser. 54, The Impact of Binary Stars on Stellar Evolution, ed. G. Beccari, & M. Boffin (Cambridge: Cambridge Univ. Press), 77

Narayan, R., Garcia, M. R., & McClintock, J. E. 1997, ApJL, 478, L79

Nelemans, G., & van Haaften, L. 2013, in ASP Conf. Ser. 470, 370 Years of Astronomy in Utrecht, ed. G. Pugliese, & M. Wijburg (San Francisco, CA: ASP), 153

Nowak, M. A., Paizis, A., Jaisawal, G. K., et al. 2019, ApJ, 874, 69

Orosz, J. A., & Bailyn, C. D. 1997, ApJ, 477, 876

Orosz, J. A., Bailyn, C. D., McClintock, J. E., & Remillard, R. A. 1996, ApJ, 468, 380

Orosz, J. A., Groot, P. J., van der Klis, M., et al. 2002, ApJ, 568, 845

Orosz, J. A., McClintock, J. E., Remillard, R. A., & Corbel, S. 2004, ApJ, 616, 376

Orosz, J. A., Steiner, J. F., McClintock, J. E., et al. 2011, ApJ, 730, 75

Paczynski, B. 1976, in IAU Symp. 73, Structure and Evolution of Close Binary Systems, ed. P. Eggleton, S. Mitton, & J. Whelan (Dordrecht: Reidel), 75

Paizis, A., Nowak, M. A., Chaty, S., et al. 2007, ApJL, 657, L109

Paizis, A., Nowak, M. A., Rodriguez, J., et al. 2015, ApJ, 808, 34

Paizis, A., Nowak, M. A., Rodriguez, J., et al. 2012, ApJ, 755, 52

Paizis, A., Nowak, M. A., Wilms, J., et al. 2011, ApJ, 738, 183

Papitto, A., Falanga, M., Hermsen, W., et al. 2020, NewAR, 91, 101544

Podsiadlowski, P., Rappaport, S., & Pfahl, E. D. 2002, ApJ, 565, 1107

Psaltis, D., Belloni, T., & van der Klis, M. 1999a, ApJ, 520, 262

Psaltis, D., Méndez, M., Wijnands, R., et al. 1998, ApJL, 501, L95

Psaltis, D., Wijnands, R., Homan, J., et al. 1999b, ApJ, 520, 763

Remillard, R. A., & McClintock, J. E. 2006, ARA&A, 44, 49

Reynolds, C. S. 2021, ARA&A, 59,

Rink, K., Caiazzo, I., & Heyl, J. 2022, arXiv:2107.06828

Romero, G. E., Vieyro, F. L., & Chaty, S. 2014, A&A, 562, L7

Russell, D. M., Fender, R. P., Hynes, R. I., et al. 2006, MNRAS, 371, 1334

Samimi, J., Share, G. H., Wood, K., et al. 1979, Natur, 278, 434

Sazonov, S., Paizis, A., Bazzano, A., et al. 2020, NewAR, 88, 101536

Shahbaz, T., Naylor, T., & Charles, P. 1997, MNRAS, 285, 607

Shahbaz, T., van der Hooft, F., Charles, P. A., & Casares, J. 1996, MNRAS, 282, L47

Shakura, N. I., & Sunyaev, R. A. 1973, A&A, 24, 337

Stella, L., & Vietri, M. 1999, PhRvL, 82, 17

Strohmayer, T., & Bildsten, L. 2006, in Compact Stellar X-ray Sources, ed. W. Lewin, & M. van der Klis, (Cambridge: Cambridge Univ. Press), 113

Sunyaev, R. A., & Titarchuk, L. G. 1985, A&A, 143, 374

Tanaka, Y., & Shibazaki, N. 1996, ARA&A, 34, 607

Tauris, T. M. 1996, A&A, 315, 453

Tauris, T. M. 2018, PhRvL, 121, 131105

Tauris, T. M., & Savonije, G. J. 1999, A&A, 350, 928

Tauris, T. M., & van den Heuvel, E. P. J. 2006, in Cambridge Astrophysics Ser. 39, Compact Stellar X-Ray Sources, ed. W. Lewin, & M. van der Klis (Cambridge: Cambridge Univ. Press), 623

Tetarenko, B. E., Sivakoff, G. R., Heinke, C. O., & Gladstone, J. C. 2016, ApJS, 222, 15

van den Eijnden, J., Degenaar, N., Russell, T. D., et al. 2021, MNRAS, 507, 3899

van der Klis, M. 1998, in NATO Advanced Science Institutes (ASI) Series C, Vol. 515, The Many Faces of Neutron Stars, ed. R. Buccheri, J. van Paradijs, & A. Alpar (Berlin: Springer), 337

van der Klis, M. 2006, in Cambridge Astrophysics Ser. 39, Compact Stellar X-Ray Sources, ed. W. Lewin, & M. van der Klis (Cambridge: Cambridge Univ. Press), 39

van der Klis, M., Jansen, F., van Paradijs, J., et al. 1985, Natur, 316, 225

van der Klis, M., Wijnands, R. A. D., Horne, K., & Chen, W. 1997, ApJL, 481, L97

van Doesburgh, M., & van der Klis, M. 2019, MNRAS, 490, 5270

Vieyro, F. L., Romero, G. E., & Chaty, S. 2016, A&A, 587, A63

Wang, J. 2016a, IJAA, 6, 82

Wang, J. 2016b, AdAst, 2016, 3424565

Warner, B., Woudt, P. A., & Pretorius, M. L. 2003, MNRAS, 344, 1193

White, N. E., & Stella, L. 1988, MNRAS, 231, 325

Wijnands, R., & van der Klis, M. 1998, Natur, 394, 344

Wijnands, R., & van der Klis, M. 1999, ApJ, 514, 939

Wijnands, R., van der Klis, M., Homan, J., et al. 2003, Natur, 424, 44

Yungelson, L. R., Kuranov, A. G., & Postnov, K. A. 2019, MNRAS, 485, 851

Yungelson, L. R., Lasota, J. P., Nelemans, G., et al. 2006, A&A, 454, 559

Zhang, C. M., Yin, H. X., Zhao, Y. H., Zhang, F., & Song, L. M. 2006, MNRAS, 366, 1373

Zhang, H., Yuan, F., & Chaty, S. 2010, ApJ, 717, 929

Zurita Heras, J. A., Chaty, S., Cadolle Bel, M., & Prat, L. 2011, MNRAS, 413, 235

Accreting Binaries
Nature, formation, and evolution
Sylvain Chaty

Chapter 6

High-mass X-ray Binaries (HMXB)

A billion years ago
—while here on Earth
multicelled life was arising and spreading—
in a galaxy far far away
two spinning black holes danced 'round one another,
rippling the fabric of space and time.

—Kip Thorne, *2020*
Reproduced with permission from Thorne & Halloran (2023).

We describe in this chapter the nature, formation and evolution of high-mass X-ray binaries (HMXB), binary systems hosting a massive star orbiting a compact object, which can be either an accreting neutron star or a black hole. We first review the nature of these binaries, their population, their Galactic distribution, the HMXB population in the Magellanic Clouds and in the galaxy M51, before reporting on the detection of cyclotron lines in accreting pulsars. We then distinguish the three main types of HMXB, located at distinct regions of the so-called Corbet diagram. We first describe the Be X-ray binaries, along with their variability, the viscous decretion disk and the rejuvenation model. We then report on multi-wavelength studies of supergiant X-ray binaries, and among them the various subclasses—obscured sgHMXB, eccentric sgHMXB, supergiant fast X-ray transients (SFXT)—along with their X-ray variability, and the process of stellar wind accretion, without forgetting the beginning atmospheric Roche lobe overflow systems. We then dive into the formation and global evolution of HMXB, describing the isolated binary evolution, the importance of stellar winds and natal kicks, along with alternative scenario of HMXB evolution, population III stars, and primordial black holes, and how to distinguish between various formation channels. We finally mention the various steps, scenario, and observables, producing binary black hole, binary neutron star, binary neutron star–black hole, and even binary white dwarf and neutron star, on their way toward merging and emission of gravitational waves. We finish this chapter by giving useful links to reviews, catalogs, database, and references.

6.1 Nature of High-Mass X-ray Binaries

We now describe *high-mass X-ray binaries* (HMXB), systems composed of a compact object, which can be either a neutron star—NS—or a black hole—BH—orbiting a luminous and massive early O or B spectral type companion star, of mass $M \gtrsim 8 M_\odot$ (see Figure 6.1). The specificity of an HMXB is mainly related to the characteristics of the companion star, with a lifetime of a few tens of millions of years. The first one is that, because it is a massive and early-type star, it usually possesses a high stellar wind, losing a substantial amount of stellar mass during the whole evolution of the binary system, with mass-loss rate potentially reaching $10^{-6} M_\odot$/year. The second one is that, because of the mass-ratio between the massive star and the compact object, the companion star usually does not fill its Roche lobe (see Section 3.3), preventing any permanent formation of accretion disk, unlike LMXB. Thus, only a small fraction of the mass lost through stellar wind is captured by the companion, resulting in a low accretion rate. Since this is an inefficient process, these systems are, contrarily to LMXB, usually brighter in optical than in X-rays, with a luminosity ratio $\frac{L_X}{L_{\rm opt}}$ ranging from $\sim 10^{-3}$ to ~ 10. High X-ray luminosities ($L_X \gtrsim 10^{35}$ erg s^{-1}) are observed in the form of strong and transient X-ray flares reaching the Eddington luminosity $L_{\rm Edd}$, in two situations: either when the compact object crosses the dense wind ejected from the companion star, or when the atmosphere of this star begins to fill its Roche lobe.

Orbital periods are spread on day-to-year timescale, some sources having short orbital period, such as Cyg X-3 hosting a WR star, with $P_{\rm orb} \sim 4.8$ hr, detected at all

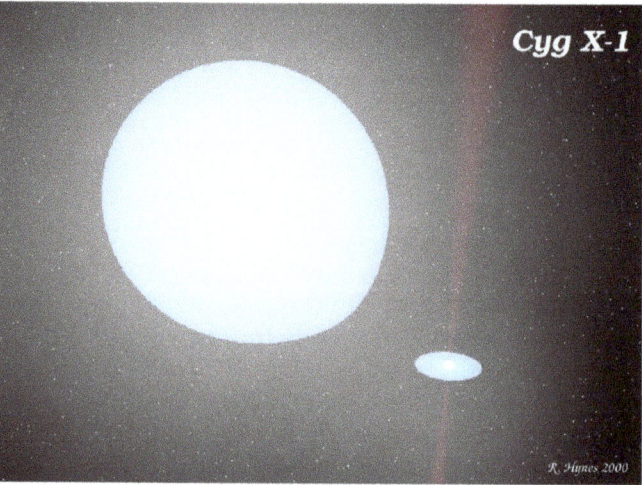

Figure 6.1. Typical image of a high-mass X-ray binary, in this particular case is represented Cyg X-1, a high-mass star in orbit around a black hole, this object being also a microquasar. The black hole accretes both through formation of an accretion disk, and through direct accretion of stellar wind. Credit: Robert Hynes.

wavelengths, from infrared to X-rays (Mason et al. 1986; Hanson et al. 2000). This HMXB has been intensively studied through orbital-phase-resolved infrared spectra, in outburst and quiescence, including Doppler tomographic analysis (see Figure 3.8), allowing astronomers to derive the dynamical mass function of the system, with a mass range for the WR He star within $5 \leqslant M \leqslant 11 M_\odot$, orbiting a compact object that is either a neutron star or a black hole (Hanson et al. 2000).

We distinguish three types of HMXB by the nature of the donor, and the process of accretion, that we order here by decreasing number of systems in each class: (i) *main sequence Oe-Be star X-ray binaries* (BeHMXB), (ii) *supergiant X-ray binaries* (sgHMXB), and (iii) *Roche-lobe overflow systems* (RLO–HMXB). We describe below each of these three different types of HMXB.

6.1.1 Population of HMXB

Our Galaxy hosts a number of 114 HMXB, as reported in Liu et al. (2006), and 117 in Bird et al. (2016a). By cross-correlating both catalogs, Fortin et al. (2018) have shown that they share 79 HMXB in common. Among these common sources, 6 sources identified as HMXB in Liu et al. (2006) have been assigned to other types by Bird et al. (2016a). Adding new identifications by Coleiro et al. (2013) and Fortin et al. (2018); the total number of HMXB currently known in our Galaxy amounts to 130. HMXB thus represent ~37% of the total number of high-energy binary systems (when adding all known LMXB and HMXB reported in Liu et al. 2007; Coleiro et al. 2013; Bird et al. 2016a; Fortin et al. 2018). Among these 130 HMXB, there are 94 *confirmed* HMXB (meaning that they are detected in X-rays, have an optical and/ or infrared counterpart, and that the spectral type of the secondary has been determined spectroscopically, following the method described in Fortin et al. 2022a), and 36 candidates (among which 10 of them do not have any sg or BeHMXB identification). These 94 confirmed HMXB are split in 56 Be/OeHMXB (+21 candidates) and 38 sgHMXB (+5 candidates), amounting to ~60% of BeHMXB and ~40% of sgHMXB, respectively. sgHMXB can be further divided in 21 *classical* persistent systems (+2 candidates), 13 SFXT (+3 candidates), and 3 RLO-HMXBs. It remains ~35% of unidentified HMXB. In total, only 4 HMXB harbor a black hole as a compact object (a list of dynamically confirmed black holes within HMXB systems is reported in Table 6.1; updated from Corral-Santana et al. 2016; and completed), and among sgHMXB there are 4 sgB[e] and 1 Wolf–Rayet secondary star.

6.1.2 Galactic Distribution

Figure 6.2 shows the Galactic distribution of all HMXB, in Galactic coordinates. Contrary to LMXB, which host old stars ($>10^9$ yr) concentrated toward the Galactic center far from their birthplace, HMXB host young stars ($<10^7$ yr) which remain close to their birthplace, leaving only little time for migration. According to the mass-age relationship of massive stars, the time between their birth and the supernova is of the order ~10 Myr (Schaller et al. 1992). Concentrated within the Galactic plane, they closely follow the gas distribution in the Galaxy, clustered

Table 6.1. Dynamically-determined Binary Parameters and Effective Spin of Black Holes within HMXB Systems.

Name	Spectral type	Porb (h)	K2 (km s^{-1})	f(M1) $M_\odot$	M1 $M_\odot$	q	i (°)	$v_{rot} \sin i$ (km s^{-1})	Spin	Ref.
HMXB										
Cyg X-1 (V1357 Cyg)	O9.7Iab	134.4	75.2	0.960	21.2 ± 2.2	0.52	27.5 ± 0.7		⩾0.983	1956
LMC X-3 (0538-641)	B3V	40.9	235	2.3 ± 0.3	6.94 ± 0.56	~0.52	69.5 ± 0.7	130	$0.25^{+0.20}_{-0.29}$	0538
LMC X-1 (0540-697)	O7III	93.8	71.6	0.12 ± 0.03	10.91 ± 1.54	~0.34	36 ± 2	~260	$0.92^{+0.05}_{-0.07}$	0540
M33 X-7 (0540-697)	O7-8III	82.9	108.9	0.777	15.65 ± 1.45	~0.22	74.6 ± 1.0	250 ± 7	0.84 ± 0.05	m33x7

Notes. Spin values are taken from Reynolds (2021). References: **1956**: Miller-Jones et al. (2021); **0538**: Orosz et al. (2014); **0540**: Orosz et al. (2009); **m33x7**: Orosz et al. (2007).

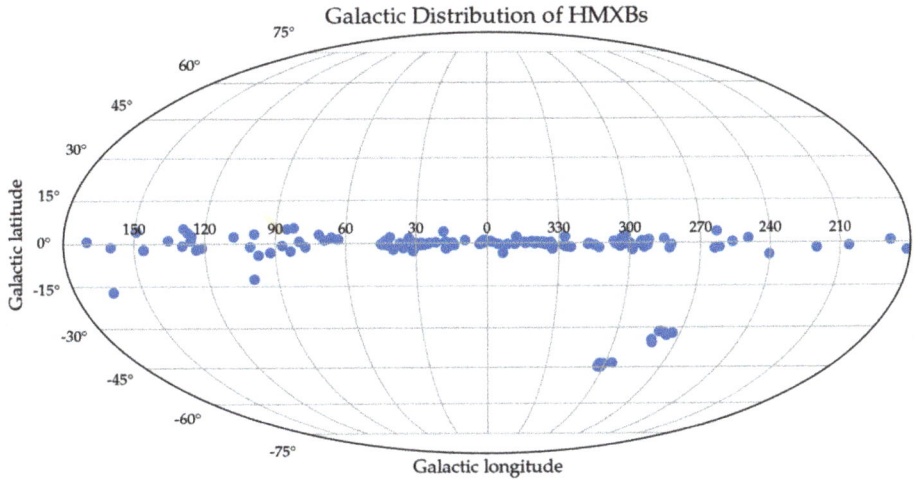

Figure 6.2. Galactic distribution, in Galactic coordinates, of all HMXB known to date, based on a catalog of 148 HMXB. They are located in the plane of the Galaxy, close to the tangential directions of the Galactic arms. Credit: Francis Fortin.

toward tangential directions of inner Galactic arms, rich in star-forming regions (Benaglia et al. 2010) and stellar formation complexes (SFC, Russeil 2003)—see also the study of open clusters within Milky Way spiral arms, from Gaia EDR3 (Castro-Ginard et al. 2021). The distribution of HMXB in our Galaxy is thus a good indicator of massive star formation, and their collective X-ray luminosity has been historically used to compute the star formation rate of the host galaxy (Grimm et al. 2002, 2003). Coleiro & Chaty (2013) derived the distance of HMXB (including both BeHMXB and sgHMXB) by adjusting the spectral type on optical/infrared magnitudes of the companion star. Then, by correlating the position of HMXB with the position of SFC, they showed that HMXB are clustered within 0.3 kpc of an SFC, with an inter-cluster distance of 1.7 kpc, remaining close to their birthplace, as reported in Figure 6.3 (see also Bodaghee et al. 2012).

Since the time to form an HMXB only takes tens of millions of years (Tauris & van den Heuvel 2006), it is an accurate marker for the passing of density waves within a given region of our Galaxy. Coleiro & Chaty (2013) computed the offset between the position of spiral arms and HMXB distribution, corresponding to a timelag of $\sim 10^7$ years, consistent with the expected delay between the birth of the massive star, and the HMXB formation. By taking into account the Galactic arm rotation, it is thus possible to derive parameters such as the age, migration distance from their birthplace, and kick velocity for BeHMXB and sgHMXB systems. The origin of this binary migration is potentially due to a combination of various effects, such as recoil due to anisotropic mass loss occurring during a supernova event (Blaauw 1961), natal kick (Brandt & Podsiadlowski 1995), and dynamical ejection from star cluster, with a final velocity up to 1.5–2 times higher than the pre-supernova velocity (Pflamm-Altenburg & Kroupa 2010). By assuming a typical kick velocity of 100km s^{-1} (Hills 1983) and a maximum migration distance due to a kick

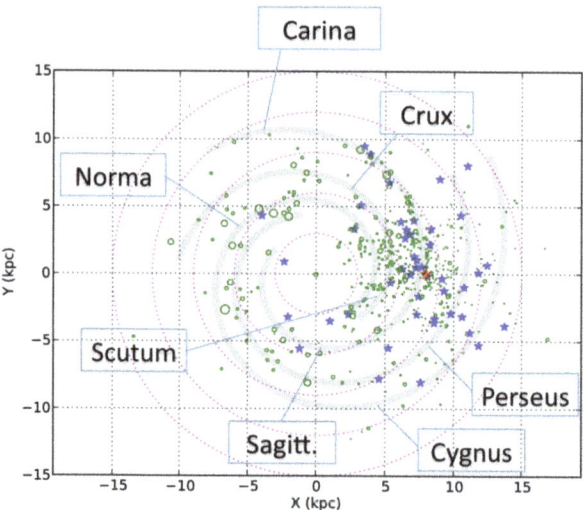

Figure 6.3. Galactic distribution of HMXB (blue stars) and stellar formation complexes SFC (green circles). The distance of HMXB has been derived by adjusting the spectral type on optical/infrared magnitudes of the companion star. The circle radius of SFC is proportional to their excitation parameter value (Russeil 2003). Galactic spiral arms are indicated, as well as the location of the Sun (red dot at position 8.5:0). Reprinted with permission of the AAS from Coleiro & Chaty (2013).

event of 0.3 kpc, Coleiro & Chaty (2013) obtain an upper limit for the time lapse between the supernova explosion and the HMXB stage of ~3 Myr (see also an updated study using Gaia EDR3 data in Fortin et al. 2022b).

As pointed out by Walter et al. (2015), there are about 20, 000 O stars in our Galaxy, with 1/3 of them located within binary systems, evolving through envelope stripping (Sana et al. 2012). Assuming that half of these systems remain bound after the natal kick received during the supernova event (see Section 6.5.3), we can estimate that one HMXB forms every ~1500 years. Then, taking into account the number of known sgHMXB in our Galaxy, they derive that their accreting phase only lasts for ~10^5 years, i.e., ~10% of the whole life of a massive star.

6.1.3 Population of HMXB in the Magellanic Clouds

A large number of BeHMXB are located within the Magellanic Clouds, which happen to be ideal laboratories to study X-ray source population, the Large Magellanic Cloud (LMC) being located at a distance of ~50 kpc, and the Small Magellanic Cloud (SMC) at ~62 kpc. In both Magellanic Clouds, BeHMXB are much more numerous than sgHMXB. This is especially true for the SMC, in which there is only one sgHMXB, SMC X-1.

From the number of confirmed BeHMXB in our Galaxy, we can estimate how many such systems are expected, based upon the mass ratio of ~50 between the Milky Way and the Magellanic Clouds. We thus expect only a few BeHMXB within the SMC, and instead there are ~120 confirmed BeHMXB, nearly half of them being X-ray pulsars (Haberl et al. 2022). ~60 HMXB are identified within the LMC, with detected pulsations in about 20 of them. Instead, sgHMXB are more numerous in the LMC than in the SMC.

This over-abundance of BeHMXB in the SMC is likely due to a different history of stellar formation in both Magellanic Clouds. The first reason is related to the presence of a strong bridge of material between the Magellanic Clouds and the Milky Way, suggesting strong tidal interactions in the past, leading to an increase of stellar formation ~100 Myr ago, triggering the birth of many new massive stars, some of them now hosted in BeHMXB (McBride et al. 2008). The second reason is that both SMC and LMC are gas-rich and metal-poor, with respectively 0.25 and 0.5 solar metallicity, thus forming more massive stars than the Milky Way (Choudhury et al. 2016, 2018).

6.1.4 Population of HMXB in the Galaxy M51

HMXB can also be studied in other galaxies, such as the Whirlpool galaxy (M51), an interacting galaxy pair NGC 5194/5195, where, using data from Chandra and Hubble Space Telescopes, Rice et al. (2021) managed to identify 334 candidate X-ray binary systems, altogether with their potential optical counterparts for half of them, with ~80% of variable sources over a 30 day timescale (see Figure 6.4).

6.1.5 Cyclotron Line in Accreting Pulsars

A large majority of HMXB systems host an *accreting pulsar*, with a rotation period that generally decreases, since accreted matter brings substantial angular momentum to the neutron star. A few systems exhibit an erratic behavior, naturally explained by the stellar wind accreted via the formation of a transient accretion disk, rotating randomly in one direction or the opposite. Accreting pulsars are usually highly magnetized neutron stars ($>10^{12}$ G), pertaining to a young stellar population, producing a periodic X-ray emission of a period of a few seconds. The electron–proton plasma falls toward the neutron star surface via an accretion column at

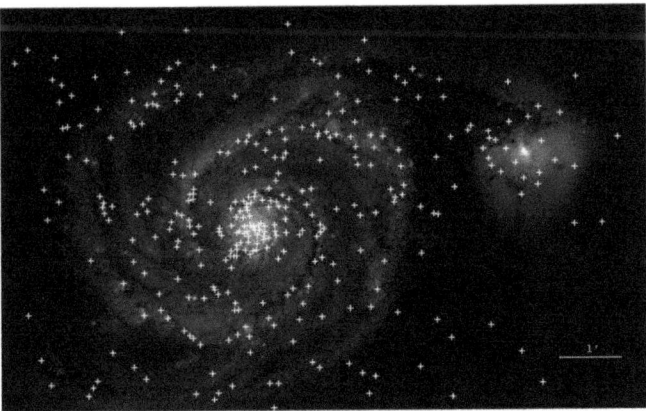

Figure 6.4. Combined HST (r: F658N, g: F555W, b: F435W) image of all candidate X-ray binary systems (white crosses) in the Whirlpool galaxy (M51). North is to the right. Reprinted with permission of the AAS from Rice et al. (2021).

almost the velocity of the light, until it is stopped by the pressure of the dipolar magnetic field, forced to move along the field lines toward magnetic poles of the neutron star, until the accreted matter heats up to a temperature of $T \sim 10^8$ K close to the NS surface, and releases its gravitational energy, emitting X-rays. X-ray spectra of accreting pulsars, emitted from regions around the magnetic poles, are characterized by a hard X-ray emission ($kT > 15$ keV), with a power law of photon index in the range $\Gamma \sim 0.3 - 2$, and a high-energy exponential cutoff between 7 and 30 keV. They also exhibit cyclotron lines (*cyclotron-resonant scattering features*, CRSF) in the range 10–100 keV. At the polar caps of the NS, a plasma, consisting of electrons, protons, and ions, moves cyclically, in a plane perpendicular to the strong magnetic field. This optically thick plasma leads to resonant scattering of hard X-ray photons. Such cyclotron lines, seen in absorption in X-ray spectra, allow direct measurements of the magnetic field strength at the neutron star surface. Their strong energy variability suggests a non-uniform accretion flow, strongly dependent on varying magnetic fields. The energy E_{cyc} of the centroid cyclotron line, created by the resonant scattering, depends on the magnetic field strength:

$$E_{cyc} = \frac{E_{Landau,n}}{1+z} = \frac{n}{1+z} \times 11.6 \text{ keV} \times B_{10^{12}G} \qquad (6.1)$$

z being the gravitational redshift of CRSF due to NS mass, and n is the number of Landau quantized levels involved: $n = 1$ for the fundamental line, corresponding to a scattering from ground level to the first excited Landau level, and $n \geqslant 2$ are the harmonics (Staubert et al. 2019).

Study of the cyclotron-resonant scattering feature allows astronomers to better understand the close environment of neutron star. Such a temporal and spectral hard X-ray study has been performed on OAO 1657-415, an accreting X-ray pulsar with a high-mass companion, observed by the NuSTAR X-ray satellite during its brightest state and in different orbital phases (Saavedra et al. 2022). The hard X-ray NuSTAR spectrum in the $3 - 79$ keV energy was modeled with a power-law continuum emission model and a high-energy cutoff, associated to pulsations and cyclotron-resonant scattering line at 35.6 ± 2.5 keV, with a positive correlation between the luminosity and the energy associated with the cyclotron line. Saavedra et al. (2022) also estimate the dipolar magnetic field at the pulsar surface of $4.0 \pm 0.2 \times 10^{12}$ G.

In the census of 36 accreting binaries hosting neutron star, based on radio observations and reported in van den Eijnden et al. (2021); there are 13 weakly magnetized ($B \leqslant 10^{10}$ G) and 23 strongly magnetized ($B \geqslant 10^{10}$ G) neutron stars (see Figure 5.23). And among the strongly magnetized category, 16 reside in HMXB. The authors point out that, while the origin of radio emission in HMXB is unclear, in all but two detected sources (Vela X-1 and 4U 1700-37), the radio emission seems more likely emanating from a jet than from the donor stellar wind.

6.2 Be X-ray Binaries

Be X-ray binaries (BeHMXB) host a main sequence donor star of spectral type O9-B3e III/IV/V, of mass ranging from 8 to $20 M_\odot$, a rapidly rotating star—close to break-up velocity—surrounded by a circumstellar equatorial (so-called *decretion*) disk of gas, formed from ejected material which has settled into a Keplerian disk and spread outwards through viscous diffusion, as revealed by the presence of a prominent Hα emission line (see Figure 6.5) and continuum free–free emission (see Section 3.5; Reig 2011). For sake of simplicity we call BeHMXB both types of systems, hosting a Be or an Oe as the donor star (Rivinius et al. 2013). This disk is created by a low velocity and high density stellar wind of $\sim 10^{-7} M_{\odot} \mathrm{yr}^{-1}$ (Lamers & Cassinelli 1999). These systems are detached (see Section 3.3.2), the Be star sitting deep inside its Roche lobe as indicated by long orbital periods, and the compact object accretes directly from the circumstellar disk, with the formation of a transient accretion disk, which has the particularity to be much smaller than in the case of LMXB.

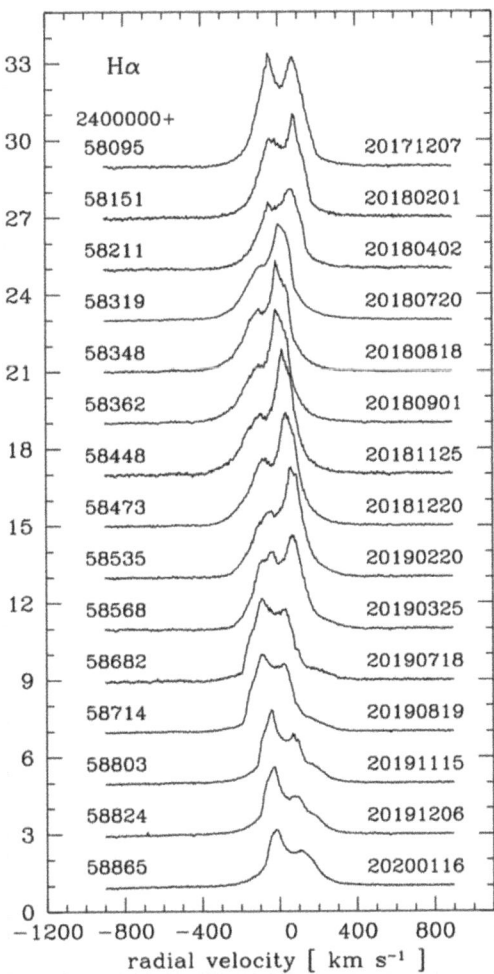

Figure 6.5. Variability of Hα emission-line profile in the BeHMXB system X Per. The date of observations is indicated in the right. Reprinted with permission from Zamanov et al. (2020).

6.2.1 Variability Properties

These systems are highly variable in X-rays, with three kinds of X-ray variability (Negueruela et al. 1998):

1. Persistent and low-luminosity X-ray emission, at a level $10^{34} \leqslant L_X \leqslant 10^{36}$ erg s^{-1}. Some classical sources, such as the BeHMXB X Per (Zamanov et al. 2020), have always been observed in this kind of variability. Some BeHMXB exhibit very low quiescent X-ray luminosity, such as IGR J01363+6610 (Tomsick et al. 2011).

2. Transient and bright Type I X-ray outbursts (originally called *Class I*, periodic transient activity; Stella et al. 1986). These periodic outbursts, with $\frac{L_X(\max)}{L_X(\min)} \geqslant 100$ take place each time the compact object, usually a NS on a wide ($P_{\mathrm{orb}} \geqslant 10$ days) and eccentric ($0.3 \leqslant e \leqslant 0.5$) orbit, approaches periastron and accretes matter from the decretion disk (see Charles & Coe 2006; Tauris & van den Heuvel 2006, and references therein). These transient HMXB are easy to detect, becoming during their maximum the brightest celestial sources for a few weeks, brighter than systems in which accretion is produced via stellar wind, with $L_X \sim 10^{36} - 10^{37}$ erg s^{-1}, before returning to quiescence for a few months or years (Bildsten et al. 1997; Coe 2000; Zurita Heras & Chaty 2008; Rodriguez et al. 2009a). Long orbital periods of ~ 100 days can be detected, by using long periods of X-ray data, for instance using the RXTE/ASM and Swift/BAT database (Yan et al. 2012). In most cases, the duration of these outbursts seems to be related to the orbital period.

3. Very luminous Type II outbursts (originally called *Class II*, irregular transient activity; Stella et al. 1986), with $\frac{L_X(\max)}{L_X(\min)}$ between 100 and 1000, which can start shortly after periastron passage, or at any other phase, since it does not show any other correlation with the orbital phase. They last for several orbital cycles, for several weeks or even months. These giant outbursts, with X-ray luminosity at a level $L_X \geqslant 10^{37}$ erg s^{-1}, are usually seen in sources which display Type I outbursts, and are probably due to a dramatic expansion of the truncated circumstellar disk (Okazaki & Negueruela 2001). Martin et al. (2014), by modeling the long-term evolution of a Be/X-ray binary system using 3D smoothed particle hydrodynamics simulations, suggest that such giant and long outbursts occur on a timescale of about 10 orbital periods, when a highly misaligned decretion disk becomes eccentric, allowing the compact object companion to capture a larger amount of material at periastron.

6.2.2 The Viscous Decretion Disk Model

The viscous decretion disk model, naturally predicting the truncation of the Be circumstellar disk, can account for most properties of Be stars. The truncation radius of the circumstellar disk mainly depends on the orbital parameters and the viscosity. On the first hand, in BeHMXB with low eccentricity, the disk is expected to be truncated at the 3:1 resonance radius, with a gap between the disk outer radius and the Roche lobe radius of the Be star. This gap is so wide that the neutron star cannot accrete enough material at periastron passage to show periodic Type I outbursts, thus displaying only occasional giant Type II outbursts. On the second hand, in BeHMXB with high eccentricity, the truncation of the decretion disk occurs at a much higher resonance radius, located close to the Roche lobe radius at periastron. In these systems, disk truncation is not efficient, allowing the neutron star to accrete material from the disk at every periastron passage, via the Lagrange point L_1. This accretion process, at low velocity relative to the neutron star, transfers substantial angular momentum, resulting in the formation of temporary accretion disks when displaying periodic Type I outbursts (see Okazaki & Negueruela 2001, and references therein).

6.2.3 BeHMXB in the Corbet Diagram

BeHMXB exhibit a correlation between the NS spin P_{spin} and orbital P_{orb} periods ($P_{spin} \propto P_{orb}^2$), as shown by their location in the Corbet diagram (Corbet 1986), an useful tool to disentangle the nature and evolution of various types of HMXB. A new version of this diagram, updated with additional sources, is reported in Figure 6.6 (see also Chaty 2013; Walter et al. 2015; Tauris et al. 2017; Kretschmar et al. 2019), where BeHMXB are scattered, roughly aligned, from the lower left to the upper right. This correlation is due to the specific process of accretion in these systems. Indeed, for accretion to occur onto the neutron star, the *Alfvén radius*, corresponding to the outer edge of the NS magnetosphere (see Section 2.2.2), and thus to the inner edge of the truncated accretion disk, must be smaller than the

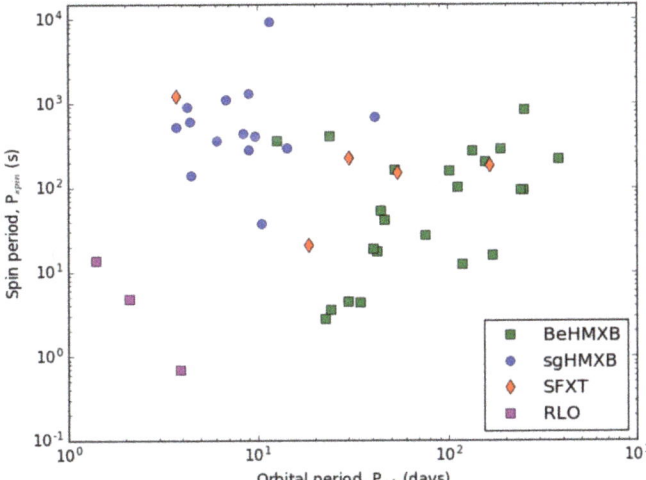

Figure 6.6. Updated Corbet diagram showing the different populations of HMXB (accreting pulsars) with measured values of both orbital P_{orb} and spin P_{spin} periods. Green squares are Be systems, blue circles are supergiant systems, red diamonds are SFXT, and purple squares are RLO systems. Reproduced with permission, Credit: Sylvain Chaty.

corotation radius of the accretion flow, rotating at Keplerian velocity (Waters & van Kerkwijk 1989). If this condition is not fulfilled, then the neutron star spins down.

Since the Alfvén radius depends on the density of the surrounding medium, and large orbits correspond to low stellar wind densities at the location of the neutron star, the correlation naturally takes place in BeHMXB, generally believed to be in spin equilibrium. A small (wide) orbit presents on average a high (low) stellar wind density, increasing (decreasing) the accretion pressure, and thus accelerating (slowing down) the NS rotation, which in turn increases (decreases) the centrifugal inhibition, preventing (allowing) additional accretion of matter. Accreted material thus efficiently transfers angular momentum at each periastron passage, rapidly spinning NS corresponding to compact orbit systems, and slowly spinning NS to wide orbit systems (as seen in Figure 6.6). Because of this interaction between the neutron star and the circumstellar disk, the orbiting neutron star gradually erodes the outer edge of the disk, preventing its growth. On average, decretion disks are smaller around Be stars within HMXB, compared to isolated Be stars (Rivinius et al. 2013).

6.2.4 The Rejuvenation Model

The formation of BeHMXB has been explained by the *rejuvenation* model (Rappaport et al. 1982), as a product of binary evolution. In this model, the mass transfer speeds up first the rotation of the outer layers of the B-spectral type secondary star, and subsequently the star itself, which was not born as fast rotator, and starts to rotate so quickly that a circumstellar disk of gas is created, giving birth to the Be phenomenon (Coe 2000; Charles & Coe 2006). The whole evolution of BeHMXB system is based upon conservative mass transfer, with two periods of mass transfer, and eventually detonation of helium star, leaving behind a neutron star. In the frame of this scenario, wide binary orbits—ranging from 100 to 600 days —must be produced before the supernova explosion. Any small asymmetry during this SN event will produce the frequently observed orbital eccentricity. Contrary to isolated Be stars (Rivinius et al. 2013), there is no Be stars of spectral type later than B2 hosted in BeHMXB (Negueruela & Coe 2002; McBride et al. 2008), likely due to the fact that wide orbits are more vulnerable to disruption during SN explosion, and this is particularly true in the case of less massive stars.

Finally, evolutionary models predict the existence of BeHMXB hosting a black hole as a compact object. However, apart from MWC 656 hosting a confirmed BH (Casares et al. 2014), most BeHMXB seem to host a neutron star.

6.3 Supergiant X-ray Binaries (sgHMXB)

sgHMXB host a supergiant star of spectral type O8-B1 I/II (typical mass $M \sim 30 M_\odot$, luminosity $L \sim 10^5 - 10^6 L_\odot$, mass loss rate $\dot{M} \sim 10^{-7} - 10^{-5} M_\odot$ yr^{-1}), characterized by an intense, slow and dense, highly supersonic and radiatively-driven stellar wind, radially outflowing from the equator. B-type supergiants typically possess slower terminal velocity winds ($v_\infty \sim 1000$km s^{-1}) than O-type supergiants ($v_\infty \sim 2000$km s^{-1}; Negueruela 2010). The so-called *classical* sgHMXB are persistent sources in the hard X-ray domain, most of them being close binary systems, with a compact object on a short ($\leqslant 10$ days) and quasi-circular orbit (low eccentricity, $e \leqslant 0.1$), directly accreting from the stellar wind, through, e.g., the Bondi–Hoyle–Lyttleton process. They frequently show eclipses, due to large companion star compared to small compact object, displaying stronger absorption close to eclipse than at inferior conjunction ($N_H \geqslant 10^{22} \mathrm{cm}^{-2}$).

Such wind-fed systems exhibit a luminous and persistent X-ray emission (from $L_X \geqslant 10^{35}$ erg s^{-1}, and up to $\sim 10^{38}$ erg s^{-1}), with superimposed large variations on short timescales (corresponding to a variability factor of $\sim 10^3$), mostly driven by hydrodynamic phenomena occurring on scales larger than the accretion radius, and a power-law X-ray spectrum with a cutoff at $\sim 10 - 30$ keV. Among the most famous sgHMXB, observed for a long time, and sometimes continuously during multi-wavelength campaigns, we can cite SMC X-1 (secondary of spectral type O9.7 Ia+), LMC X-4 (O8 III), Vela X-1 (B0.5 Ia), Cen X-3 (O9 III/Veq), GX 301-2 (B1.5 Iaeq), 4U 1538-522 (B0.2 Ia), IGR J16207-5129 (B8 IIIe), OAO 1657-415 (Ofpe/WNL), 4U 1700-377 (O6 Iafcp), EXO 1722-363 (B0-1 Ia), SAX J1802.7-2017 (B Ib), XTE J1855-026 (B0 Iaep), 4U 1907+09 (O8.5 Iab), and 4U 2206+54 (O9.5 Vep) (see, e.g., Lutovinov et al. 2013; Kretschmar et al. 2019).

6.3.1 sgHMXB in the Corbet Diagram

sgHMXB are located in the upper left part of the Corbet diagram (Figure 6.6), characterized by small orbital periods $P_{orb} \sim 3 - 10$ days and long spin periods $P_{spin} \sim 100 - 10,000$ s. They do not show any correlation between P_{orb} and P_{spin}, the spin periods varying erratically in time, due to the fact that there is no net transfer of angular momentum via wind accretion, alternately co- and counter-rotating with the neutron star. For typical OB supergiants, the spin period is given by the formula:

$$P_{spin} \propto \dot{M}^{-3/7} v_\infty^{12/7} \tag{6.2}$$

where v_∞ is the wind terminal velocity of the supergiant (Waters & van Kerkwijk 1989). The detection of long pulsations, and of cyclotron-resonant scattering features (see Section 6.1.5), imply that these sgHMXB host young NS with magnetic fields $B \sim 10^{11} - 10^{12}$ G.

6.3.2 Multi-wavelength Studies

The INTernational Gamma-Ray Astrophysics Laboratory (INTEGRAL) satellite, a medium size mission from the European Space Agency (Winkler & Courvoisier 2003; Kuulkers et al. 2021), was launched in 2002. With its wide field of view ($\sim 30°$), unique coverage in the hard X-ray and soft γ-ray band, and frequent scans of the Galactic plane (reaching 92% of completeness), INTEGRAL has revolutionized our view (and understanding) of sgHMXB in our Galaxy, by tripling their number (Chaty 2013; Walter et al. 2015; van den Heuvel et al. 2021). Since the source localization is not accurate enough with INTEGRAL ($\sim 2 - 5'$), follow-up multi-wavelength observation campaigns are required, first in the soft ($\leqslant 10$ keV) X-rays (with satellites such as Swift/XRT, XMM-Newton, Chandra) to get a better

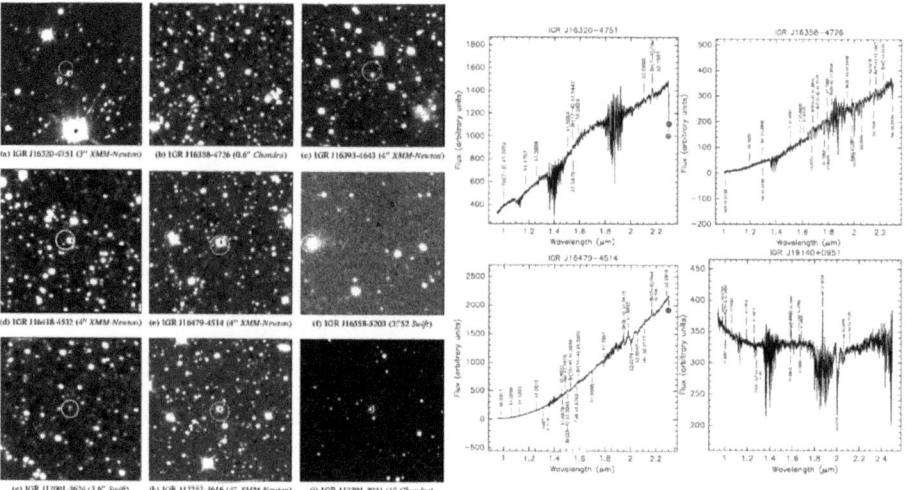

Figure 6.7. Study of HMXB detected by the INTEGRAL satellite. *Left:* identification of the infrared counterpart on ESO NTT images, thanks to its localization by X-ray satellite. *Right:* infrared spectra used to identify the nature of the secondary star. Reproduced with permission from Astronomy & Astrophysics, Credit ESO (Chaty et al. 2008).

localization (sub-arcsec), and then in optical and infrared (ESO/VLT) domains to get the photometry, and then spectroscopy, to unambiguously identify the spectral type of the companion star, allowing us to pinpoint the real nature of the binary system (see Figure 6.7). Examples of multi-wavelength observing campaigns can be found in Butler et al. (2009), Chaty et al. (2008), Coleiro et al. (2013), Fortin et al. (2018), Ozbey Arabaci et al. (2012), Rahoui et al. (2008), Rodriguez et al. (2008, 2009b), and Tomsick et al. (2006, 2008, 2009a, 2012, 2020, 2021). These campaigns have revealed three new subclasses of sgHMXB with specific properties, that we now describe. These observations, aided of modeling, helped us to probe the variations and geometry of the accretion column and emission regions. For reviews on HMXB, see, e.g., Chaty (2013), Lutovinov et al. (2013), Walter et al. (2015), Kretschmar et al. (2019), Krivonos et al. (2021), and references therein.

6.3.3 Obscured sgHMXB

A small subclass of sgHMXB ($\sim$5 systems) exhibits a permanently high intrinsic and local extinction in the range $10^{23} \leqslant N_H \leqslant 10^{24} \mathrm{cm}^{-2}$, with a compact object deeply embedded in the dense stellar wind. These highly absorbed systems are characterized by short orbital periods ranging from 3.7 to 9.7 days, the neutron star orbiting close to the surface of their O8–B1 stellar type companion stars (see Figure 6.8). Their luminosity lies in the range $10^{36} \leqslant L_X \leqslant 10^{38}$ erg s^{-1}. They can exhibit a complex temporal lightcurve, such as IGR J16207-5129 (Tomsick et al. 2009b; Bodaghee et al. 2010). Near and mid-infrared excess have been detected toward many of these sources, with the specificity that this infrared excess is neither due to stellar emission, nor emanating from interstellar dust on the line-of-sight, but instead is created by the

Figure 6.8. Scenario illustrating a neutron star orbiting a supergiant star on a circular orbit, accreting from the clumpy stellar wind of the supergiant (in this particular case, the system IGR J16318-4848 is represented). The accretion of matter is persistent in the case of the obscured sources, where the compact object orbits inside the cocoon of dust enshrouding the whole system. Reprinted with permission from Elsevier, Chaty (2013).

presence of a large amount of circumstellar dust, surrounding the close vicinity of the supergiant star, explaining a substantial fraction of the X-ray obscuration (Chaty et al. 2008; Rahoui et al. 2008). Some of these sources, such as IGR J16283-4838, can be detected far away, located beyond the Galactic center (Pellizza et al. 2011). It can be sometimes difficult to exactly evaluate the true nature of such a system, like the obscured accreting X-ray pulsar IGR J18179-1621, studied through X-ray and NIR observations, which combines properties usually associated to sources of distinct nature: a long outburst lasting for several weeks, a high intrinsic absorption, and a short $P_{\rm spin} = 11.82$ s (Nowak et al. 2012). It also frequently happens that obscured sgHMXB share X-ray properties similar to obscured AGN, especially Seyfert galaxies (Tomsick et al. 2015, 2016).

6.3.3.1 Stellar Wind Structure in IGR J16320-4751

A successful technique, to quantitatively investigate the geometrical and physical properties of stellar wind structures within these obscured binaries, aimed at measuring the temporal and spectral variation of their intrinsic absorption along the orbit, for instance using long (more than 10 years) lightcurves of XMM-Newton and Swift/BAT X-ray satellites. This has been done on the obscured sgHMXB IGR J16320-4751, characterized by a column density $N_{\rm H} \sim 10^{23}$cm^{-2}, more than an order of magnitude higher than the interstellar absorption along its line of sight. This system, composed of a neutron star of spin period $P_{\rm spin} = 1300$ s, and an O8 I supergiant star, with an orbital period $P_{\rm orb} \sim 9$ days, displays high-variability and flaring activity in X-rays on several timescales, with short and bright flares (flux increase of a factor 10 in 300 s). By fitting the highly absorbed continuum and Fe absorption edge at 7 keV, with a thermally comptonized model, García et al. (2018)

reveal a clear modulation of the column density $N_{\rm H}$ with the orbital phase, the stellar wind being the main contributor to both continuum absorption and Fe K line emission.

6.3.3.2 Eclipses in IGR J18027-2016

This method, measuring the temporal and spectral variation of the intrinsic absorption along the orbit, has also been used to reveal stellar wind structures in the eclipsing obscured sgHMXB IGR J18027-2016. This system, composed of a neutron star of spin period $P_{\rm spin} \sim 140$ s, and a late OB supergiant star, with an orbital period $P_{\rm orb} \sim 4.57$ days, shows an asymmetric eclipse profile stable on the long term. Including two absorption components—the first one for the interstellar absorption, and the second one for the intrinsic absorption—Fogantini et al. (2021) show that the local column density outside the eclipse is at a low-level value of $N_{\rm H} \sim 6 \times 10^{22}{\rm cm}^{-2}$, while when the neutron star enters and leaves the eclipse (ingress and egress respectively), the spectrum hardens, and the column density increases by a factor $\geqslant 3$, reaching maximum values up to $\sim 35 \times 10^{22}{\rm cm}^{-2}$. This is explained by a scenario in which both a photo-ionization wake and an accretion wake are responsible for the variation of column density along the orbit of the neutron star around the supergiant, in agreement with the variability of the Fe-line complex.

6.3.3.3 The Extreme Case IGR J16318-4848

One of the most extreme case is the highly obscured sgHMXB hosting a sgB[e] star IGR J16318-4848, which was in 2003 the first source detected by the INTEGRAL satellite, characterized by a high intrinsic absorption $N_{\rm H} \sim 10^{24}{\rm cm}^{-2}$. This source is not only highly absorbed in X-rays, but also in infrared, the infrared emission being due to a surprising environment surrounding the sgB[e] companion star,[1] a supergiant star surrounded by a complex and stratified envelope, at different densities and temperatures (Filliatre & Chaty 2004; Kaplan et al. 2006; Moon et al. 2007). Multiwavelength studies putting together ESO/VLT, Spitzer and Herschel optical to midinfrared spectroscopic observations, combined with modeling of stellar atmosphere and wind, confirm the presence of absorbing material (dust and cold gas) enshrouding the whole binary system. The sgB[e] star is surrounded by an irradiated inner edge (the *rim*) heated to a temperature of $\sim 3800 - 5500$ K, along with a viscous circumbinary dust disk component at an inner temperature of ~ 750 K, at $\sim 0.7 R_{\rm star}$, thus excluding a spherical geometry for the dust component (Chaty & Rahoui 2012).

Additional broad-band spectroscopic observations with the X-shooter instrument on the VLT, combined with a modeling of stellar atmosphere and wind using the Potsdam Wolf–Rayet code, suggest that the compact object orbits within the cavity present between the star and the disk, and show that the sgB[e] star likely hosts an helium-enhanced atmosphere, due to its intense wind shedding part of its hydrogen envelope. Finally, this study also shows that the high absorption of the system is explained by its high inclination, with the system seen nearly edge-on (Fortin et al. 2020; see Figure 6.9). IGR J16318-4848 shares many similar properties with another

---

[1] The [e] of sgB[e] means that its spectrum exhibits forbidden lines in emission.

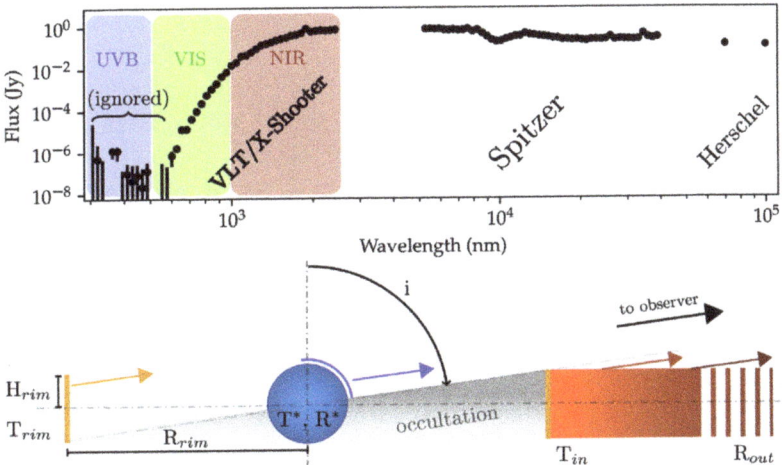

Figure 6.9. Spectral energy distribution, from visible to mid-infrared domain (VLT/X-Shooter, Spitzer, and Herschel), and scenario of the obscured HMXB IGR J16318-4848, showing the supergiant star, surrounded by a cavity, a rim and then a disk of matter. Reprinted with permission of the AAS from Fortin et al. (2020).

sgB[e] source, CI Camelopardalis (CI Cam), exhibiting intense X-ray variability on timescales of days, likely due to an accreting neutron star immersed within a dense and stratified circumstellar envelope. This source also exhibits differences in the accretion flux and circumstellar extinction, which according to Bartlett et al. (2019) represent either changes in the environment due to variable mass loss from the star, or to orbital motion of the accretor.

6.3.3.4 The Hypergiant Case GX 301-2

Another extreme example is the highly obscured HMXB GX 301-2 located at the distance of ~3 kpc, hosting a neutron star orbiting an *hypergiant* star characterized by a strong stellar wind, and surrounded by circumstellar material, detected at 70 and 100 μm with the Herschel far-infrared telescope (Servillat et al. 2014). This circumstellar material is mainly concentrated within a rimmed viscous disk enshrouding the whole binary system, similarly to IGR J16318-4848, extending up to a radius of ~8 au, with a temperature going from $T_{max} \sim 1500$ K at the inner rim, corresponding to dust sublimation, to $T_{min} \sim 200$ K at the edge, allowing for dust formation. Sharing many similarities with other highly obscured X-ray binaries, and B[e] supergiants such as IGR J16318-4848, GX 301-2 might represent a transitional stage in the evolution of massive stars in binary systems, linking supergiant B[e] systems to *luminous blue variables* (LBV), massive supergiant stars characterized by significant outbursts and mass loss (Mennekens & Vanbeveren 2014).

It is likely that these persistent sgHMXB, with $P_{orb} < 10$ days, become obscured when transiting toward Roche lobe overflow, like RLO–HMXB systems. They are characterized by slow winds, mostly because stellar wind is slower in binaries than in single stars, the neutron star cutting off wind acceleration via ionization. Since these obscured sgHMXB are detected ~20% of the observing time in X-rays, one can estimate that they remain, on average, X-ray active during $\sim$20, 000 years, when they are close to overflowing their Roche lobe. When the companion star becomes close to Roche lobe overflow, a deep spiral-in of the neutron star initiates, along with a parallel shrinking of the orbit (van den Heuvel 1983), corresponding to the *Common Envelope* Phase (see Section 3.4 and Paczynski 1976). There exists also the possibility for some systems to accrete both through Roche lobe overflow and stellar wind accretion, such as Cyg X-1, a particular sgHMXB hosting a black hole.

6.3.4 Eccentric sgHMXB

Another subclass of sgHMXB is constituted of eccentric systems, systems which have the particularity to generate variability amplitude reaching even higher factors —up to 100—thus appearing as transient systems, at the periodic time when they are crossing zones of various wind density, traveling along their orbit.

6.3.5 Supergiant Fast X-ray Transients (SFXT)

A significant subclass of sgHMXB is constituted of 12 (+5 candidates) *supergiant fast X-ray transients* (SFXT, Negueruela et al. 2006; Chaty 2013; Walter et al. 2015; and references therein). Contrary to classical sgHMXB, these systems are not persistent, but instead exhibit short and intense X-ray outbursts with low *duty cycle* (fraction of time spent in a high luminous state) of $\leqslant$10%, an unusual characteristic among HMXB, rising in tens of minutes up to a peak luminosity $L_X \sim 10^{36} - 10^{37}$ erg s^{-1}, lasting for a few hours, and alternating with long ($\sim$70 days) periods of quiescence at $L_X \sim 10^{31} - 10^{33}$ erg s^{-1}, with a huge variability factor $10^4 \leqslant \frac{L_{max}}{L_{min}} \leqslant 10^5$.

6.3.5.1 Typical Examples of SFXT
Among SFXT, IGR J17544-2619 represents an archetypical source, displaying rich variability (see Figure 6.10, and Pellizza et al. 2006; Romano et al. 2015). This source has been extensively studied through multi-wavelength simultaneous observational campaign, using the X-ray satellites Chandra, XMM-Newton, NuSTAR and ESO facilities, from faint X-ray quiescent state to bright X-ray outbursts. It exhibits thermal and bulk Comptonization, similarly to young pulsars in HMXB. Like other SFXT, it is significantly sub-luminous compared to other sgHMXB. Optical and infrared observations reveal micro-variability compatible with supergiant stars (Bozzo et al. 2016).

The SFXT SAX J1818.6-1703, a flaring transient X-ray source serendipitously discovered by BeppoSAX in 1998 during an observation of the Galactic center, displays short and bright flares and an unusually very low quiescent level, implying a dynamical range in intensity as large as $10^3 - 10^4$. It also exhibits an unusually long orbital period of $P_{orb} \sim 30$ days, and an elapsed accretion phase of $\sim$6 days, implying an elliptical orbit and an early B0.5-1 I supergiant spectral type, with eccentricities $e \sim 0.3 - 0.4$. Huge variations of the X-ray flux can be explained through accretion of macro-clumps formed within the stellar wind, SFXT behaving like sgHMXB, but with different orbital parameters (Zurita Heras & Chaty 2009). Another SFXT, IGR

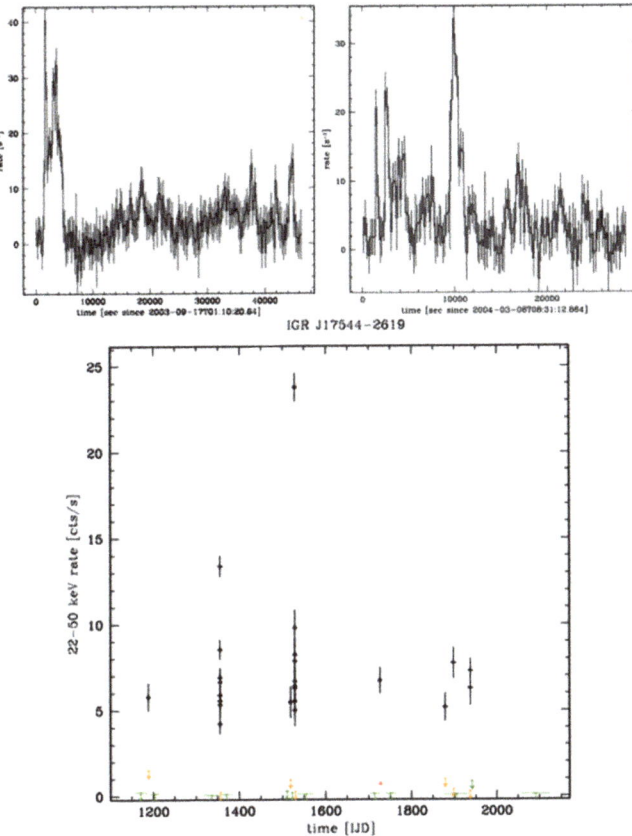

Figure 6.10. X-ray typical behavior of the archetypical SFXT IGR J17544-2619: INTEGRAL pointed (top panels, revolutions 113 and 171, respectively) and 200 s binned (lower panel) lightcurves. Reprinted with permission from Chaty (2013).

J17391-3021, exhibit weak flares observed by the X-ray satellite Suzaku (Bodaghee et al. 2011), which might represent a low-activity state.

6.3.5.2 *SFXT in the Corbet Diagram*

These systems, characterized by a compact object, orbiting either on a circular or eccentric orbit with $P_{orb} \sim 3 - 100$ days and $P_{spin} \sim 100 - 1000$ s, span a vast location in the Corbet diagram, mostly inbetween BeHMXB and sgHMXB (see Figure 6.6). To explain the scattered distribution of SFXT in the Corbet diagram, Liu et al. (2011) proposed that the majority of the systems evolved directly from HMXB hosting OB-type main sequence star (i.e., without a significant accretion history since the formation of the neutron star), whereas a minority of the systems evolved from BeHMXB (i.e., SFXT with an accretion history and thus located within the area of BeHMXB). Among these *intermediate* SFXT, we can cite the sources IGR J18483-0311 (Rahoui & Chaty 2008) and IGR J16465-4507, both hosting an early B

(B0.5–B1) supergiant star, the latter exhibiting a high rotation velocity of $v = 320 \pm 8$km s^{-1} as shown by fitting its spectra on a stellar spectral model (Chaty et al. 2016).

SFXT, characterized by short orbital period but reaching low luminosities, require a mechanism quenching the accretion of stellar wind onto the neutron star surface. Various processes have been invoked to explain such flares and variability, such as stellar wind inhomogeneities (*clumps*), magnetic gating and centrifugal accretion barrier due to spinning neutron star, transitory accretion disk, eccentricity, etc.

6.3.6 X-ray Variability

Variability of the X-ray luminosity of wind-fed sgHMXB is complex, even if most of the behavior seems reproducible by invoking hydrodynamical effects. The X-ray luminosity is mainly determined by two parameters: the density and velocity of the stellar wind surrounding the accreting neutron star. By assuming a smooth stellar wind and a mass to luminosity conversion factor of $0.1mc^2$, X-ray luminosities reached by a wind-fed sgHMXB mostly depend on its orbital period and eccentricity, ranging from $L \sim 10^{33}$ erg s^{-1} for an SFXT with P_{orb} of a few days, to $L \sim 10^{37}$ erg s^{-1} for an RLO–HMXB with P_{orb} of ~10 days, and reaching even up to $L \sim 10^{38}$ erg s^{-1} for an sgHMXB with P_{orb} of a few days (Castor et al. 1975). We now review some of the mechanisms allowing to reproduce at the same time the high range of X-ray luminosity and variability, following the discussion in Walter et al. (2015).

6.3.6.1 Hydrodynamical Effects

The neutron star itself produces hydrodynamical effects on the stellar wind, increasing the variability of the accretion rate in wind-fed systems with circular orbits, by a factor 10–100 (Blondin et al. 1990). When the system is close to Roche lobe overflow, the tidal stream increases the wind density toward the compact object (Blondin et al. 1991), further increasing the variability, particularly in eccentric systems such as GX 301-2. In addition, photo-ionization of the wind by the neutron star, due to the collision between the primary stellar wind, and a gas stream flowing toward the surface of the neutron star (Blondin 1994), generates a shock front with additional wind inhomogeneities, leading to the emergence of low-density bubbles expanding to $\sim 10^{11}$ cm before crashing on the accretion radius of the magneto-sphere. This mechanism generates X-ray luminosity variations by a factor of 10^3 and transient modulations with a characteristic timescale of ~6500 s, similar to the ones detected in Vela X-1 (Manousakis & Walter 2015).

6.3.6.2 Stellar Wind Macro-clumping

Line-driven instability can generate huge density variations in the stellar wind of massive stars, suggesting the existence of dense clumps in the wind of OB secondary stars. In the *macro-clumping* scenario, each SFXT flare is explained by the accretion of a single clump, in the Bondi–Hoyle–Lyttleton approximation, the strong variability in X-rays directly resulting from the strong density and velocity fluctuations in the wind. The typical parameters in this scenario are: a compact object with large orbital radius: $10R_\star$, a clump size of a few tenths of $R_\star$, a clump mass of $10^{19} - 10^{22}$ g (for $N_H = 10^{22} - 10^{23} cm^{-2}$), a mass loss rate of $10^{-5} - 10^{-6} M_\odot$ yr^{-1}, a clump separation of the order of $R_\star$ at the orbital radius, and a volume filling factor in the range $0.02 - 0.1$ (Oskinova et al. 2012). The flare-to-quiescent count rate ratio is directly related to the $\frac{clump}{inter-clump}$ density ratio, which ranges between 15 and 50 for intermediate SFXT, and between 10^2 and 10^4 for classical SFXT. A very high degree of porosity is required to reproduce the observed outburst frequency in SFXT, in good agreement with UV line profiles and line-driven instabilities at large radii (Runacres & Owocki 2005; Owocki 2007; Walter & Zurita Heras 2007). The number of clumps versus radius in a ring of given width and height, for a velocity law with $\beta = 0.8$ and porosity parameter $L_0 = 0.35$, is given in Oskinova et al. (2007). While observations provide clear evidence for clumpy winds (see, e.g., Oskinova et al. 2006), both the details and geometry of such inhomogeneities remain poorly constrained. Wind clumping does not seem to be the unique cause, to explain variability factors as large as 10^3 in classical sgHMXB such as Vela X-1, requiring alternative scenario. However, it is clear that the neutron star can perturb the stellar wind, and even more if it is a massive one, with $M \geqslant 1.8 M_\odot$ (Negueruela et al. 2008).

6.3.6.3 Magnetic Gating
An alternative scenario invokes the possibility that the X-ray variability is generated by Kelvin–Helmholtz instability at high flux level at the magnetospheric boundary, close to the top of the accretion column, leading to *magnetic gating*, similar to keeping a barrier closed during quiescence to prevent accretion, and opening it up during outbursts, to allow direct accretion. The required magnetic field is within the range $\sim 2 - 10 \times 10^{13}$ G (Stella et al. 1986; Bozzo et al. 2008).

6.3.7 Basics of Stellar Wind Accretion

Assuming that the spin period of a neutron star is nearly in equilibrium, the magnetic and corotation radii have a similar value, and the mass accretion rate $\dot{M}_{acc}$ in a smooth stellar wind can be obtained in the frame of the Bondi–Hoyle–Lyttleton accretion scenario (Bondi 1952):

$$\dot{M}_{acc} = \dot{m}_w \times \frac{G^2 M_{NS}^2}{a^2 v_w v_{rel}^3} \qquad (6.3)$$

where $\dot{m}_w$ is the wind mass loss rate from the secondary, a the orbital separation, v_w the wind velocity, and v_{rel} is the relative velocity of the stellar wind, in the frame of the neutron star:

$$v_{rel} = (v_{orb}^2 + v_w^2)^{1/2} \qquad (6.4)$$

where v_{orb} is the orbital velocity of the neutron star. The wind velocity v_w, typically in the range $100 - 1000 km$ s^{-1}, is usually larger than the orbital velocity v_{orb}, for $P_{orb} \sim 10$ days wind-fed systems (Shakura et al. 2012), so that v_{orb} can be neglected:

$=> v_{\mathrm{rel}} \sim v_w$. v_w is given by the so-called $\beta-$ law, β being a parameter usually taken as $\beta = 1$ (Puls et al. 2008):

$$v_w = v_\infty \left(1 - \frac{R_2}{a}\right)^\beta \tag{6.5}$$

where v_∞ is the terminal velocity of the line-driven stellar wind, and R_2 the radius of the secondary. We note that wind parameters such as $\dot{m}_w$ and v_∞ remain largely uncertain.

6.3.7.1 Accretion, Magnetospheric, and Corotation Radii

When the spin period P_{spin} is too high, the accreted plasma cannot reach the neutron star surface, and in some conditions can even be expelled out. This inhibition of accretion due to rotation is the *propeller effect* (Illarionov & Sunyaev 1975). To better understand this effect, we consider three typical radii:

1. The *accretion (Bondi) radius* r_{acc}, at which the inflowing material is gravitationally bound to fall onto the surface of the neutron star (Bondi 1952):

$$r_{\mathrm{acc}} = \frac{2GM_{\mathrm{NS}}}{v_w^2} = 3.7 \times 10^{10} \, (v_{10^3 \mathrm{km\ s}^{-1}})^{-2} \ \mathrm{cm}. \tag{6.6}$$

2. The *magnetospheric radius* r_{mag}, at which the magnetic pressure of the neutron star balances the ram pressure of the inflowing matter. When $r_{\mathrm{mag}} > r_{\mathrm{acc}}$ (Davies & Pringle 1981):

$$r_{\mathrm{mag}} = 3.3 \times 10^{10} \left(\dot{M}_{10^{-6} M_\odot \mathrm{yr}^{-1}} \times v_{10^3 \mathrm{km\ s}^{-1}} \right)^{-1/6}$$
$$(a_{P_{\mathrm{orb}}=10d} \times \mu_{10^{33} \mathrm{G\ cm}^{-3}})^{1/3} \ \mathrm{cm}. \tag{6.7}$$

3. The *corotation radius* r_{cor}, at which the neutron star angular velocity equals the Keplerian angular velocity:

$$r_{\mathrm{cor}} = 1.7 \times 10^{10} (P_{\mathrm{spin}_{10^3 s}})^{2/3} \mathrm{cm}. \tag{6.8}$$

6.3.7.2 Various Regimes of Accretion

A magnetosphere extending beyond the accretion radius prevents most of the inflowing matter from being accreted (see Section 2.2.2 and Bondi 1952). Determined mainly by its spin, the corotation radius can change only over long timescales related to secular evolution of the neutron star. Thus, according to various radius configurations, different regimes take place, as described below, and reported in Figure 6.11 (we follow here the discussion in Bozzo et al. 2008):

1. *Super-Keplerian inhibition regime* (inflowing matter outside r_{acc}): r_{acc} and $r_{\mathrm{cor}} < r_{\mathrm{mag}}$, thus with a magnetospheric radius rotating more slowly that the inner regions close to the neutron star. In this case the inflowing matter is shocked and halted close to r_{mag}, in direct interaction with the neutron star magnetosphere which is locally super-Keplerian, not able to proceed further

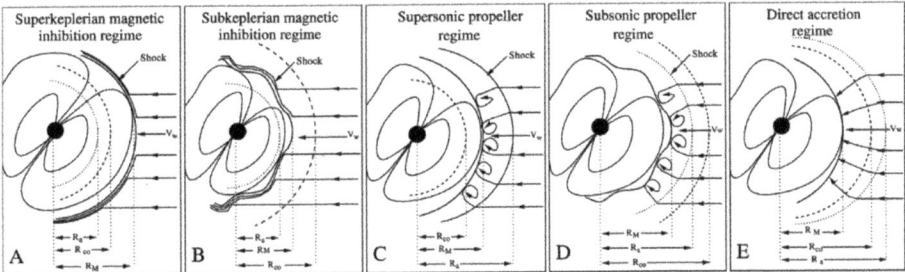

Figure 6.11. Schematic view of a magnetized neutron star interacting with the inflowing matter from its supergiant companion. The different regimes are shown, based on the relative position of the accretion, magnetospheric and corotation radii. Reprinted with permission of the AAS from Bozzo et al. (2008).

toward the neutron star surface. This prevents accretion, dissipates the rotational energy, and finally spins down the neutron star.

2. *Sub-Keplerian inhibition regime* (inflowing matter outside r_{acc}): $r_{acc} < r_{mag} < r_{cor}$, matter penetrates the neutron star magnetosphere, the boundary between inflowing matter and magnetosphere is subject to Kelvin–Helmholtz instability (Harding & Leventhal 1992), all kinetic energy of the wind is converted into thermal energy, and the shear velocity is dominated by the magnetosphere rotation. The diffusion-induced mass accretion rate is orders of magnitude smaller than due to Kelvin–Helmholtz instability, leading to negligible contribution of the total X-ray luminosity.

3. *Supersonic propeller regime* (inflowing matter inside r_{acc}): $r_{cor} < r_{mag} < r_{acc}$, inflowing matter is shocked and halted at the magnetosphere, redistributed into a quasi-spherical *atmospheric* configuration, under hydrostatic equilibrium. Similarly to super-Keplerian inhibition regime, this prevents any accretion, dissipates the rotational energy, and finally spins down the neutron star.

4. *Subsonic propeller regime* (inflowing matter inside r_{acc}): $r_{mag} < r_{acc} < r_{cor}$, the magnetosphere rotation is now subsonic compared to the surrounding material, the centrifugal barrier does not operate anymore, but the neutron star rotational energy dissipation is too high for inflowing matter to penetrate the magnetosphere, and only a fraction of this matter can be accreted, mainly due to Kelvin–Helmholtz instability and Bohm diffusion. If $\dot{M}_{acc} < \dot{M}_C$, where $\dot{M}_C$ is the critical limit when gas radiative cooling takes place, direct accretion onto the neutron star surface occurs, at a rate corresponding to the inflowing matter.

5. *Direct accretion regime*: $r_{mag} < r_{cor} < r_{acc}$, if matter outside the magnetosphere cools down efficiently, then accretion onto the neutron star takes place at the same capture rate $\dot{M}_{capt}$ at which it flows toward the magnetosphere, at a standard accretion regime in which the system achieves the highest mass to luminosity conversion efficiency:

$$L_{acc} = \frac{GM\dot{M}_{capt}}{R} = 2 \times 10^{35} \dot{M}_{capt} \left(10^{15} \text{g s}^{-1}\right) \text{erg s}^{-1}. \quad (6.9)$$

6.4 Beginning Atmospheric Roche-lobe Overflow Systems

Beginning atmospheric RLO–HMXB host a massive ($>10M_\odot$) giant or sub-giant companion star ($L \geqslant 10^5 L_\odot$, $R \sim 10-30 R_\odot$), almost filling its Roche lobe, where accreted matter flows via inner Lagrangian point L1 to form an accretion disk, thus with an accretion process similar to *disk-fed* LMXB (Savonije 1978). They constitute the *classical–bright* HMXB. Among them, Cen X-3, one of the first detected HMXB (Giacconi et al. 1971), and the only RLO–HMXB identified in our Galaxy, is an X-ray pulsar accreting from a massive ($M \geqslant 15M_\odot$) companion star. We can also mention SMC X-1 and LMC X-4 within the Magellanic Clouds, with accretion of matter occurring through the formation of a permanent accretion disk, leading to a high X-ray luminosity, ranging from $L_X \sim 10^{38}$, up to 10^{39} erg s^{-1} during outbursts, and even amounting up to $L_X \sim 10^{40}$ erg s^{-1} when the donor star nearly fills the Roche lobe, the accretion becoming dominated by a tidal stream. Accretion disk instabilities might cause the emission of regular X-ray pulses, as proposed by Dubus et al. (2001) in the frame of the disk instability model, initially developed for X-ray transients.

Roche-lobe overflow is rarely observed, since the consequence is that the compact object quickly becomes enshrouded. The lifetime of these systems, before they become dynamically unstable, is thus expected to be relatively short (10^3 to 10^5 yr; Tauris & van den Heuvel 2006). These intense X-ray sources are characterized by regular X-ray eclipses and ellipsoidal variations produced by tidally deformed companion star.

6.4.1 RLO–HMXB in the Corbet Diagram

There are only a few sources, located in the lower left of the Corbet diagram (Figure 6.6), characterized by short orbital ($P_{orb} \sim 1-5$ days) and spin periods ($P_{spin} \sim 10^{-1}-10$ s), suggesting that neutron stars within these systems have spun up in the course of their evolution.

6.5 Formation and Evolution of HMXB

6.5.1 Isolated Binary Evolution of an HMXB

The overall evolution of an HMXB is represented in Figure 6.12 in the frame of the *isolated binary evolution* scenario, also called *in the field*. The evolution of such a binary is also represented in the Hertzsprung–Russell (HR) diagram in Figure 6.13, to illustrate the evolution of each star within the binary system, through the exchange of matter and angular momentum. In addition to all single-star evolution aspects, features such as mass transfer through accretion, common envelope evolution, collision, supernova kick and angular momentum loss mechanism have to be included, to fully reproduce the overall evolution of an HMXB (Hurley et al. 2002).

6.5.1.1 Formation

The whole story begins with two relatively massive *zero-age main sequence* (ZAMS) stars, of mass $M_{ZAMS} \geqslant 12M_\odot$, born in an interstellar molecular cloud. One of the star can be less massive than $12M_\odot$, but it has attracted enough matter from its

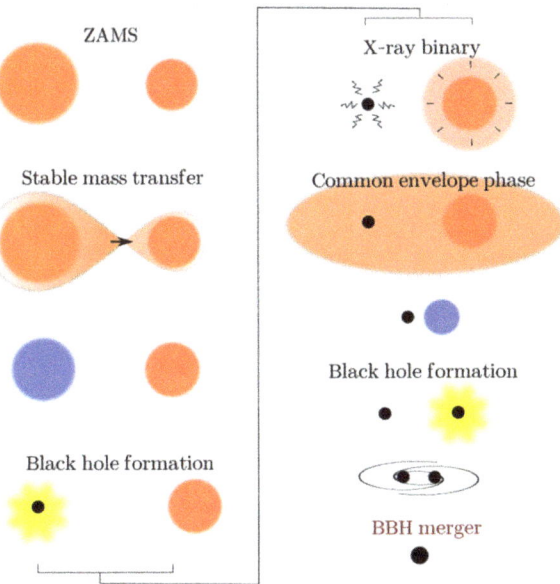

Figure 6.12. Schematic scenario of the isolated binary evolution channel of HMXB, through common envelope. Reproduced with permission from Astronomy & Astrophysics, Credit ESO (García et al. 2021).

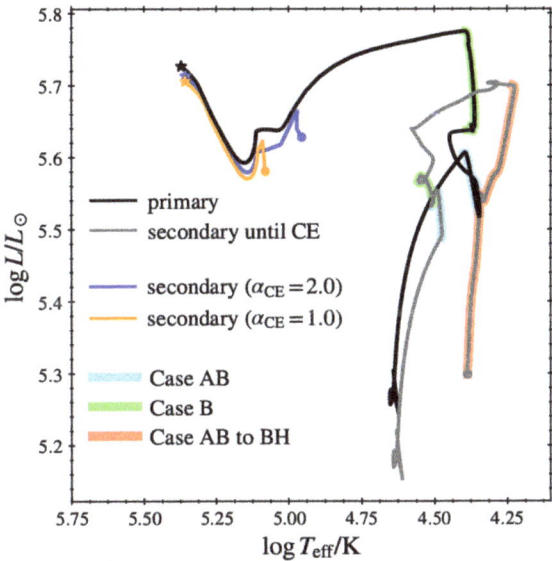

Figure 6.13. Full evolution, in the HR diagram, of a binary system constituted of two massive stars born at ZAMS (bottom right). Case AB stable mass transfer phase is indicated in blue, case B in green. After the mass transfer phase, the primary moves to the left and collapses to form a black hole, indicated by a black star. The secondary then expands, and a case AB stable mass transfer occurs (drawn in salmon), until an unstable CE phase is triggered (gray circle). The detach of the secondary occurs at the blue circle for CE efficiency $\alpha_{CE} = 2.0$ and at the orange circle for $\alpha_{CE} = 1.0$. A second BH is formed at the top left corner (colored stars). Reproduced with permission from Astronomy & Astrophysics, Credit ESO (García et al. 2021).

companion, to end up above the threshold mass to trigger a supernova. The more massive star, called star #1, reaches first the end of its main sequence phase, having completed the fusion of hydrogen into helium, the core contracts, and begins to burn helium, leading to an expansion of its hydrogen envelope. If the binary is tight enough, then it begins to experience its first dynamically stable and semi-conservative mass transfer phase (case B, see Section 3.3), during which star #1 fills its Roche lobe and transfers mass to star #2, on a thermal timescale for the primary, and non-conservative mass transfer since star #2 cannot accrete all of this incoming mass. Since the binary loses mass, in addition to stellar wind, the binary orbit widens. Star #1 loses its entire envelope, and leaves behind a naked helium-burning *Wolf–Rayet* star (which typically accounts for ~30% to 50% of its initial mass).

6.5.1.2 First Supernova

The first supernova event occurs, when star #1 collapses into a compact object (a neutron star if the mass of the helium star is $M \geqslant 2.8 M_\odot$ as in Figure 5.24, right panel, and a black hole if $M \geqslant 8 M_\odot$ as in Figure 6.12). In case the first supernova event forms a neutron star, it has accreted at this point a maximum of ~$0.02 M_\odot$ (Tauris et al. 2017). After this collapse, in case the binary system is not disrupted by a natal kick due to mass ejection during asymmetric supernova event (Atri et al. 2019), star #2 continues to evolve, leaving the main sequence toward red giant phase, expanding until its radius, which can become as large as several thousand $R_\odot$, begins to fill its Roche lobe (via beginning atmospheric Roche-lobe overflow). At this stage the system enters the accreting phase, becomes an HMXB, with a timescale of accretion of ~10^5 yr. Star #2 continues to evolve, within a binary system of orbital separation between a few hundred to a few thousand $R_\odot$.

6.5.1.3 Common Envelope

As a result of star #2 evolution, its extended envelope begins to enshroud the whole binary system, now entering the common envelope phase, characterized by a runaway process of dynamically unstable mass transfer (see Section 3.4, and a review in Ivanova et al. 2013). Since the neutron star or black hole now orbits inside the envelope of the red giant, it causes a strong drag and heavy loss of angular momentum, quickly leading to a rapidly shrinking orbit. The following evolution of the system depends on whether the common envelope will be ejected, and on the fate of the compact object inside the massive envelope.

- If the orbital period of the HMXB at the time of the Roche-Lobe overflow is relatively short (typically $\leqslant 1$ yr; Tauris et al. 2017), then the binding energy of the donor star envelope is too large for the envelope to be successfully ejected via orbital energy. In this case, the compact object (usually a neutron star) is dragged, leading to fast spiral-in, until sitting down right at the center of the massive star, becoming a *Thorne–Zytkow* (TZO) object (Thorne & Zytkow 1975). This exotic stellar object, formed of a neutron core surrounded by an extended envelope, externally looks like a cool red supergiant, with anomalous enhancements of rubydium, lithium, and mobedium. They are difficult to distinguish from normal red supergiants, and there is currently only one candidate, still unconfirmed: HV 2112, in the SMC (Levesque et al. 2014). The story ends here for these prematurely merged systems, in which the compact object resulting from star #1, now resides *forever* at the heart of star #2. We point out here that there exists an alternative pathway to form such TZO object, via the *stellar core-merger-induced collapse* model, involving a white dwarf instead of a neutron star (Ablimit et al. 2022).

- For systems having a longer orbital period at the time of the Roche-Lobe overflow ($\geqslant 1$ yr), then the binding energy of the donor star envelope is small enough to be successfully ejected via deposition of dissipated orbital and thermal energy. This common envelope phase thus ends up with the ejection of the envelope, producing a tight binary system, with a compact object now orbiting in a few hours (corresponding to an orbital separation a of a few $R_\odot$, typically $\leqslant 10 R_\odot$, much smaller than the initial separation) around star #2, which has entirely lost its envelope, thus corresponding to a WR star (see Figure 6.14).

It is interesting to notice here that the main difference between LMXB and HMXB systems—in the frame of this isolated binary evolution scenario—is that, LMXB are observed in their accreting, and thus X-ray emitting phase, *after* the CE phase, while in contrast HMXB are observed during their accreting phase, *before* the CE phase (see Figure 5.24). This means that while we only see LMXB that have survived the CE phase, we do not know how many HMXB will survive this phase of their evolution.

6.5.1.4 Second Supernova
It is then the turn to the naked core of the giant star #2 to explode, forming the second compact object. When star #2 collapses into a compact object, either neutron

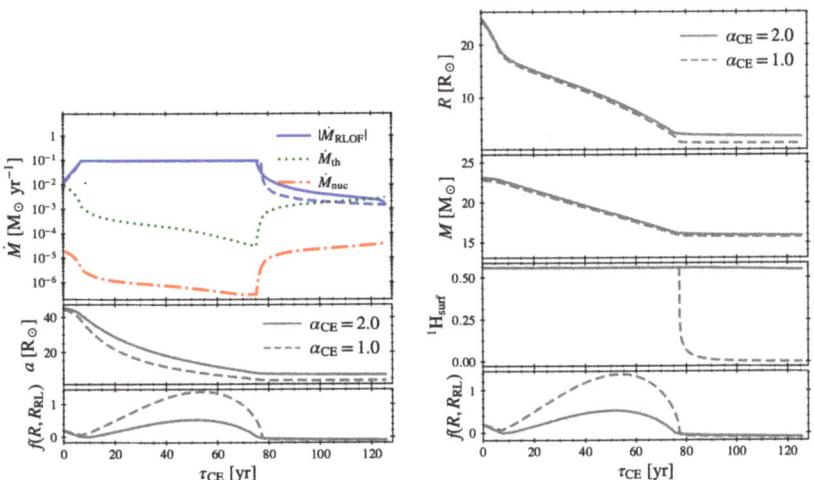

Figure 6.14. HMXB evolution during the common envelope phase. *Left:* mass transfer rate $\dot{M}$, orbital separation a and stellar radius compared to Roche lobe R/R_L, versus time τ_{CE}. *Right:* radius R, mass M and R/R_L, versus τ_{CE}, for different values of common envelope efficiencies α_{CE}. Reproduced with permission from Astronomy & Astrophysics, Credit ESO (García et al. 2021).

star or black hole (as in the case of Figure 5.24 right panel or Figure 6.12, respectively), without receiving a strong natal kick (see below) the system enters the last—and long—phase of its binary evolution, both compact objects now orbiting around each other with a short orbital period, radiating away gravitational energy at each orbit. The compact binary becomes more compact with time, until eventually merging. After a time of the order of a few $\sim 10^9$ years (the delay time τ_{GW}, between formation and merger, is proportional to the separation a^4; Peters 1964), both components collapse, finally merging in a product which likely is a black hole. One of the main uncertainty of this isolated scenario, is the fraction of binaries that will initiate and survive the common envelope phase, avoiding a premature merger and yet coming close enough to interact and eventually merge in less than the Hubble time.

6.5.2 Stellar Winds

As already mentioned in Section 6.3.7; hot ($\geqslant 10^4$ K) and massive ($M_{ZAMS} \geqslant 30 M_\odot$) stars lose a substantial fraction of their mass through the process of line-driven stellar winds, which heavily depends on the absolute metallicity Z (Castor et al. 1975). The mass-loss rate due to stellar winds is written as $\dot{M} \propto Z^\alpha$ (Vink 2001; Chen et al. 2015). The most recent models suggest that the exponent $\alpha \sim 0.85$ is not constant, but instead depends on the luminosity of the star (Puls et al. 2008; Vink 2011). The mass loss is higher when the stellar luminosity L is closer to the Eddington value L_{Edd}, quenching the dependence with metallicity when $L \geqslant L_{Edd}$. A star with $M_{ZAMS} = 90 M_\odot$ will end up with a final mass $M \sim 25 M_\odot$ for solar metallicity $Z = 10^{-2}$, and $M \sim 35 M_\odot$ for metallicity $Z \leqslant 2 \times 10^{-4}$ (see Figure 6.15; Spera et al. 2015).

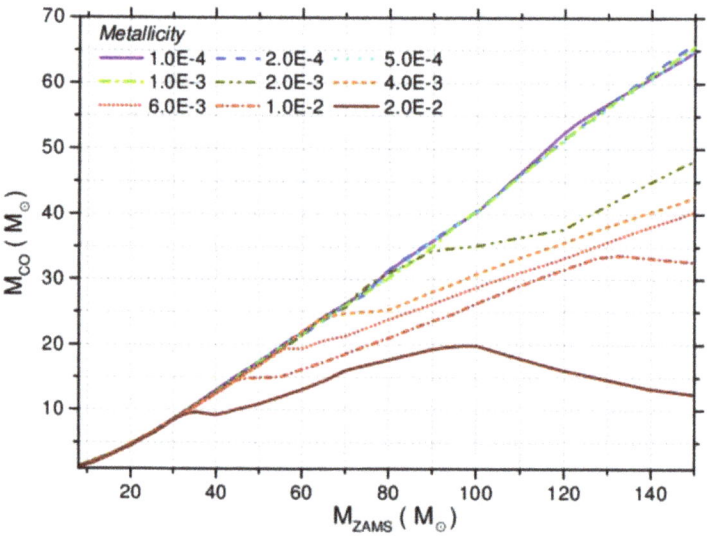

Figure 6.15. Dependence of final stellar mass as a function of the initial ZAMS mass, for different values of metallicity. Reprinted with permission from Spera et al. (2015).

While in single stars, stellar winds uniquely determine the final mass of the star, this is obviously not the case for massive stars within binaries. Since low-metallicity stars retain a large fraction of their mass and develop larger CO-core, they end up their evolution with larger pre-collapse mass, and thus are more likely to directly collapse into large mass black holes (Belczynski et al. 2010). One of the main uncertainty of most HMXB evolution is the mass loss rate through stellar winds, particularly during specific stellar evolutionary phases (see for instance the extreme case Eta Carinae; Abdo et al. 2010a), such as *luminous blue variables* (LBV)—massive supergiant stars characterized by significant outbursts and mass loss (Mennekens & Vanbeveren 2014)—and naked helium cores (*Wolf–Rayet* stars). For reviews on winds in massive stars, particularly within binary systems, see, e.g., Martínez-Núñez et al. (2017), Kretschmar et al. (2019).

Hirai & Mandel (2021) have explored the effect of anisotropic line-driven winds on the properties of accretion onto black holes in HMXB, in which the tidal force from the companion star can modify the wind structure, reducing either the wind terminal velocity due to weaker effective surface gravity, or the mass flux due to gravity darkening. Their study shows that a focused accretion stream naturally forms when the Roche-lobe filling factor reaches values $\geqslant 0.8 - 0.9$, close to beginning atmospheric Roche-lobe overflow, bringing enough angular momentum to form an accretion disk around the black hole, which suggests that only systems reaching this filling factor will become observable.

6.5.3 Natal Kick

Along with common envelope and stellar winds, natal kicks constitute at the same time a key mechanism, and one the main sources of uncertainty, within the isolated evolution scenario (see Figure 6.16). The natal kick is a crucial event in the evolution of a binary system, having an effect both on its orbital parameters and systemic velocity (Brandt & Podsiadlowski 1995). Natal kicks are directly related to the fundamental mechanisms occurring during the SN event, either ejecta-driven or neutrino-driven (Fryer & Kusenko 2006). They are also connected to other fields of astrophysics, since observed NS-kick distribution in HMXB better constrains the

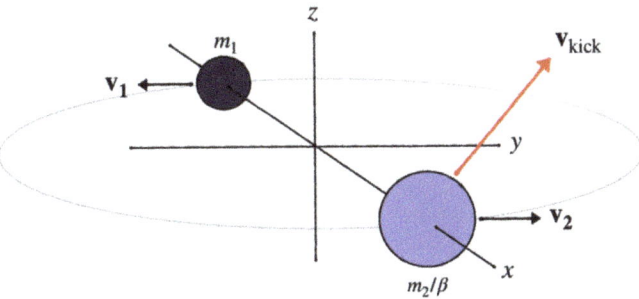

Figure 6.16. Schematic of the natal kick received by a component of a binary system (orbit situated in the $x - y$ plane), at the instant of the secondary stellar collapse. Reprinted with permission of the AAS from Callister et al. (2021).

population synthesis models (Baibhav et al. 2019), used to estimate the compact merger rates detected by LIGO-Virgo (see Shao & Li 2018; Abbott et al. 2021a, for binary NSBH mergers). A better understanding of the kick effect thus allows astronomers to answer to fundamental questions related to the formation of compact objects, and survival rate of binaries, that we will describe in the following.

6.5.3.1 Blaauw Kick

The natal kick potentially transfers to the binary a significant velocity in case the supernova event forming the NS is asymmetrical (Shklovskii 1970). It is thus necessary to accurately assess the impact of such kicks, in order to quantify the survival rate of binaries, disrupted by natal kicks stronger than typically $\sim 100 \mathrm{km\ s^{-1}}$ (but depending on various binary parameters, see Hills 1983). In particular, a consequence of the virial theorem is that a binary is disrupted by a supernova event, if more than half of the total mass of the binary ($M_1 + M_2$) is suddenly ejected, during a spherically symmetric mass loss, known as *Blaauw kick* (Blaauw 1961). Thus, binaries exhibiting large systemic velocities and low eccentricities, as well as those with low systemic velocities and high eccentricities, are hardly explained without invoking an asymmetric SN kick (in a random direction) at birth. The parameters to take into account, in case of kicks received at the NS formation, are the mass ratio of the binary, the metallicity, and potentially the magnetic field.

6.5.3.2 Simulations of Natal Kick

An analytical equation describes the impact of kick, linking the pre-supernova to post-supernova orbital parameters, assuming a Maxwellian distribution of kick velocities and an isotropic probability of kick direction (Kalogera 1996). For each binary, and a sufficiently high value of kick velocity, there exists a critical angle for which the direction of the kick received at the supernova event, relative to the orientation of the pre-supernova velocity, will disrupt the binary system. Many binaries survive the two supernova events occurring in the course of their evolution, first because of large-scale mass transfer prior to supernova event, and also because of likely low kicks. Kalogera (2000) computed the expected spin–orbit misalignment angle after the second core-collapse event, during the formation of close compact binaries, deriving that between 30 and 80% of binaries containing a black hole and a neutron star which coalesce within 10^{10} years, exhibit misalignment angles $\geqslant 30°$, constraining the kick magnitudes required for binaries to remain bound.

Based on Chandra X-ray observations of pulsar jets, in particular of Crab and Vela pulsars, theoretical models were initiated, in order to explain the apparent alignment of the spin axes, proper motion directions, and polarization vectors of pulsars, by effects of rotation and natal kick received during the formation of the proto-neutron star, by asymmetric mass ejection and/or neutrino emission. These initial models (Lai 2001; Lai et al. 2001) suggest that, in order to produce a kick velocity of a few hundred km s^{-1}, the neutron star must be born with $P_{\mathrm{spin}} \sim 1$ ms, without efficient spin-down effect due to gravitational radiation, compared to standard magnetic braking.

In the context of large natal supernova kicks in the observed distribution of pulsar velocities, Podsiadlowski et al. (2004) examined the effects of binary evolution on the final core structure of a massive star, involving its rotation rate, helium core size, carbon burning strength, and the final iron core mass. These parameters strongly influence the collapse during the supernova event, and subsequently the neutron star kicks. They show that stars in relatively close binaries, with initial masses in the range $8 \leqslant M \leqslant 11 M_\odot$, generally finish up with relatively smaller final iron core mass, after having lost their envelope during binary interaction, and produce neutron stars with low-velocity kicks (see Figure 6.17).

The first 3D-SN simulations, based on asymmetries growing in the core of a massive star just prior to collapse, showed that the newly-formed neutron star is not only kicked by ejected matter, but also by neutrinos, carrying away momentum (Fryer 2004). However, these simulations could not reproduce NS velocities in excess of 200km s^{-1}, while some NS are observed at velocities in excess of 500km s^{-1}. Further simulations, based on young isolated pulsar within pulsar wind nebula, and radio polarization observations, have shown that a correlation exists between the pulsar motion due to natal kick, and the spin direction (Ng & Romani 2007).

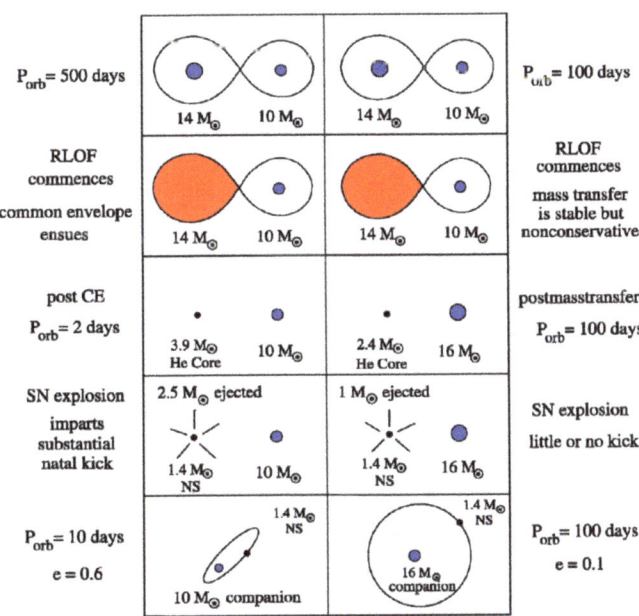

Figure 6.17. Dichotomy of natal kick scenario for HMXB. *Left:* the binary, wide enough at the onset of mass transfer, forms a 3.9$M_\odot$ helium core primary, leading to a large natal kick, and then to a final binary with a short orbital period and large eccentricity. *Right:* the primary envelope is lost while the helium core has a low mass (2.4$M_\odot$), resulting in a prompt SN event and a small natal kick, forming a wide binary with small eccentricity. Reprinted with permission of the AAS from Podsiadlowski et al. (2004).

Observation and modeling of natal kicks have been performed on various sources, such as isolated radio pulsars (Lyne & Lorimer 1994), and recycled pulsars associated with globular clusters or supernova remnants (Hobbs et al. 2005).

6.5.3.3 Observations of NS Natal Kick

Observations of pulsars, exhibiting high velocity proper motion of $\sim$400km s^{-1}, suggest that most neutron stars receive a momentum kick at their formation, with a large spread in kick magnitude, depending on the past interaction history of the exploding star, and thus correlating with the NS mass (Tauris et al. 2017).

Observations of the binary LS I +61°303 show that it is running away at $\sim$27km s^{-1} from its birthplace, invoking an asymmetric natal kick received at the supernova that formed the compact object, and an ejected mass $\leqslant 2M_\odot$, shooting out the binary with a linear momentum of $\sim$430$M_\odot$km s^{-1}, comparable to values found in runaway neutron stars and millisecond pulsars (Mirabel et al. 2004; Papitto et al. 2020). Monte-Carlo simulations, combined with observations, of Cir X-1, an LMXB hosting a NS, showed that an important kick velocity of $\sim$740km s^{-1} (with a minimum of $\sim$500km s^{-1}) is required to account not only for its radial velocity, but also for its orbit and eccentricity of 0.94, thus identified as a *survivor* of a highly asymmetric supernova (Tauris et al. 1999). Precise interferometric data from the Australian Long Baseline Array allowed to determine the position, velocity and orbit of PSR B1259-63, a binary which traveled $\sim$8 pc from its birthplace, due to a moderate kick received at the first SN event (Miller-Jones et al. 2018). A statistical study by Bodaghee et al. (2021) of HMXB distribution in the Small Magellanic Cloud showed that they remain close to their likely birthplace, with an average kick velocity value of $\leqslant$34km s^{-1}, less than their counterparts in the Milky Way (see Section 6.1.2 and Bodaghee et al. 2012; Coleiro & Chaty 2013), suggesting that the Galactic environment plays a fundamental role in stellar evolution, and particularly in what concerns kick velocity of HMXB.

Binary population synthesis models have been explored by varying prescription for natal kick, remnant mass, and mass accretion efficiency, in order to constrain different models, comparing with parallax and proper motion obtained for young isolated radio pulsars and Be-X-ray binaries (Igoshev et al. 2021). They obtain a proportion of 20% of weak natal kicks with $v \sim 45$km s^{-1}, the remaining natal kicks being higher, with $v = 336$km s^{-1}, and suggest a new model of natal kicks for electron capture SN, to satisfy observations both of isolated radio pulsars and Be X-ray binaries.

6.5.3.4 Observations of BH Natal Kick

Stellar-mass black holes also receive natal kick during their formation, due to asymmetric mass ejection during fallback supernova (Janka 2013). Population synthesis computations performed on black hole kicks allow astronomers to compare the expected vertical distribution of both BH and NS–HMXB in our Galaxy (Repetto et al. 2012; Repetto & Nelemans 2015; Repetto et al. 2017). Hubble Space Telescope observations of the black hole LMXB GRO J1655-40 revealed a high runaway velocity, explained by the natal kick imparted during the formation of the compact object (Mirabel et al. 2002).

Proper motions of 16 black hole X-ray binaries, obtained through Very Long Baseline Interferometry (VLBI) and Gaia DR2 data (Atri et al. 2019), show that most of these systems exhibit potential kick velocity $\geqslant 70$km s^{-1}, interpreted as a majority of BH acquiring strong kicks at their formation. The distribution of natal kick of this BH population presents a mean velocity of ~ 107km s^{-1}, without any correlation with the black hole mass, thus excluding any link between the formation mechanism of stellar mass black holes and their mass. This study also suggests that strong kicks in BH–HMXB systems are in agreement with the presence of low-frequency QPO in their PDS (see Section 5.3.5 on X-ray variability of BH systems), and the spin–orbit misalignment.

To conclude, natal kicks and mass loss are more likely to disrupt binaries hosting a black hole than a neutron star, in the course of the asymmetric supernova event. And in the case that the binary remains bound, the sudden mass loss during the supernova event still affects the orbit, by making it highly eccentric (Giacobbo & Mapelli 2018, 2020).

6.5.3.5 *Dependence of the Natal Kick on the Companion Mass*

The dependence of peculiar velocity v_{pec} (defined by the proper motion minus Galactic orbital motion) on the mass of the companion star shows a slight positive correlation, as reported in Figure 6.18 from a sample of 35 Galactic NS–HMXB, selected as described in Fortin et al. (2022a). They all possess (i) a referenced high-energy detection (γ and/or X-rays), (ii) a single optical/infrared counterpart other

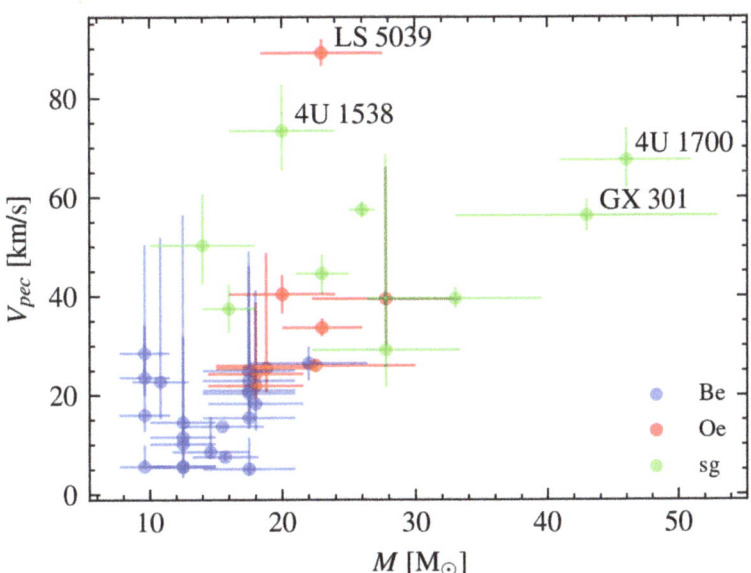

Figure 6.18. Peculiar velocity versus secondary mass in a selected HMXB sample. Image Credit: Francis Fortin; Reproduced with permission from Astronomy & Astrophysics, Credit ESO (Fortin et al. 2022a).

than Gaia, (iii) a Gaia EDR3 counterpart, and (iv) their companion spectral type was derived with spectroscopy. This study shows that the peculiar (assumed to be of same order than natal kick) velocity versus non-degenerate companion mass plane hints at a similar trend, with a median peculiar velocity of $<v_{pec}> \gtrsim 116 \pm 16 km\ s^{-1}$. More specifically, BeHMXB exhibit relatively low peculiar velocity ($<v_{pec}> \gtrsim 91 \pm 16 km\ s^{-1}$), while the slightly more massive OeHMXB exhibit relatively high peculiar velocity ($<v_{pec}> \gtrsim 126 \pm 35 km\ s^{-1}$), and the even more massive sgHMXB systematically higher peculiar velocity ($<v_{pec}> \gtrsim 147 \pm 38 km\ s^{-1}$), with little, or null, dependence on the companion mass. For instance, Fortin et al. (2022a) point out that all BeHMXB systems stay beyond $v_{pec} \leqslant 40 km\ s^{-1}$, while only 2 out of 9 supergiants are found below this peculiar velocity (see Figure 6.18).

6.5.4 Alternative Scenario of HMXB Evolution

In parallel to this isolated binary evolution scenario, there are alternative possibilities, in particular avoiding the common envelope phase, to form a binary which will experience an accreting phase at some stage of their evolution (see discussion in Mandel & Farmer 2017).

6.5.4.1 Dynamical Formation Channel

An alternative scenario to form a massive binary is through a dynamical process (see Figure 6.19), in a dense stellar cluster embedded in a molecular cloud (Lada & Lada 2003): either a young massive star cluster (age $\leqslant 100$ Myr, mass $10^4 - 10^5 M_\odot$; Portegies Zwart et al. 2010), a globular cluster (age ~ 12 Gyr, mass $10^4 - 10^6 M_\odot$; Abdo et al. 2009; Gratton et al. 2019), or even a nuclear star cluster (dense and massive structure $10^5 - 10^8 M_\odot$, at the center of most galaxies; Neumayer et al. 2020). The particularity of these star clusters is to have at the same time a high density ($\geqslant 10^3$ stars/pc) and a low-velocity dispersion (from a few to a few tens of km s^{-1}— apart from the nuclear cluster where it can reach a few hundred km s^{-1}), so that some stars draw closer together, forming a deeper potential well, while other stars move outwards and may even entirely escape from the system (Spitzer 1984, 1986). For instance, according to the scenario of Gieles et al. (2021), the globular cluster Palomar 5, located in the Galactic halo, formed with an usual black hole mass fraction of a few percent, before stars were lost at a rate higher than black holes. This process gradually increased the black hole fraction, now amounting to $\sim 20\%$ of the actual mass cluster, which now hosts a substantial population of stellar-mass black holes, enhancing tidal stripping and tail formation. This scenario even predicts that, while such a cluster will be eventually constituted of black holes at 100% in nearly 1 Gyr, this is not the fate of initially denser clusters, which eventually end up with lower black hole fractions.

In such star clusters, black holes rapidly become the most massive objects within a few tens of millions of years, dynamical relaxation causing them to sink toward the cluster core, where they form new black hole binaries. In this case, the two components of such a binary have not been formed within the same molecular cloud, but have met during the course of their evolution. Due to mass segregation,

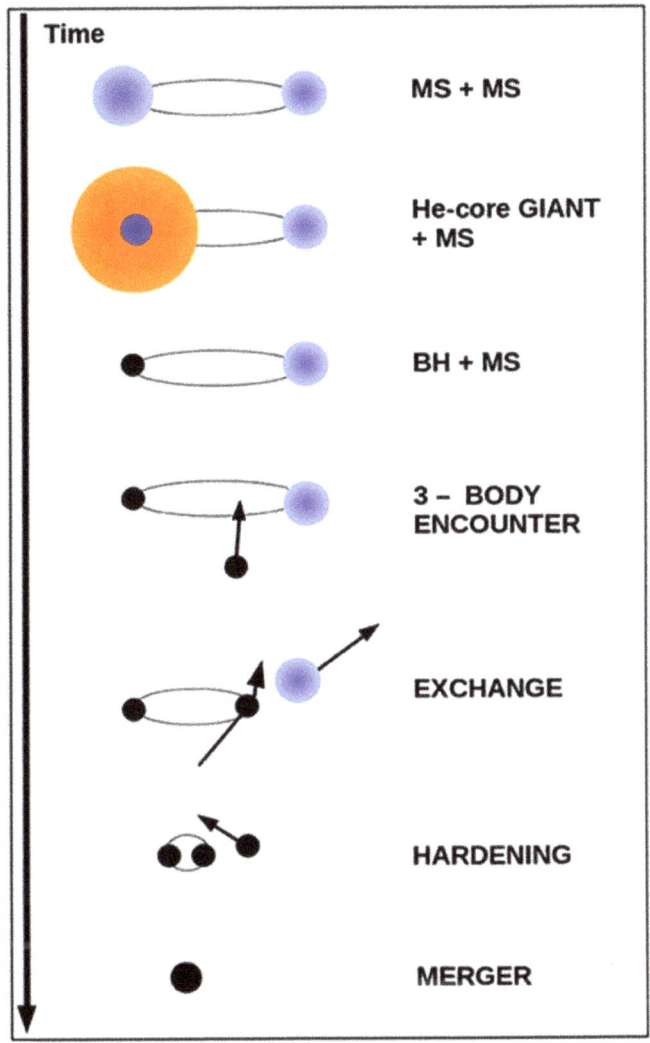

Figure 6.19. Dynamical capture in dense stellar cluster, an alternative scenario to form a binary black hole. Reprinted with permission from Mapelli (2020), Creative Commons Attribution License (CC BY).

the most massive stars in the cluster tend to sink toward the center, where they may form binaries through three-body interactions, in which the lightest object tends to be ejected, and the two heavier objects form a new binary. Thus, after a close three-body encounter between a single object and a binary (Hills & Fullerton 1980), a dynamical exchange of the lightest component of the binary takes place, ejected by heavier ones, this scenario of repeated mergers favoring the formation of binaries of increasing total mass. Stellar-mass black holes in young star clusters are more efficient in this process of forming new binaries, because of their higher mass (10 times more black hole than neutron star binaries; Ziosi et al. 2014).

We distinguish *soft* (wide) and *hard* (tight) binaries, by the orbital speeds of the binary components. Soft binaries, with orbital speeds smaller than the velocity dispersion inside the cluster, tend to be disrupted after an interaction with an interloper. On the contrary, hard binaries, with higher orbital speeds, tend to become more tightly bound, reducing their orbital separation after a super-elastic interaction, the interloper leaving the binary with higher speed by removing some of its internal energy (*dynamical hardening*; Portegies Zwart & McMillan 2000). When the density of the cluster is high enough, this process ensures numerous interactions to harden the binary, ultimately ejected from the cluster, until it merges in a Hubble time, after reducing its orbital separation, due to gravitational radiation (Mapelli 2016; Mapelli & Giacobbo 2018).

Such mergers in dense star clusters, in which dynamics is primordial with $n \geqslant 10^3$ stars/pc^3, eventually lead to the formation of black holes of mass up to $M \sim 80 M_\odot$, thus more massive than in the isolated scenario, where we expect black holes of mass up to $\sim 40 M_\odot$ with the same metallicity (García et al. 2021), leading to a final black hole of mass up to ~ 160 and $80 M_\odot$ for a binary black hole in stellar cluster and in the field, respectively. In addition, a few percent of merging black holes in young stellar clusters would reach a mass within the pair-instability mass gap ($\sim 60 \leqslant M \leqslant 120 M_\odot$), as estimated by Di Carlo et al. (2019, 2020), Rodriguez et al. (2019), Fragione et al. (2022).

Finally, successive episodes of repeated mergers of binary BH, i.e., *runaway collisions* within a dense (nuclear) stellar cluster at the center of galaxies, might even form *intermediate-mass black holes* (IMBH) with mass $M \geqslant 100 M_\odot$, especially at low metallicity (Portegies Zwart & McMillan 2002; Portegies Zwart et al. 2004; Giersz et al. 2015; Mapelli 2016). Such dense star clusters require a deep potential well, in order to retain most of the BH merger products, after receiving significant recoil kicks, due to anisotropic emission of gravitational radiation. By using simulations including full stellar evolution, Fragione et al. (2022) show that a massive stellar BH seed can grow up to $10^3 - 10^4 M_\odot$, after a series of repeated mergers of lower mass BH. Lowering the cluster metallicity even facilitates the process, leading to larger final mass of the resulting BH. They also show that the growing BH spin decreases in magnitude after successive mergers, leading to a negative correlation between BH final mass and spin. This is of prime importance for the interpretation of results obtained by LIGO-Virgo, as soon as the population of detected IMBH will be statistically significant.

To conclude, this dynamical formation of massive black hole binaries within young massive star clusters (see Figure 6.19), naturally leaves several imprints on the mass, spin and orbital properties of BBH and resulting BH, characterized not only by a higher mass than BBH formed in the field, but also by higher initial eccentricity and isotropically oriented spins.

6.5.4.2 *Chemically Homogeneous Evolution (CHE)*

An alternative scenario of HMXB evolution, called *chemically homogeneous evolution* (CHE), is based on the possibility that some massive and low-metallicity stars do not experience any expansion phase, and convert most of their mass into a black hole. In this case, we begin with a tight binary constituted of two such stars nearly filling their Roche lobe, with an orbital period of only a few days, causing strong tidal energy dissipation, accelerating the rotation of each star up to a few tens of percent of their break-up velocities, until they both become tidally locked, with the stellar rotation period equal to the orbital period. Because of this tidally–induced high spin, such fast rotating stars efficiently mix internal chemical species, draining hydrogen into the core, and helium into the envelope, until eventually all the hydrogen is fused into helium (see Figure 6.20). Then, when they reach the main sequence, both stars behave like a contracting Wolf–Rayet naked helium star. Since the metallicity is low, the small wind-driven mass loss does not significantly widen the binary orbit, remaining small enough to simultaneously avoid both mass transfer and premature merger, even by overfilling their Roche lobe. Both stars then evolve into a black hole, forming a tight binary black hole, and radiate away gravitational

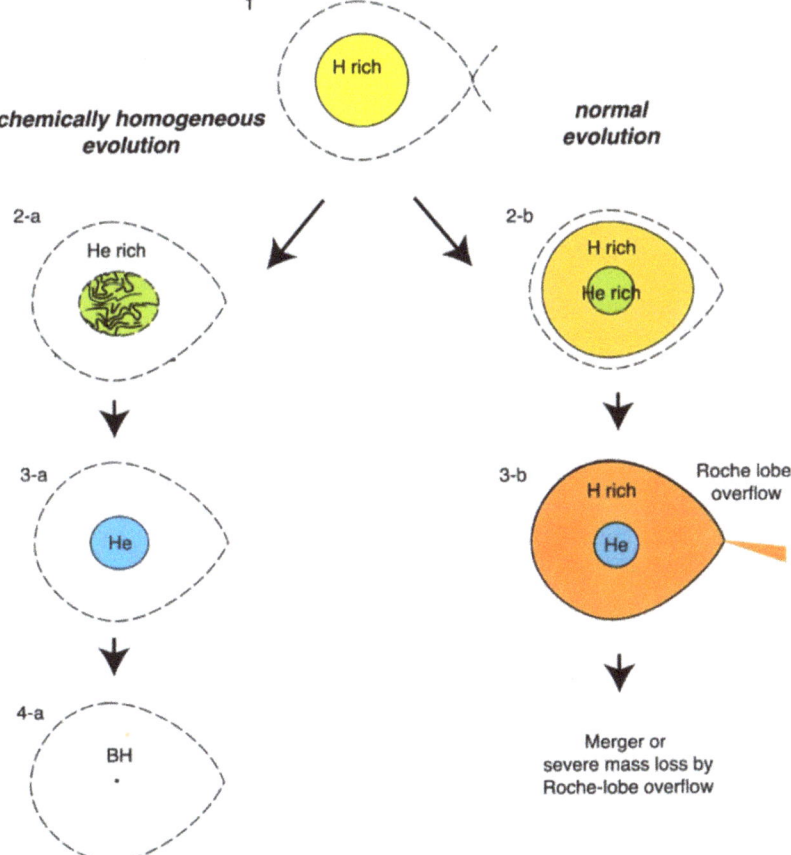

Figure 6.20. Comparison between normal evolution (*right*), versus chemically homogeneous evolution (CHE) scenario of HMXB (*left*). The enhanced mixing causes the star to shrink inside its Roche lobe, instead of expanding. This prevents low-metallicity star to experience a large mass loss, thus potentially leading to the most massive stellar-mass black hole binaries. Reprinted with permission from Mandel & de Mink (2016).

energy while orbiting around each other, until they merge in less than the Hubble time. This alternative—over-contact binary evolution—scenario can potentially explain the most massive stellar-mass black hole binaries, with total black hole mass above $\sim 50 M_\odot$, detected with gravitational waves, such as GW150914, with a predicted rate of $\sim 10 \text{Gpc}^{-3}\text{yr}^{-1}$ (de Mink et al. 2009; de Mink & Mandel 2016; Mandel & de Mink 2016; Mandel & Farmer 2017; Marchant et al. 2016).

6.5.4.3 Population III Stars

In addition to the alternative scenario described above, to form massive close binary black holes that can merge in the Hubble time, there is a pathway which is similar to classical isolated *(in the field)* evolution, but involving very low metallicity and massive binaries, or even early metal-free first-generation—*Population III*—stars, formed from the primordial hydrogen and helium gas left-over from the Big Bang (see Figure 7.4). Such stars, characterized by a weak, or even null, stellar wind, are thought to grow to enormous mass possibly up to $10^3 M_\odot$, thus consuming their hydrogen very quickly, leading to stellar lifetime of a few million years, and in the course of their evolution these stellar progenitors keep most of their mass (Postnov & Kuranov 2017).

While even telescopes like the James Webb Space Telescope might not resolve such stars individually, galaxy spectra showing only hydrogen and helium lines might reveal their presence, allowing to study early-star formation, and how these stellar giants, after exploding into supernova events, contributed to enrichment in heavy atoms during the evolution of the Universe. Ultraviolet radiation emitted by these massive and hot stars might also be among the source of re-ionization of the Universe (the *epoch of re-ionization* occurred from $\sim 4 \times 10^5$ years, up to ~ 1 Gyr after the Big Bang), in addition to radiation emanating from supermassive black holes at the heart of most galaxies. At the end of their life, these Population III stars eventually collapse into black holes, providing *seeds* that likely merge into super-massive black holes.

6.5.4.4 Primordial Black Holes

To be complete, we also have to mention the possibility to form massive close binary black holes, directly from primordial stellar-mass black holes. This has been initially proposed by Nakamura et al. (1997) in the context of the MACHO experiment. Primordial black holes, of mass $\sim 0.5 M_\odot$, would be formed in the early universe at the radiation-dominated stage, when the temperature was ~ 1 GeV, and potentially represent a substantial part of dark matter in the Universe (Bird et al. 2016b).

6.5.5 How to Distinguish Between These Formation Channels?

Many parameters—some still very much unconstrained—enter in every formation channels. Concerning the isolated scenario evolution, we have already described the common envelope phase, probably the most uncertain phase of stellar life (see Sections 3.4 and 6.5.1.3; Fragos et al. 2019; García et al. 2021), but we have to add the physics of core-collapse supernovae (see Section 2.1.5), linked to the existence of natal kicks (see Section 6.5.3), and the strength of stellar winds (see Section 6.5.2), depending on the metallicity, among the main sources of uncertainty about the isolated binary formation channel. It might be possible to distinguish between various evolution scenario, and in particular between isolated and dynamical formation, by comparing fundamental parameters measured in black hole binaries:

1. *Mass* distribution of formed black holes: the isolated *in the field* scenario tends to produce binaries made of two black holes of similar masses (see Section 6.6.2.2; Abbott et al. 2021c; Vitale 2021);

2. *Eccentricity*: while isolated binary evolution scenario produce low or even null-eccentricity binaries, dynamical formation leads to binaries with significant eccentricity, potentially leading to eccentric binary black hole coalescences (Rodriguez et al. 2018; Abbott et al. 2021d);

3. *Spin*: most physical processes occurring during the various steps of isolated scenario tend to produce individual spin vectors of black holes preferentially aligned with the global orbital angular momentum direction. In contrast, other scenario, and in particular dynamical formation, tend to produce an isotropic distribution of black hole spin vectors, randomly oriented by subsequent encounters, and thus generally misaligned with the orbital angular momentum (Rodriguez et al. 2016; Farr et al. 2017; Reynolds 2021; Franciolini et al. 2022; Zevin et al. 2021).

6.6 Final Evolution of HMXB: Toward GW Emission

In the previous section we described the main physical processes occurring in the frame of the isolated binary evolution scenario via the common envelope phase, from the formation of massive stars to the merging of binary stellar-mass black holes. The initial separation of a binary must be of the order of a few tens of $R_\odot$ for both compact objects to merge within a Hubble time, after a long approach due to gravitational-wave release, as computed by Peters (1964). The detection of the phase and amplitude evolution of these events gives access to unique information on many physical properties of the source, such as mass, spin, distance, and also equation of state, in particular for neutron stars. While no electromagnetic radiation is generally expected during the merging of a *binary black hole* (BBH), it is expected for a *binary neutron star* (BNS) and for a *neutron star–black hole* (NSBH) binary. We summarize in the following the main results of gravitational wave (GW) detections obtained by LIGO and Virgo observatories, with more than 50 GW events detected so far in O1, O2, and O3 campaigns. Up to now, all detections perfectly match the waveform of two compact objects accelerated in extreme gravitational fields, deforming the spacetime and emitting gravitational waves, as predicted by the general theory of relativity. Upgraded existing, and future GW detectors, are expected to detect ~100 BBH mergers per year, as well as numerous BNS and NSBH, in the mass range from ~1 to a few hundred $M_\odot$ (see the review by Barack et al. 2019).

Gravitational waves corresponding to the merger of compact object binaries are emitted at the frequency:

$$\nu_{GW} \sim \frac{c}{2\pi R_s} \sim \frac{4400}{\dfrac{M}{M_\odot}}\text{Hz} \qquad (6.10)$$

with R_S the Schwarzschild radius (see Section 2.2.1). The resulting *strain* (dimensionless spatial deformation) of a compact object binary merger is of the order:

$$h = \frac{\Delta L}{L} \sim 10^{-21}. \qquad (6.11)$$

Thus the LIGO-Virgo detectors, with interferometric arms of L 3–4 km length, measure $\Delta L \sim 3-4 \times 10^{-18}$ m, corresponding to tiny relative changes in the separation of mirrors, of $\sim 1/1000$ the diameter of the proton.[2] The science targets of LIGO-Virgo detectors are in the $10^2 - 10^3$ Hz frequency range, corresponding to a BH mass range of $1 - 10^3 M_\odot$, typically $\sim 150 M_\odot$, with inspiral and merger phases, as described below (see Figures 6.21 and 9.3). Before describing the first GW results, let us point out that GW interferometers are already at a very advanced stage, when

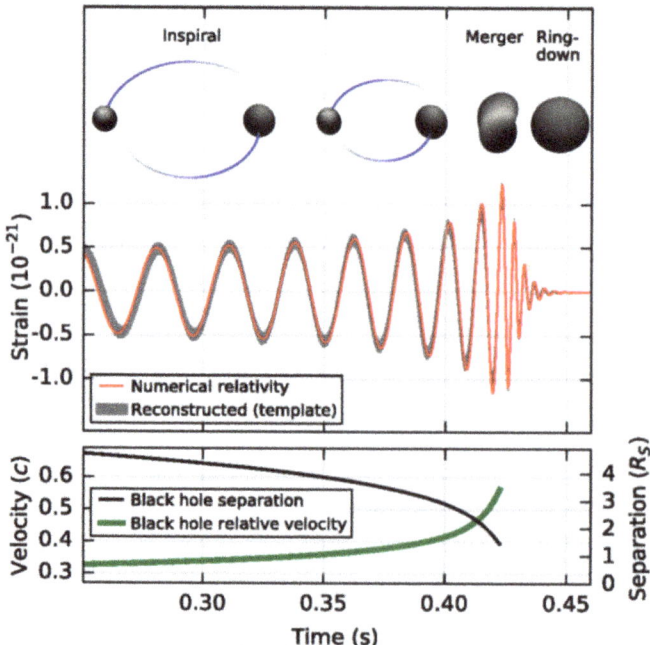

Figure 6.21. *Top:* gravitational wave strain amplitude from the binary black hole GW150914 (only the final 0.5 s is shown here). The inset images show numerical relativity models of the black hole horizons, as they merge. *Bottom:* effective relative velocity and Keplerian separation in units of Schwarzschild radii of both black holes. Reprinted with permission from the ASP from Abbott et al. (2016a), Creative Commons Attribution 3.0 License.

---

[2] A useful analogy is the one of a GW passing between the Sun and Proxima Centauri—the closest star— changing the distance (4.2 l.y., or 4×10^{13} km) by the width of one hair.

comparing with optical, radio and X-ray satellites, a few years only after these telescopes obtained their first light.

6.6.1 Quantitative Results

Before describing the events in detail, we first enumerate here quantitatively the detections of gravitational wave events, by the LIGO-Virgo interferometers, reported in Figure 6.22.

- 11 merger events during the O1 and O2 campaigns from September 2015 to August 2017 (3 BBH during O1, 7 BBH and 1 BNS—see Figure 6.23—during O2 respectively; see catalog GWTC-1, Abbott et al. 2019a, 2019b);
- 39 merger events during the O3a campaign from April to September 2019 (35 BBH, 3 NSBH, and 1 BNS; see catalog GWTC-2, Abbott et al. 2021b, 2021c);
- 8 new merger events, and a retraction of 3 of them, following a reanalysis of the O3a campaign (catalog GWTC-2.1, The LIGO Scientific Collaboration 2021);
- 35 merger events (32 BBH and 3 NSBH) during the O3b campaign from November 2019 to March 2020 (catalog GWTC-3, Abbott et al. 2021e).

6.6.2 Binary black hole (BBH)

The LIGO, Virgo and KAGRA (LVK) GW interferometers are mostly sensitive to the merger phase of BBH (see Figures 6.21 and 9.3), with a detected total BH mass range ranging from ~20 to $\sim140 M_\odot$, and low effective spins measured for most of them. All BBH detected up to now, during the first three observing runs of LIGO–Virgo, behave as expected for Kerr black holes in the frame of general relativity.

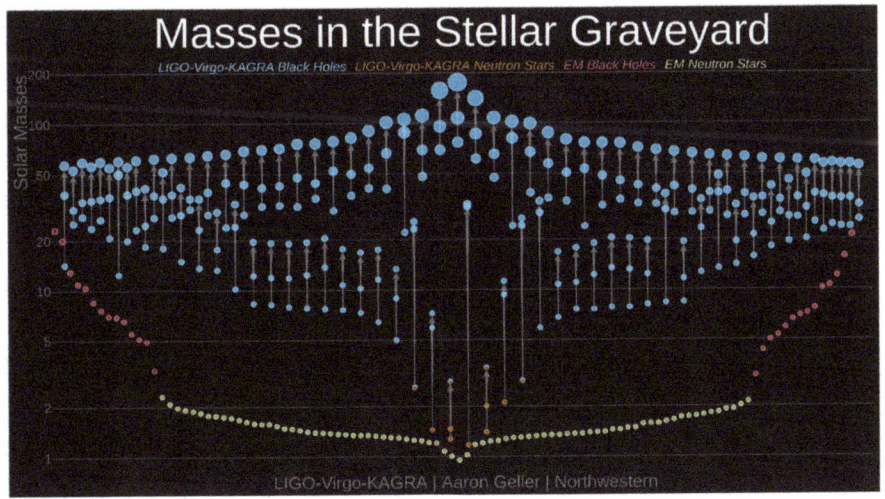

Figure 6.22. Population of compact objects, detected with GW detectors (black holes in blue and neutron stars in orange) and electromagnetic (EM) observatories (black holes in red and neutron stars in yellow). Credit: LIGO–Virgo–KAGRA/Aaron Geller/Northwestern.

BBH, when merging, are currently detectable up to an horizon of 214 Mpc for O2, 475 Mpc for O3a, 574 Mpc for O3b, 900 Mpc for O4, and 1800 Mpc for O5.[3]

6.6.2.1 The Merging

The subsequent merger phases are (see Figure 6.21):

1. early inspiral phase with amplitude modulation resulting from spin-induced precession of the orbital plane—characterized by a low velocity and weak gravitational field;
2. late inspiral/plunge phase—high velocity and strong gravitational field—(post-Newtonian theory);
3. merger phase—non-linear, non-perturbative effects, rapidly varying gravitational field—(numerical relativity);
4. ringdown phase where the newly-formed black hole rings down to an equilibrium state—excitation of quasi-normal modes/spacetime vibration—(BH perturbation theory).

6.6.2.2 Mass Distribution of Black Holes

The ~85 BBH GW events are consistent with a mass distribution of BH that differs from a single power law, thus unlike the mass distribution of stellar progenitors. Instead, the mass distribution of the most massive black hole is consistent with a two-component distribution, a Gaussian component and a power-law component, which would suggest at least two distinct formation scenario (Abbott et al. 2021c; Vitale 2021). On one hand, the Gaussian component—centered at $M \sim 33 M_\odot$—arises from black holes piling up below the pair-instability mass gap. On the other hand, the power-law component—with a maximum BH mass $M_{max} \sim 52 M_\odot$, close to the lower end of the pair instability (PISN) mass gap—emanates from black holes generated by the collapse of massive stars forming the parent stellar population (see Section 2.1.5).

By comparing the properties of black holes within BBH (detected through their GW emission when merging), and within LMXB or HMXB (studied with electromagnetic ways) it appears that the first ones tend to be more massive than the latter ones (Fishbach & Kalogera 2022). While this discrepancy could be due to observational selection effect in the case of HMXB (with a small sample of only 3 HMXB), this does not seem to be the case for LMXB, in which BH are significantly lighter than their counterparts in BBH population. However, a possibility is that it only reflects a correlation between masses of the compact object and its stellar companion, constituting the binary system.

In the context of the isolated binary evolution scenario, García et al. (2021) computed the properties of progenitors of low-mass binary black-hole mergers (of the order of $M \sim 7 - 10 M_\odot$), such as the two merger events GW151226 and GW170608 detected by LIGO-Virgo, using the 1D hydrodynamical stellar-evolution code $MESA$,[4] adapted to include the black-hole formation and the unstable mass transfer occurring during the common envelope phase. All progenitors, from binaries with

---

[3] http://public.virgo-gw.eu/the-gravitational-waves-horizon
[4] http://mesa.sourceforge.net

initial separations in the $30-200R_\odot$ range, experience a stable mass transfer interaction before the formation of the first black hole. This is followed by a second unstable mass-transfer episode, leading to a common envelope ejection, occurring either when the secondary star crosses the Hertzsprung gap, or when it is burning He in its core. Only progenitors passing through the common envelope phase are able to merge in less than a Hubble time. The masses of both black holes at the end of the binary evolution range from $\sim5M_\odot$ up to $\sim45M_\odot$, with mass ratio preferentially close to 1 (although other mass ratios are possible; Giacobbo & Mapelli 2018; García et al. 2021). Integrated merger-rate densities are computed in the range $0.2-5.0\mathrm{Gpc}^{-3}\mathrm{yr}^{-1}$ in the local universe, leading to detection rates of $1.2-3.3\ \mathrm{yr}^{-1}$, compatible with rates observed by LIGO-Virgo (García et al. 2021).

In contrast, scenario proposing alternatives to common envelope have a preferred BH mass range on a higher range, between 25 and $60M_\odot$ (Marchant et al. 2016), with high and aligned spins, null eccentricity at the end of their evolution, and long delay times of $\geqslant3$ Gyr (de Mink & Mandel 2016).

6.6.2.3 Spin Distribution of Black Holes

All GW events observed up to now by LIGO-Virgo, during the first three observing runs, indicate that the measured effective spins of merged BH are preferentially small or positive, i.e., aligned with the angular orbital momentum, but not always equal to zero, like for instance GW151226 and GW191204_171526. Some BBH exhibit effective spins oriented along the negative (i.e., opposite) direction of the orbital angular momentum, like for instance the event GW191109_010717. These *GW*-BH thus seem distinct from the Galactic BH within LMXB or HMXB, studied with electromagnetic ways, for which high effective spins have been measured (see Section 5.3.5). The fact that *electromagnetic*-BH spin significantly faster than *GW*-BH (Reynolds 2021; Fishbach & Kalogera 2022) is likely related to different formation channels, due to mass dichotomy of the donor star, with higher mass for HMXB than for LMXB. It is likely that BH are born with negligible spin, and later spun up during accreting phase. However, while BH are efficiently spun up by accretion within LMXB (most *electromagnetic*-BH), it is usually not the case for HMXB (natural progenitors of *GW*-BH), in which stellar accretion takes place, without net transfer of angular momentum (see Section 6.3.7).

Based on the GWTC-2 catalog of BBH gravitational wave detections, Callister et al. (2021) analyzed their spin magnitudes and orientations, suggesting the presence of binaries exhibiting component spins significantly misaligned with respect to their orbital angular momenta. This is a signature of a formation process distinct from isolated *in the field* formation scenario, for instance via standard common envelope (CE) evolution, thus suggesting instead a dynamical binary formation, within dense stellar clusters. By studying the impact of natal kick received at the black hole formation, in the context of CE scenario, they find that, for isolated black holes born with small natal spins, BBH formed through CE phase require extreme natal kicks to match the observed BBH population detected by LIGO-Virgo, with a velocity dispersion $\sigma\sim970\mathrm{km\ s}^{-1}$. Callister et al. (2021) thus suggest that, to avoid such extreme kicks, alternative channels have to contribute to the BBH population.

6.6.2.4 Uncertainties and Local Merger Rates

Taking into account the uncertainties mentioned above, rough calculations in the frame of the isolated scenario predict local merger-rate densities from a few to a few thousand events $Gpc^{-3}yr^{-1}$, based on the star formation rate of the local universe (Belczynski et al. 2016; Mandel & Farmer 2017; Mapelli & Giacobbo 2018; Giacobbo & Mapelli 2018, 2020; Tang et al. 2020; García et al. 2021). For alternative scenarios, local merger-rate densities are expected to be $\sim 10 Gpc^{-3}yr^{-1}$ (Mandel & de Mink 2016), though with large uncertainties (see Gallegos-Garcia et al. 2021).

6.6.2.5 Detections of BBH

We now describe the most striking GW events involving BH, which are the most common among the 90 events detected so far by LIGO-Virgo, and illustrated in Figure 6.22.

- **GW150914** is the first direct and unambiguous detection of gravitational waves, associated with the merger of two black holes of masses $M_1 = 36$ and $M_2 = 29 M_\odot$ (Abbott et al. 2016b, 2016c). The energy involved in this process is huge, with a black hole radius $R_S = 90$ km, at a separation of $a = 350$ km, making 75 orbits per second before colliding at half the speed of light, and forming a $M = 62 M_\odot$ black hole (thus losing $3 M_\odot$ in the process), at a distance of $D \sim 410$ Mpc (see Figure 6.21). This event, marking the dawn of multimessenger astronomy (even if in the case of BBH no electromagnetic signal is expected), constituting the first detection of a binary black hole, made of heavy stellar-mass black holes, which were not expected, well above the mass range of black holes discovered via classical X-ray studies, in the range $\sim 5 - 15 M_\odot$. Such heavy stellar-mass black holes had to be formed in a very poor metallic environment.

- **GW190412** is a merger of two black holes at a distance $D = 740$ Mpc, with asymmetric masses $M_1 \sim 30 M_\odot$ and $M_2 \sim 8 M_\odot$, thus with a large mass ratio $\sim 4:1$ and unusual spin parameters, with some evidence for orbital precession, both facts suggesting alternative formation scenario, such as dynamical formation, likely in a globular cluster (Abbott et al. 2020a; Mandel & Fragos 2020).

- **GW190521** is the heaviest BBH (consisting of two black holes of $M_1 = 85$ and $M_2 = 66 M_\odot$, with a total mass of $M = 151 M_\odot$), more massive than expected from standard stellar evolution, within the (pulsational) pair-instability supernova (PISN) mass gap, and thus that cannot be formed directly because of pair instability (see Section 2.1.5). The resulting black hole has a mass of $\sim 142 M_\odot$, consistent with an IMBH (Abbott et al. 2020b, 2020c). This detection raises fundamental questions about black hole formation, with the first indication of a possible alternative way of black hole formation, with one or both components formed during a previous merger event, perhaps through a dynamical formation, explaining the unusually high mass, and opening the way toward the formation of IMBH via successive mergers, nuclear star clusters being a likely efficient site of such formation.

- **GW190924_021846** ($M_1 \sim 9$ and $M_2 \sim 5M_\odot$ at a distance of $D = 0.6$ Gpc) and **GW191129_134029** ($M_1 = 10.7$ and $M_2 = 6.7M_\odot$, with a resulting merger of mass $16.8M_\odot$), are the lightest BBH (Abbott et al. 2021e).

Assuming that compact binary formation follows cosmic star formation, the detection rate of binary black holes is $R_{\mathrm{BBH}}(z = 0) \sim 24^{+14}_{-9}\mathrm{Gpc}^{-3}\mathrm{yr}^{-1}$ (Abbott et al. 2021c). More than a hundred GW signals are expected every year with future improvements on the observatories, which should allow a good characterization of the BBH population and mass distribution, up to a redshift $z \sim 1$, better constrain their formation channel, and in particular the formation of intermediate-mass black hole. BBH further away than redshift $z = 1$ would be too far to be directly detected by the LIGO-Virgo interferometers, but would still contribute to the gravitational wave background. Various models of binary black holes, merging through a variety of evolutionary channels, theoretically predict a range of local merging rates, summarized in Mandel & Broekgaarden (2022). Finally, Wagg et al. (2021) estimate that $\sim 6 - 154$ such binaries should be detected on the 4-year lifetime of LISA.

6.6.3 Binary Neutron Star (BNS)

Binary neutron stars (BNSs) are systems that will be detected not only by LIGO-Virgo interferometers when merging and emitting high-frequency GW, but also in large number by the Square-Kilometre-Array (SKA), as radio pulsars (Tauris et al. 2017). A BNS is treated as a 2-body problem in general relativity. General relativistic effects allow to constrain mass more precisely than in classical mechanics using Kepler's third law, because of uncertainty due to system inclination, using for instance the pericenter precession of orbital motion:

$$\dot{\omega} = \frac{3}{1 - e^2}\left(\frac{2\pi}{T}\right)^{5/3}\frac{G^{2/3}}{c^2}(M_1 + M_2)^{2/3}. \tag{6.12}$$

6.6.3.1 The Hulse and Taylor Binary Pulsar
With gravitational wave measurements, it is possible to extract mass parameters directly from the gravitational wave signal itself. When period change due to gravitational wave emission can be measured, then the mass and orbital parameter measurements become very accurate, allowing for instance to predict the orbital evolution, such as the Hulse and Taylor binary pulsar PSR B1913+16, located at D=7.1 kpc (Hulse & Taylor 1975). Discovered in 1974, this binary neutron star, containing a pulsar, allowed an accurate measurement of the orbital period change, due to emission of gravitational waves. The orbital period, now of 7.75 hr, and decreasing of 76.5 μs per year as predicted due to GW emission, will eventually lead to the merging of the binary neutron star (of masses respectively $M_1 = 1.441\,4 \pm 0.000\,2M_\odot$ and $M_2 = 1.386\,7 \pm 0.000\,2M_\odot$) in 302 Myr (see Figure 6.24; Weisberg & Huang 2016). The eccentricity of this system is currently $e = 0.617$, in case its orbit were circular, the binary would merge after a much longer delay, in 1.65 Gyr (see Figure 6.25; Peters 1964). In fact, mass transfer episodes and

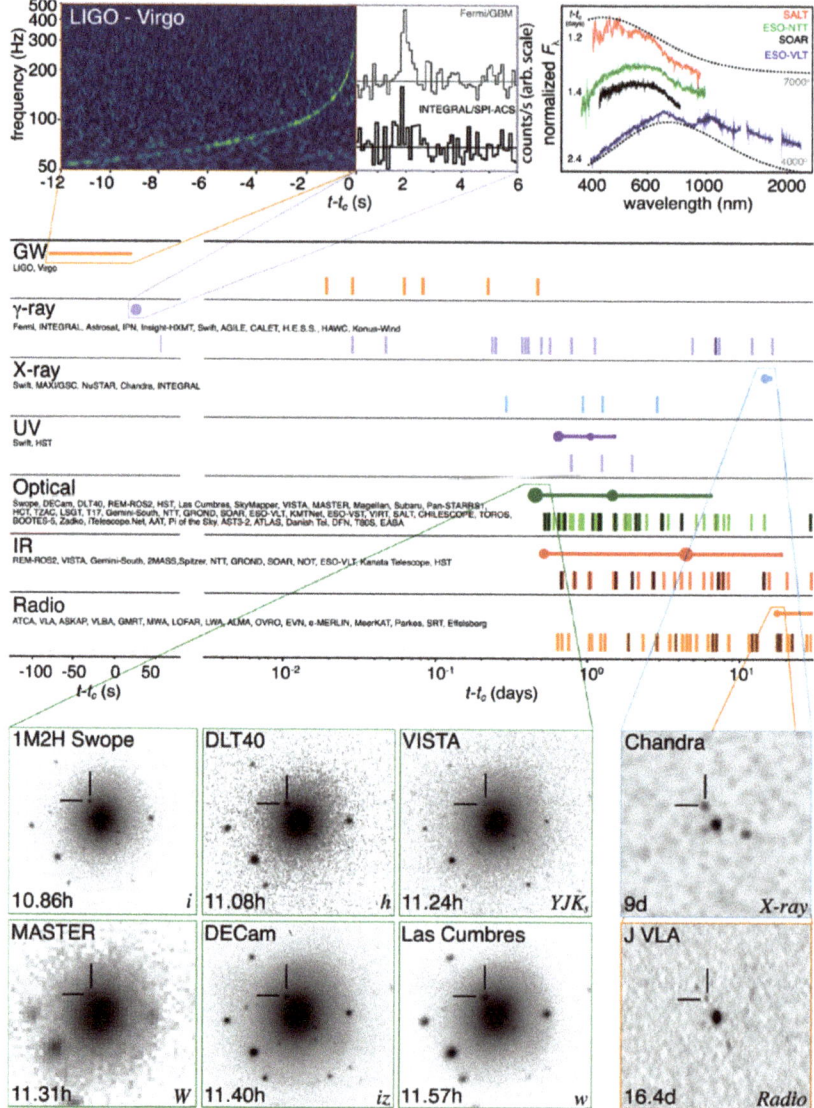

Figure 6.23. Timeline of the discovery of GW170817, GRB 170817A, SSS17a/AT 2017gfo, and the wealth of follow-up observations. Insets at the top show the first detections in the gravitational wave, γ-ray, and optical to infrared spectra. Insets at the bottom show the first detections in optical, X-ray, and radio bands. Reprinted with permission of the AAS from Abbott et al. (2017b), Creative Commons Attribution 3.0 License.

gravitational-wave decay are expected to efficiently reduce eccentricity, so that almost all isolated binaries have nearly null eccentricity at the end of their evolution.

6.6.3.2 Detections of BNS

BNS, of mass range between $1.0 - 2.5 M_{\odot}$, are systems that when merging are currently detectable, by interferometers of LIGO, Virgo and **KAGRA (LVK)**

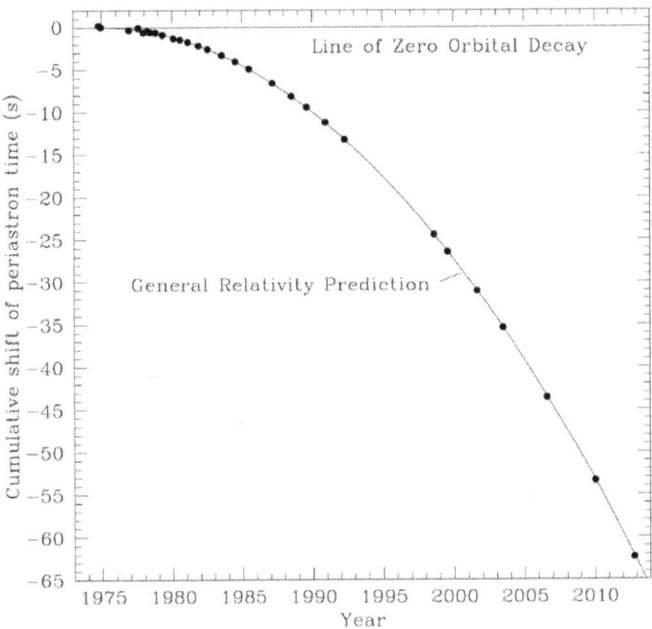

Figure 6.24. Variation of the orbital period of the binary pulsar PSR B1913+16. Reprinted with permission of the AAS from Weisberg & Huang (2016).

collaborations, up to an horizon of 28 Mpc for O2, 47 Mpc for O3a, 60 Mpc for O3b, 100 Mpc for O4, and 200 Mpc for O5,[5] with a rate $R_{BNS}(z = 0) = 320^{+490}_{-240}$Gpc^{-3}yr^{-1}. More specifically, LIGO aims for BNS at a sensitivity of 160–190 Mpc, Virgo of 80–115 Mpc, and KAGRA greater than 1 Mpc sensitivity. In the late phase of the inspiral, just before the collision, the deformation and disruption of both neutron stars generate a tidal tail of debris traveling at 0.2 c, and the formation of an accretion disk. Such a merger produces gravitational waves, eventually detected by gravitational wave detectors in the minutes prior to the collision, accompanied by the emission of a narrow beam of $\gamma-$ rays, via a short ($\leqslant$2 s) and energetic ($\sim10^{52}$ erg s^{-1}) γ-ray burst (GRB[6]), due to matter accreted onto the newly-formed compact object, forming a jet along the direction of the magnetic field (see Figure 6.23). Meanwhile, in the center the merging either produces an unstable neutron star, which in turn collapses into a black hole in ms to s timescale, or directly collapses into a black hole. With the energy loss due to γ-ray jet interacting with the ISM, the beaming angle increases, allowing the detection of an electromagnetic counterpart at lower energies, from X-ray to radio domain. In parallel, the neutron-rich material surrounding the newly-formed object produces large amounts of electromagnetic radiation, generated by the decay of heavy nuclei formed during the *r-process* (explosive nucleosynthesis,

---

[5] http://public.virgo-gw.eu/the-gravitational-waves-horizon

[6] Instead, long GRB are more likely due to the collapse of poorly metallic and very massive stars (Vreeswijk et al. 1999).

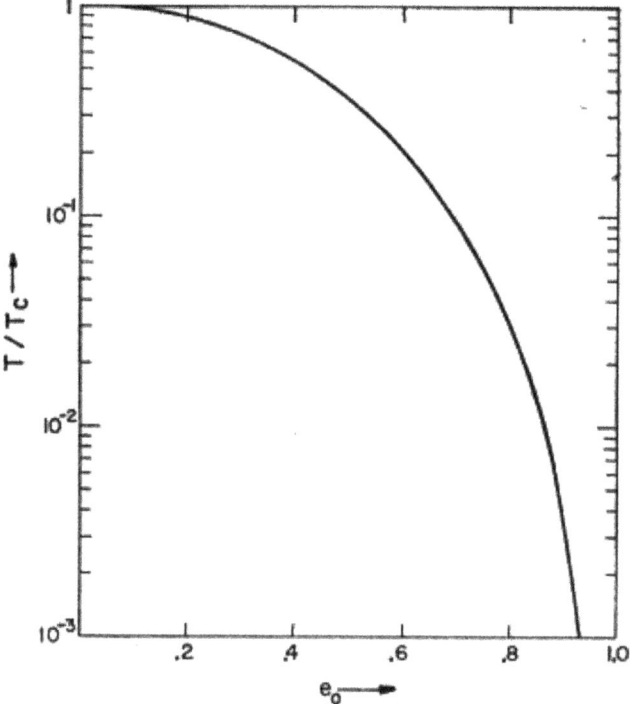

Figure 6.25. Merging time versus eccentricity of a binary. Reprinted with permission from Peters (1964).

Mennekens & Vanbeveren 2014), and the merger is usually associated to a *kilonova*, detectable during hours to years post-event.

We now describe the two GW events involving NS, detected so far by LIGO-Virgo, and illustrated in Figure 6.22.

- **GW170817** was detected as a gravitational wave event by LIGO-Virgo as a loud GW signal (SNR = 32.4) during ∼100 s and ∼3000 orbiting cycles, associated to the merging of two neutron stars, with a chirp mass $M_{\text{chirp}} = \mu^{3/5} M^{2/5} = 1.98 M_{\odot}$ (Abbott et al. 2017a). The merging also generated an emission of a short γ-ray burst, GRB 170817A, detected by the Fermi γ-ray Space Telescope ∼1.7 s post-merger time (see a wealth of multimessenger observations in Figure 6.23). Less than 11 hours after the event, the galaxy NGC 4993, located at a distance $D \sim 40\text{Mpc}$, was identified in the optical domain as the host of the merger, which occurred outside the center (AT 2017gfo). At this distance, it was the closest known short GRB, probably seen off-axis with respect to relativistic jet. An extensive multi-wavelength observing campaign, involving numerous ground and space-based telescopes, acquired unique data in X-ray, ultraviolet, optical, infrared, and radio domain, and reported in Figure 6.23 (Abbott et al. 2017b; Evans et al. 2017). The joint analysis of GW170817 and GRB170817A allowed to constrain a bit more the yet unknown equation of state in the core of NS, and in particular to derive a mass between 1 and $1.9 M_\odot$, with a total mass $2.73 M_\odot$ (Abbott et al. 2017c, 2019c), and a radius of ~12 km for both components pre-merger (Abbott et al. 2018). Multi-wavelength observations have also shown that this BNS merger produced $\geqslant 0.05 M_\odot$ of atomic elements heavier than iron (Drout et al. 2017; Pian et al. 2017), demonstrating that neutron star mergers play a key role in the rapid neutron capture (*r-process*) explosive nucleosynthesis in the Universe (although Galactic chemical evolution models tend to show that such events alone cannot reproduce the observed element abundance patterns of extremely metal-poor stars, suggesting the existence of alternative sites of r-process nucleosynthesis, such as supernova or even hypernova, associated to long-duration γ-ray bursts; Yong et al. 2021).This event also allowed an independent measure of the Hubble constant $H_0 = 70.0^{+12.0}_{-8.0}$ km/s/Mpc, simultaneously using both the redshift (measured in optical) and the luminosity distance (obtained by GW; Abbott et al. 2017d). In addition, the observed time delay between GW170817 and GRB170817A led to strongly constrain the speed of GW: $|c_g - c_{em}| \leqslant 10^{-15}$ c, thus very close to the speed of light, allowing to eliminate a series of alternative theories (Abbott et al. 2017e). This event remains the only one associated to an electromagnetic counterpart, marking the advent of combined gravitational and electromagnetic astronomy: *multimessenger astrophysics*.

- **GW190425**, a second BNS merger, has been discovered during O3 campaign (Abbott et al. 2021b), with $M_1 = 2$, $M_2 = 1.4 M_\odot$, and a total mass $M = 3.4 M_\odot$, thus heavier than any other known binary pulsar within an HMXB measured via X-ray observations (maximum total mass of $M \sim 2.88 M_\odot$; Lazarus et al. 2016). It is difficult to explain how such a massive binary has formed: in the isolated binary channel, the natal kick which occurs during supernova event is more disruptive when the newly-formed compact object is more massive (see Section 6.5.3), and the dynamical channel tends to form more massive black holes clustering toward the center of stellar cluster, but not more massive neutron stars, which tend to remain off nucleus of the cluster, in locations with poorer density. Another possibility remains: that GW190425 was in fact a NS–BH merger, which would explain its high mass.

Various models of binary neutron stars, merging through a variety of evolutionary channels, theoretically predict a range of local merging rates, summarized in Mandel & Broekgaarden (2022). While the rate of binary neutron star detected by LIGO-Virgo lies in the lower end of the anticipated range, we can expect a few BNS mergers every year. BNS population is among the favorite candidate sources to be detected by the LISA interferometer. Lau et al. (2020) and Wagg et al. (2021) estimate that ~3 – 35 such binaries will be detected at an orbital frequency ~0.8 mHz with a S/N ratio $\geqslant 8$, during the 4-year lifetime of LISA, mainly within the Galaxy, but also a few systems from the Magellanic Clouds. These detections will deliver information such as orbital separation, eccentricities, and chirp masses, allowing astronomers to constrain progenitors, and formation channels. In addition, their localization at the arc-minute level will be useful to derive NS natal kicks, and birthplace of these systems (Coleiro & Chaty 2013).

6.6.4 Binary Neutron Star–Black Hole (NSBH)

6.6.4.1 Detections of NSBH

We now describe the few GW events involving NS and BH, detected so far by LIGO-Virgo, and illustrated in Figure 6.22. It is expected from these events to detect some electromagnetic signal coming from the neutron star, just before or during the merger with the black hole.

- **GW190814** was detected on 2019 August 14 by the LIGO-Virgo interferometers (Abbott et al. 2021b), an event also labeled S190814bv. This event is potentially the first merger of a compact neutron star–black hole binary (NSBH), hosting a black hole of $\sim 23 M_\odot$, and a compact object of mass $M \sim 2.6 M_\odot$, falling right within the lower mass gap, and thus of unknown nature: it is either the lightest known BH, or the heaviest NS. Although an extensive observing campaign was conducted to search in optical and infrared domain, no electromagnetic counterpart—such as a kilonova, a γ-ray burst, or an afterglow, produced by an outflow during and after the merger—was detected, up to an ejecta mass limit of $M \geqslant 0.1 M_\odot$, which makes unlikely the tidal disruption of the neutron star during the merger (Ackley et al. 2020; Kilpatrick et al. 2021). The nature of this event thus remains unconstrained. If the object is indeed a neutron star, then the minimum BH mass would be $M_{min} = 6 M_\odot$, consistent with observations of X-ray binaries. However, since the mass is likely too large to be supported by the equation of state of a neutron star, this object might as well be a black hole, in this case pushing the minimum BH mass to $M_{min} = 2.5 M_\odot$, still narrowing the neutron star–black hole gap mentioned in Section 2.1.5.
- **GW191219_163120** is a very asymmetrical merging between a neutron star of mass $M_1 = 1.2 M_\odot$ and a black hole of mass $M_2 = 31 M_\odot$. The mass of the neutron star makes it one of the lightest ever observed (Abbott et al. 2021e).
- **GW200105_162426 and GW200115_042309** are two additional compact binary coalescences of a neutron star and a black hole (Abbott et al. 2021a; The LIGO Scientific Collaboration 2021). GW200105_162426 (a marginal candidate, see Abbott et al. 2021e) has the following characteristics: $M_1 = 8.9 M_\odot$, $M_2 = 1.9 M_\odot$, luminosity distance $D = 280$ Mpc and primary spin $\leqslant 0.23$, and GW200115_042309: $M_1 = 5.7 M_\odot$, $M_2 = 1.5 M_\odot$, $D = 300$ Mpc, with the particularity of a large and negative spin projection (~ -0.2), i.e., misaligned with the orbital angular momentum (note however the alternative interpretation of a non-spinning binary; Mandel & Smith 2021). These events allow to infer an NSBH merger-rate density in a range $45 - 130 \mathrm{Gpc}^{-3}\mathrm{yr}^{-1}$ (Abbott et al. 2021a). However, they do not allow to constrain the spin or tidal deformation of the secondary component.

Progenitors of NSBH binaries are expected to be mainly massive stars evolving in the field, thus depending on the natal kick distribution for BH and NS, and on the common envelope phase. They are unlikely to form through dynamical formation in dense star clusters, which instead tend to form higher mass components. Under the assumption that initial stellar spins are aligned with the binary angular momentum, Fragione et al. (2021) show that both large natal kicks for NS ($\geqslant$150km s^{-1}) and high efficiencies for common envelope ejection are required to simultaneously explain both the high merger rate and large spin–orbit misalignment of events such as GW200115.

One of the interest of NSBH mergers is directly linked to their multimessenger potential, with a GW detection followed by an electromagnetic counterpart, the NS being tidally disrupted, leaving behind debris accreted by the BH. Such a detection would reveal fundamental clues on both the equation of state (EoS) of NS and the BH spin. Fragione (2021) computing the rate of merger events followed by an EM counterpart, find that $\geqslant$50% of such mergers would lead to an EM counterpart, only in the case that BH are born with high spin, which for the moment is disfavored by current GW detections, thus suggesting that it is unlikely for NSBH mergers to become multimessenger sources.

Various models of binary neutron star–black hole, merging through a variety of evolutionary channels, theoretically predict a range of local merging rates, summarized in Mandel & Broekgaarden (2022). Finally, Wagg et al. (2021) estimate that ~2 – 198 such binaries should be detected during the 4-year lifetime of LISA.

6.6.5 Binary White Dwarf and Neutron Star

No merging of such system has been observed yet, with a predicted merging rate of 10–100 times less than for binary neutron stars, but which would be potentially accompanied by a higher emission, thus leading to a detection rate which potentially could be comparable to the one of binary neutron stars.

6.7 Reviews, Catalogs, Database and References

6.7.1 Reviews

- Reviews on HMXB: Tauris & van den Heuvel (2006); Chaty (2013); Lutovinov et al. (2013); Walter et al. (2015); Kretschmar et al. (2019); Krivonos et al. (2021);
- Review on the common envelope phase: Ivanova et al. (2013);
- Reviews on stellar winds: Puls et al. (2008); Martínez-Núñez et al. (2017); Kretschmar et al. (2019);
- Review on double neutron star formation and evolution: Tauris et al. (2017);
- Review on rates of compact object coalescences: Mandel & Broekgaarden (2022);
- Review on gravitational waves emitted by black holes and link with fundamental physics: Barack et al. (2019).

6.7.2 Catalogs

- Catalogs of HMXB: Liu et al. (2006); Lutovinov et al. (2013); Walter et al. (2015);
- Catalog of BeHMXB: Rivinius et al. (2013);

- List of NS–HMXB sources, along with their Gaia EDR3 counterparts: Fortin et al. (2022a) and https://apc.u-paris.fr/~fortin/HMXB_catalogue.html;
- Catalog of gravitational wave events detected by LIGO, Virgo, and KAGRA collaboration, along with a list of general detection resources tools: https://www.ligo.org/detections.php;
- List of papers from the LIGO, Virgo, and KAGRA collaboration: https://pnp.ligo.org/ppcomm/Papers.html.

6.7.3 Database

We give in Table 6.1 the list of dynamically-determined binary parameters and effective spin of black hole systems within HMXB.

References

Abbott, B. P., Abbott, R., Abbott, T. D., et al. 2016a, PhRvL, 116, 061102

Abbott, B. P., Abbott, R., Abbott, T. D., et al. 2016b, ApJL, 818, L22

Abbott, B. P., Abbott, R., Abbott, T. D., et al. 2016c, ApJL, 826, L13

Abbott, B. P., Abbott, R., Abbott, T. D., et al. 2019a, PhRvX, 9, 031040

Abbott, B. P., Abbott, R., Abbott, T. D., et al. 2019b, PhRvD, 100, 104036

Abbott, B. P., Abbott, R., Abbott, T. D., et al. 2017a, PhRvL, 119, 161101

Abbott, B. P., Abbott, R., Abbott, T. D., et al. 2017b, ApJL, 848, L12

Abbott, B. P., Abbott, R., Abbott, T. D., et al. 2017c, ApJL, 850, L40

Abbott, B. P., Abbott, R., Abbott, T. D., et al. 2017d, Natur, 551, 85

Abbott, B. P., Abbott, R., Abbott, T. D., et al. 2017e, ApJL, 848, L13

Abbott, B. P., Abbott, R., Abbott, T. D., et al. 2019c, PhRvX, 9, 011001

Abbott, B. P., Abbott, R., Abbott, T. D., et al. 2018, PhRvL, 121, 161101

Abbott, R., Abbott, T. D., Abraham, S., et al. 2021a, ApJL, 915, L5

Abbott, R., Abbott, T. D., Abraham, S., et al. 2021b, PhRvX, 11, 021053

Abbott, R., Abbott, T. D., Abraham, S., et al. 2021c, ApJL, 913, L7

Abbott, R., Abbott, T. D., Abraham, S., et al. 2020a, PhRvD, 102, 043015

Abbott, R., Abbott, T. D., Abraham, S., et al. 2020b, PhRvL, 125, 101102

Abbott, R., Abbott, T. D., Abraham, S., et al. 2020c, ApJL, 900, L13

Abbott, R., Abbott, T. D., Acernese, F., et al. 2021d, PhRvD, 104, 102001

Abbott, R., Abbott, T. D., Acernese, F., et al. 2021e, arXiv:2111.03606

Abdo, A. A., Ackermann, M., Ajello, M., et al. 2010a, ApJ, 723, 649

Abdo, A. A., Ackermann, M., Ajello, M., et al. 2009, Sci, 325, 845

Abdo, A. A., Ackermann, M., Ajello, M., et al. 2010b, A&A, 524, A75

Ablimit, I., Podsiadlowski, P., Hirai, R., & Wicker, J. 2022, MNRAS, 513, 4802

Ackley, K., Amati, L., Barbieri, C., et al. 2020, A&A, 643, A113

Atri, P., Miller-Jones, J. C. A., Bahramian, A., et al. 2019, MNRAS, 489, 3116

Baibhav, V., Berti, E., Gerosa, D., et al. 2019, PhRvD, 100, 064060

Barack, L., Cardoso, V., Nissanke, S., et al. 2019, CQGra, 36, 143001

Bartlett, E. S., Clark, J. S., & Negueruela, I. 2019, A&A, 622, A93

Belczynski, K., Bulik, T., Fryer, C. L., et al. 2010, ApJ, 714, 1217

Belczynski, K., Holz, D. E., Bulik, T., & O'Shaughnessy, R. 2016, Natur, 534, 512

Benaglia, P., Ribó, M., Combi, J. A., et al. 2010, A&A, 523, A62

Bildsten, L., Chakrabarty, D., Chiu, J., et al. 1997, ApJS, 113, 367

Bird, A. J., Bazzano, A., Malizia, A., et al. 2016a, ApJS, 223, 15

Bird, S., Cholis, I., Muñoz, J. B., et al. 2016b, PhRvL, 116, 201301

Blaauw, A. 1961, BAN, 15, 265

Blondin, J. M. 1994, ApJ, 435, 756

Blondin, J. M., Kallman, T. R., Fryxell, B. A., & Taam, R. E. 1990, ApJ, 356, 591

Blondin, J. M., Stevens, I. R., & Kallman, T. R. 1991, ApJ, 371, 684

Bodaghee, A., Antoniou, V., Zezas, A., et al. 2021, ApJ, 919, 81

Bodaghee, A., Tomsick, J. A., Rodriguez, J., et al. 2010, ApJ, 719, 451

Bodaghee, A., Tomsick, J. A., Rodriguez, J., et al. 2011, ApJ, 727, 59

Bodaghee, A., Tomsick, J. A., Rodriguez, J., & James, J. B. 2012, ApJ, 744, 108

Bondi, H. 1952, MNRAS, 112, 195

Bozzo, E., Bhalerao, V., Pradhan, P., et al. 2016, A&A, 596, A16

Bozzo, E., Falanga, M., & Stella, L. 2008, ApJ, 683, 1031

Brandt, N., & Podsiadlowski, P. 1995, MNRAS, 274, 461

Butler, S. C., Tomsick, J. A., Chaty, S., et al. 2009, ApJ, 698, 502

Callister, T. A., Farr, W. M., & Renzo, M. 2021, ApJ, 920, 157

Casares, J., Negueruela, I., Ribó, M., et al. 2014, Natur, 505, 378

Castor, J. I., Abbott, D. C., & Klein, R. I. 1975, ApJ, 195, 157

Castro-Ginard, A., McMillan, P. J., Luri, X., et al. 2021, A&A, 652, A162

Charles, P. A., & Coe, M. J. 2006, in Cambridge Astrophysics Ser. 39, Compact Stellar X-ray Sources, ed. W. Lewin, & M. van der Klis (Cambridge: Cambridge Univ. Press), 215

Chaty, S. 2013, AdSpR, 52, 2132

Chaty, S., LeReun, A., Negueruela, I., et al. 2016, A&A, 591, A87

Chaty, S., & Rahoui, F. 2012, ApJ, 751, 150

Chaty, S., Rahoui, F., Foellmi, C., et al. 2008, A&A, 484, 783

Chen, Y., Bressan, A., Girardi, L., et al. 2015, MNRAS, 452, 1068

Choudhury, S., Subramaniam, A., & Cole, A. A. 2016, MNRAS, 455, 1855

Choudhury, S., Subramaniam, A., Cole, A. A., & Sohn, Y. J. 2018, MNRAS, 475, 4279

Coe, M. J. 2000, in ASP Conf. Ser. 214: IAU Colloq. 175: The Be Phenomenon in Early-Type Stars, ed. M. A. Smith, H. F. Henrichs, & J. Fabregat (San Francisco, CA: ASP), 656

Coleiro, A., & Chaty, S. 2013, ApJ, 764, 185

Coleiro, A., Chaty, S., Zurita Heras, J. A., Rahoui, F., & Tomsick, J. A. 2013, A&A, 560, A108

Corbet, R. H. D. 1986, MNRAS, 220, 1047

Corral-Santana, J. M., Casares, J., Muñoz-Darias, T., et al. 2016, A&A, 587, A61

Davies, R. E., & Pringle, J. E. 1981, MNRAS, 196, 209

de Mink, S. E., Cantiello, M., Langer, N., et al. 2009, A&A, 497, 243

de Mink, S. E., & Mandel, I. 2016, MNRAS, 460, 3545

Di Carlo, U. N., Giacobbo, N., Mapelli, M., et al. 2019, MNRAS, 487, 2947

Di Carlo, U. N., Mapelli, M., Bouffanais, Y., et al. 2020, MNRAS, 497, 1043

Drout, M. R., Piro, A. L., Shappee, B. J., et al. 2017, Sci, 358, 1570

Dubus, G., Hameury, J. M., & Lasota, J. P. 2001, A&A, 373, 251

Evans, P. A., Cenko, S. B., Kennea, J. A., et al. 2017, Sci, 358, 1565

Farr, W. M., Stevenson, S., Miller, M. C., et al. 2017, Natur, 548, 426

Filliatre, P., & Chaty, S. 2004, ApJ, 616, 469

Fishbach, M., & Kalogera, V. 2022, ApJL, 929, L26

Fogantini, F. A., García, F., Combi, J. A., & Chaty, S. 2021, A&A, 647, A75

Fortin, F., Chaty, S., Coleiro, A., Tomsick, J. A., & Nitschelm, C. H. R. 2018, A&A, 618, A150

Fortin, F., Chaty, S., & Sander, A. 2020, ApJ, 894, 86

Fortin, F., García, F., Chaty, S., Chassande-Mottin, E., & Simaz Bunzel, A. 2022a, arXiv:2206.03904

Fortin, F., García, F., & Chaty, S. 2022b, arXiv:2207.02114

Fragione, G. 2021, ApJL, 923, L2

Fragione, G., Kocsis, B., Rasio, F. A., & Silk, J. 2022, ApJ, 927, 231

Fragione, G., Loeb, A., & Rasio, F. A. 2021, ApJL, 918, L38

Fragos, T., Andrews, J. J., Ramirez-Ruiz, E., et al. 2019, ApJL, 883, L45

Franciolini, G., Baibhav, V., De Luca, V., et al. 2022, PhRvD, 105, 083526

Fryer, C. L. 2004, ApJL, 601, L175

Fryer, C. L., & Kusenko, A. 2006, ApJS, 163, 335

Gallegos-Garcia, M., Berry, C. P. L., Marchant, P., & Kalogera, V. 2021, ApJ, 922, 110

García, F., Fogantini, F. A., Chaty, S., & Combi, J. A. 2018, A&A, 618, A61

García, F., Simaz Bunzel, A., Chaty, S., Porter, E., & Chassande-Mottin, E. 2021, A&A, 649, A114

Giacconi, R., Gursky, H., Kellogg, E., Schreier, E., & Tananbaum, H. 1971, ApJL, 167, L67

Giacobbo, N., & Mapelli, M. 2018, MNRAS, 480, 2011

Giacobbo, N., & Mapelli, M. 2020, ApJ, 891, 141

Gieles, M., Erkal, D., Antonini, F., Balbinot, E., & Peñarrubia, J. 2021, NatAs, 5, 957

Giersz, M., Leigh, N., Hypki, A., Lützgendorf, N., & Askar, A. 2015, MNRAS, 454, 3150

Gratton, R., Bragaglia, A., Carretta, E., et al. 2019, A&ARv, 27, 8

Grimm, H., Gilfanov, M., & Sunyaev, R. 2003, MNRAS, 339, 793

Grimm, H-J., Gilfanov, M., & Sunyaev, R. 2002, A&A, 391, 923

Haberl, F., Maitra, C., Carpano, S., et al. 2022, A&A, 661, A25

Hanson, M. M., Still, M. D., & Fender, R. P. 2000, ApJ, 541, 308

Harding, A. K., & Leventhal, M. 1992, Natur, 357, 388

Hills, J. G. 1983, ApJ, 267, 322

Hills, J. G., & Fullerton, L. W. 1980, AJ, 85, 1281

Hirai, R., & Mandel, I. 2021, PASA, 38, e056

Hobbs, G., Lorimer, D. R., Lyne, A. G., & Kramer, M. 2005, MNRAS, 360, 974

Hulse, R. A., & Taylor, J. H. 1975, ApJL, 195, L51

Hurley, J. R., Tout, C. A., & Pols, O. R. 2002, MNRAS, 329, 897

Igoshev, A. P., Chruslinska, M., Dorozsmai, A., & Toonen, S. 2021, MNRAS, 508, 3345

Illarionov, A. F., & Sunyaev, R. A. 1975, A&A, 39, 185

Ivanova, N., Justham, S., Chen, X., et al. 2013, A&ARv, 21, 59

Janka, H-T. 2013, MNRAS, 434, 1355

Kalogera, V. 1996, ApJ, 471, 352

Kalogera, V. 2000, ApJ, 541, 319

Kaplan, D. L., Moon, D-S, & Reach, W. T. 2006, ApJ, 649, L107

Kilpatrick, C. D., Coulter, D. A., Arcavi, I., et al. 2021, ApJ, 923, 258

Kretschmar, P., Fürst, F., Sidoli, L., et al. 2019, NewAR, 86, 101546

Krivonos, R. A., Bird, A. J., Churazov, E. M., et al. 2021, NewAR, 92, 101612

Kuulkers, E., Ferrigno, C., Kretschmar, P., et al. 2021, NewAR, 93, 101629

Lada, C. J., & Lada, E. A. 2003, ARA&A, 41, 57

Lai, D. 2001, in Physics of Neutron Star Interiors, ed. D. Blaschke, N. K. Glendenning, & A. Sedrakian (Berlin: Springer), 424

Lai, D., Chernoff, D. F., & Cordes, J. M. 2001, ApJ, 549, 1111

Lamers, H. J. G. L. M., & Cassinelli, J. P. 1999, Introduction to Stellar Winds (Cambridge: Cambridge Univ. Press)

Lau, M. Y. M., Mandel, I., Vigna-Gómez, A., et al. 2020, MNRAS, 492, 3061

Lazarus, P., Freire, P. C. C., Allen, B., et al. 2016, ApJ, 831, 150

Levesque, E. M., Massey, P., Zytkow, A. N., & Morrell, N. 2014, MNRAS, 443, L94

Liu, Q. Z., Chaty, S., & Yan, J. Z. 2011, MNRAS, 415, 3349

Liu, Q. Z., van Paradijs, J., & van den Heuvel, E. P. J. 2006, A&A, 455, 1165

Liu, Q. Z., van Paradijs, J., & van den Heuvel, E. P. J. 2007, A&A, 469, 807

Lutovinov, A. A., Revnivtsev, M. G., Tsygankov, S. S., & Krivonos, R. A. 2013, MNRAS, 431, 327

Lyne, A. G., & Lorimer, D. R. 1994, Natur, 369, 127

Mandel, I., & Broekgaarden, F. S. 2022, LRR, 25, 1

Mandel, I., & de Mink, S. E. 2016, MNRAS, 458, 2634

Mandel, I., & Farmer, A. 2017, Natur, 547, 284

Mandel, I., & Fragos, T. 2020, ApJL, 895, L28

Mandel, I., & Smith, R. J. E. 2021, ApJL, 922, L14

Manousakis, A., & Walter, R. 2015, A&A, 575, A58

Mapelli, M. 2016, MNRAS, 459, 3432

Mapelli, M. 2020, FrASS, 7, 38

Mapelli, M., & Giacobbo, N. 2018, MNRAS, 479, 4391

Marchant, P., Langer, N., Podsiadlowski, P., Tauris, T. M., & Moriya, T. J. 2016, A&A, 588, A50

Martin, R. G., Nixon, C., Armitage, P. J., Lubow, S. H., & Price, D. J. 2014, ApJL, 790, L34

Martínez-Núñez, S., Kretschmar, P., Bozzo, E., et al. 2017, SSRv, 212, 59

Mason, K. O., Cordova, F. A., & White, N. E. 1986, ApJ, 309, 700

McBride, V. A., Coe, M. J., Negueruela, I., Schurch, M. P. E., & McGowan, K. E. 2008, MNRAS, 388, 1198

Mennekens, N., & Vanbeveren, D. 2014, A&A, 564, A134

Miller-Jones, J. C. A., Bahramian, A., Orosz, J. A., et al. 2021, Sci, 371, 1046

Miller-Jones, J. C. A., Deller, A. T., Shannon, R. M., et al. 2018, MNRAS, 479, 4849

Mirabel, I. F., Mignani, R., Rodrigues, I., et al. 2002, A&A, 395, 595

Mirabel, I. F., Rodrigues, I., & Liu, Q. Z. 2004, A&A, 422, L29

Moon, D., Kaplan, D. L., Reach, W. T., et al. 2007, ApJL, 671, L53

Nakamura, T., Sasaki, M., Tanaka, T., & Thorne, K. S. 1997, ApJL, 487, L139

Negueruela, I. 2010, in ASP Conf. Ser. 422, High Energy Phenomena in Massive Stars, ed. J. Martí, P. L. Luque-Escamilla, & J. A. Combi (San Francisco, CA: ASP), 57

Negueruela, I., & Coe, M. J. 2002, A&A, 385, 517

Negueruela, I., Reig, P., Coe, M. J., & Fabregat, J. 1998, A&A, 336, 251

Negueruela, I., Smith, D. M., Reig, P., Chaty, S., & Torrejón, J. M. 2006, in ESA Special Publication SP-604, Proc. of the X-ray Universe 2005, ed. A. Wilson (Paris: European Space Agency), 165

Negueruela, I., Torrejón, J. M., Reig, P., Ribó, M., & Smith, D. M. 2008, in AIP Conf. Ser. 1010, A Population Explosion:The Nature & Evolution of X-ray Binaries in Diverse

Environments, ed. R. M. Bandyopadhyay, S. Wachter, D. Gelino, & C. R. Gelino (New York: AIP), 252

Neumayer, N., Seth, A., & Böker, T. 2020, A&ARv, 28, 4

Ng, C. Y., & Romani, R. W. 2007, ApJ, 660, 1357

Nowak, M. A., Paizis, A., Rodriguez, J., et al. 2012, ApJ, 757, 143

Okazaki, A. T., & Negueruela, I. 2001, A&A, 377, 161

Orosz, J. A., McClintock, J. E., Narayan, R., et al. 2007, Natur, 449, 872

Orosz, J. A., Steeghs, D., McClintock, J. E., et al. 2009, ApJ, 697, 573

Orosz, J. A., Steiner, J. F., McClintock, J. E., et al. 2014, ApJ, 794, 154

Oskinova, L. M., Feldmeier, A., & Hamann, W. R. 2006, MNRAS, 372, 313

Oskinova, L. M., Feldmeier, A., & Kretschmar, P. 2012, MNRAS, 421, 2820

Oskinova, L. M., Hamann, W-R., & Feldmeier, A. 2007, A&A, 476, 1331

Owocki, S. P. 2007, in ASP Conf. Ser. 361, Active OB-Stars: Laboratories for Stellar and Circumstellar Physics, ed. A. T. Okazaki, S. P. Owocki, & S. Stefl (San Francisco, CA: ASP), 3

Ozbey Arabaci, M., Kalemci, E., Tomsick, J. A., et al. 2012, ApJ, 761, 4

Paczynski, B. 1976, in IAU Symp. 73, Structure and Evolution of Close Binary Systems, ed. P. Eggleton, S. Mitton, & J. Whelan (Dordrecht: Reidel), 75

Papitto, A., Falanga, M., Hermsen, W., et al. 2020, NewAR, 91, 101544

Pellizza, L. J., Chaty, S., & Chisari, N. E. 2011, A&A, 526, A15

Pellizza, L. J., Chaty, S., & Negueruela, I. 2006, A&A, 455, 653

Peters, P. C. 1964, PhRv, 136, 1224

Pflamm-Altenburg, J., & Kroupa, P. 2010, MNRAS, 404, 1564

Pian, E., D'Avanzo, P., Benetti, S., et al. 2017, Natur, 551, 67

Podsiadlowski, P., Langer, N., Poelarends, A. J. T., et al. 2004, ApJ, 612, 1044

Portegies Zwart, S. F., Baumgardt, H., Hut, P., Makino, J., & McMillan, S. L. W. 2004, Natur, 428, 724

Portegies Zwart, S. F., & McMillan, S. L. W. 2000, ApJL, 528, L17

Portegies Zwart, S. F., & McMillan, S. L. W. 2002, ApJ, 576, 899

Portegies Zwart, S. F., McMillan, S. L. W., & Gieles, M. 2010, ARA&A, 48, 431

Postnov, K., & Kuranov, A. 2017, in IAU Symp. 329, The Lives and Death-Throes of Massive Stars, ed. J. J. Eldridge, J. C. Bray, L. A. S. McClelland, & L. Xiao (Cambridge: Cambridge Univ. Press), 118

Puls, J., Vink, J. S., & Najarro, F. 2008, A&ARv, 16, 209

Rahoui, F., & Chaty, S. 2008, A&A, 492, 163

Rahoui, F., Chaty, S., Lagage, P. O., & Pantin, E. 2008, A&A, 484, 801

Rappaport, S., Joss, P. C., & Webbink, R. F. 1982, ApJ, 254, 616

Reig, P. 2011, Ap&SS, 332, 1

Repetto, S., Davies, M. B., & Sigurdsson, S. 2012, MNRAS, 425, 2799

Repetto, S., Igoshev, A. P., & Nelemans, G. 2017, MNRAS, 467, 298

Repetto, S., & Nelemans, G. 2015, MNRAS, 453, 3341

Reynolds, C. S. 2021, ARA&A, 59,

Rice, J. R., Rangelov, B., Prestwich, A., et al. 2021, ApJ, 922, 178

Rivinius, T., Carciofi, A. C., & Martayan, C. 2013, A&ARv, 21, 69

Rodriguez, C. L., Amaro-Seoane, P., Chatterjee, S., & Rasio, F. A. 2018, PhRvL, 120, 151101

Rodriguez, C. L., Zevin, M., Amaro-Seoane, P., et al. 2019, PhRvD, 100, 043027

Rodriguez, C. L., Zevin, M., Pankow, C., Kalogera, V., & Rasio, F. A. 2016, ApJL, 832, L2

Rodriguez, J., Tomsick, J. A., Bodaghee, A., et al. 2009a, A&A, 508, 889

Rodriguez, J., Tomsick, J. A., & Chaty, S. 2008, A&A, 482, 731

Rodriguez, J., Tomsick, J. A., & Chaty, S. 2009b, A&A, 494, 417

Romano, P., Bozzo, E., Mangano, V., et al. 2015, A&A, 576, L4

Runacres, M. C., & Owocki, S. P. 2005, A&A, 429, 323

Russeil, D. 2003, A&A, 397, 133

Saavedra, E. A., Fogantini, F. A., Combi, J. A., García, F., & Chaty, S. 2022, A&A, 659, A48

Sana, H., de Mink, S. E., de Koter, A., et al. 2012, Sci, 337, 444

Savonije, G. J. 1978, A&A, 62, 317

Schaller, G., Schaerer, D., Meynet, G., & Maeder, A. 1992, A&AS, 96, 269

Servillat, M., Coleiro, A., Chaty, S., Rahoui, F., & Zurita Heras, J. A. 2014, ApJ, 797, 114

Shakura, N., Postnov, K., Kochetkova, A., & Hjalmarsdotter, L. 2012, MNRAS, 420, 216

Shao, Y., & Li, X-D. 2018, ApJ, 867, 124

Shklovskii, I. S. 1970, SvA, 13, 562

Spera, M., Mapelli, M., & Bressan, A. 2015, MNRAS, 451, 4086

Spitzer, L. Jr 1984, Sci, 225, 465

Spitzer, L. Jr 1986, in The Use of Supercomputers in Stellar Dynamics, ed. P. Hut, & L. W. McMillan (Berlin: Springer), 3

Staubert, R., Trümper, J., Kendziorra, E., et al. 2019, A&A, 622, A61

Stella, L., White, N. E., & Rosner, R. 1986, ApJ, 308, 669

Tang, P. N., Eldridge, J. J., Stanway, E. R., & Bray, J. C. 2020, MNRAS, 493, L6

Tauris, T. M., Fender, R. P., van den Heuvel, E. P. J., Johnston, H. M., & Wu, K. 1999, MNRAS, 310, 1165

Tauris, T. M., Kramer, M., Freire, P. C. C., et al. 2017, ApJ, 846, 170

Tauris, T. M., & van den Heuvel, E. P. J. 2006, in Cambridge Astrophysics Ser. 39, Compact Stellar X-ray Sources, ed. W. Lewin, & M. van der Klis (Cambridge: Cambridge Univ. Press), 623

The LIGO Scientific Collaboration,, the Virgo Collaboration, & Abbott, R. 2021, arXiv: 2108.01045

Thorne, K., & Halloran, L. 2023, The Warped Side of our Universe (New York: Liveright/Norton)

Thorne, K. S., & Zytkow, A. N. 1975, ApJL, 199, L19

Tomsick, J. A., Bodaghee, A., Chaty, S., et al. 2020, ApJ, 889, 53

Tomsick, J. A., Bodaghee, A., Rodriguez, J., et al. 2012, ApJL, 750, L39

Tomsick, J. A., Chaty, S., Rodriguez, J., et al. 2006, ApJ, 647, 1309

Tomsick, J. A., Chaty, S., Rodriguez, J., Walter, R., & Kaaret, P. 2008, ApJ, 685, 1143

Tomsick, J. A., Chaty, S., Rodriguez, J., Walter, R., & Kaaret, P. 2009a, ApJ, 701, 811

Tomsick, J. A., Chaty, S., Rodriguez, J., et al. 2009b, ApJ, 694, 344

Tomsick, J. A., Coughenour, B. M., Hare, J., et al. 2021, ApJ, 914, 48

Tomsick, J. A., Heinke, C., Halpern, J., et al. 2011, ApJ, 728, 86

Tomsick, J. A., Krivonos, R., Rahoui, F., et al. 2015, MNRAS, 449, 597

Tomsick, J. A., Krivonos, R., Wang, Q., et al. 2016, ApJ, 816, 38

van den Eijnden, J., Degenaar, N., Russell, T. D., et al. 2021, MNRAS, 507, 3899

van den Heuvel, E., Bélanger, G., Hanlon, L., & Kuulkers, E. 2021, NewAR, 93, 101633

van den Heuvel, E. P. J. 1983, in Accretion-Driven Stellar X-ray Sources, ed. W. H. G. Lewin, & E. P. J. van den Heuvel (Cambridge: Cambridge Univ. Press), 303

Vink, J. S. 2011, Ap&SS, 336, 163

Vink, J. S., de Koter, A., & Lamers, H. J. G. L. M. 2001, A&A, 369, 574

Vitale, S. 2021, Sci, 372, eabc7397

Vreeswijk, P. M., Galama, T. J., Owens, A., et al. 1999, ApJ, 523, 171

Wagg, T., Broekgaarden, F. S., de Mink, S. E., et al. 2021, arXiv:2111.13704

Walter, R., Lutovinov, A. A., Bozzo, E., & Tsygankov, S. S. 2015, A&ARv, 23, 2

Walter, R., & Zurita Heras, J. 2007, A&A, 476, 335

Waters, L. B. F. M., & van Kerkwijk, M. H. 1989, A&A, 223, 196

Weisberg, J. M., & Huang, Y. 2016, ApJ, 829, 55

Winkler, C., Courvoisier, T. J-L., & Di Cocco, G. 2003, A&A, 411, L1

Yan, J., Zurita Heras, J. A., Chaty, S., Li, H., & Liu, Q. 2012, ApJ, 753, 73

Yong, D., Kobayashi, C., Da Costa, G. S., et al. 2021, Natur, 595, 223

Zamanov, R. K., Stoyanov, K. A., Wolter, U., et al. 2020, MNRAS, 499, 3650

Zevin, M., Bavera, S. S., Berry, C. P. L., et al. 2021, ApJ, 910, 152

Ziosi, B. M., Mapelli, M., Branchesi, M., & Tormen, G. 2014, MNRAS, 441, 3703

Zurita Heras, J. A., & Chaty, S. 2008, A&A, 489, 657

Zurita Heras, J. A., & Chaty, S. 2009, A&A, 493, L1

Chapter 7

Other Accreting Binaries

They will tear and be absorbed by the black hole
I'll still be a part of the Universe

—Melvin Tao, 2019

We describe in this chapter the nature, formation, and evolution of other types of accreting binaries. We begin with the intermediate-mass X-ray binaries hosting an intermediate-mass star orbiting a compact object such as a neutron star or black hole, describing their observations and their evolution. We go on with the ultra-luminous X-ray sources, which can be explained by invoking either some anisotropic emission from an accreting compact object, either neutron star or stellar-mass black hole, or some isotropic emission from an intermediate-mass black hole. We finally describe the γ-ray binaries, which are peculiar HMXB mainly emitting in γ-rays. We then finish this chapter by giving useful links to reviews, catalogs, database, and references.

7.1 Intermediate-mass X-ray Binaries (IMXB)

In the previous sections we have described binary systems with companion (secondary) star of mass less than $\sim 1 M_\odot$ (low-mass X-ray binaries, LMXB) and above $\sim 10 M_\odot$ (high-mass X-ray binaries, HMXB). Obviously there exists many stars in the range $1 - 10 M_\odot$, and we classify under the generic name of *intermediate-mass X-ray binaries* (IMXB) the binary systems hosting such companion stars of intermediate mass.

7.1.1 Observations of IMXB

These systems are not easily detected by X-ray instruments, because they usually are weak X-ray sources, in-between LMXB experiencing efficient accretion process via

Roche-lobe overflow (RLO), and HMXB via stellar wind accretion. In addition, secondary stars in IMXB are not massive enough to produce sufficiently high stellar wind mass-loss rates to power a strong X-ray source. Then, when it evolves toward RLO, this accreting phase only lasts for a short time, probably of the order of 10^3 years, because of a large mass ratio between the secondary star and the compact object, particularly in the case of a neutron star (Tauris & van den Heuvel 2006). In addition, the high-mass transfer rate $\dot{M} > 10^{-4} M_\odot$/year in place during this phase, greater than $\dot{M}_{Edd}$, prevents any emitted X-rays from piercing through the dense gas surrounding the accreting compact object. Thus, contrary to LMXB and HMXB systems, IMXB cannot be persistently strong X-ray sources.

Despite this selection effect, a few IMXB are detected, like the pulsating source Her X-1 (HZ Her/X0115+63, $P_{orb} = 1.7$ d), which hosts a $1.8 - 2.0 M_\odot$ evolved sub-giant star filling its Roche lobe, and orbiting a highly magnetized neutron star ($B \sim 5 \times 10^{12}$ G, $P_{spin} = 1.24$ s). It is the first X-ray pulsar in which the magnetic field of the NS was measured (Trümper et al. 1978), directly from the cyclotron resonance scattering feature (CRSF, see Section 6.1.5), seen as absorption/emission line in the X-ray spectrum. On one hand, the optical light-curve of the companion star HZ Her is modulated with respect to the orbital period, due to X-ray illumination. On the other hand, the X-ray light-curve of Her X-1 is modulated with a ~ 35 d period, explained by the retrograde orbital precession of the accretion disk (see Figure 7.1; Klochkov et al. 2006).

We also mention the interesting neutron star-IMXB system Cyg X-2 ($P_{orb} = 9.8$ d), whose evolution suggests that the secondary star had an initial mass of $\sim 3.5 M_\odot$, transferring mass near the end of its main-sequence phase, until it became a sub-giant of mass $0.5 M_\odot$, with a non-degenerate helium core (Podsiadlowski & Rappaport 2000).

Other IMXB host black holes and a donor star less massive than the black hole, such a ratio allowing a more stable RLO, like the systems GS 1354-64, 4U 1543-47, GRO J1655-40, and V4641 Sgr. In all these IMXB, observers derived the mass of the black hole thanks to dynamical measurements (see Section 5.1.4): a list of dynamically confirmed black holes within IMXB systems is reported in Table 7.1.

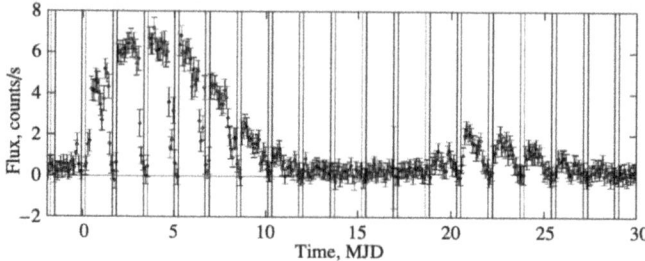

Figure 7.1. X-ray light-curve (RXTE/ASM $2 - 12$ keV) of Her X-1 obtained with the Rossi X-ray Timing Explorer satellite, constructed for cycles started at the orbital phase 0.75. The vertical lines indicate the time of orbital eclipses ($P_{orb} = 1.7$ d), superimposed on the 35 d precessional orbital period. Reprinted with permission from Klochkov et al. (2006).

Table 7.1. Dynamically-determined Binary Parameters and Effective Spin of Black Holes within IMXB Systems.

Name	Spectral type	Porb (h)	K2 (km s⁻¹)	f(M1) $M_\odot$	M1 $M_\odot$	q	i (°)	$v_{rot} \sin i$ (km s⁻¹)	Spin	Ref.
IMXB										
GS 1354-64 (BW Cir)	G5III	$61.068 \pm 2\,10^{-3}$	279 ± 5	5.7 ± 0.3	$\geqslant 6.9$	$0.12^{+0.03}_{-0.04}$	$\leqslant 79$	69 ± 8	$\geqslant 0.98$	1354
4U 1543-475 (IL Lup)	A2V	$26.793\,77 \pm 7\,10^{-5}$	124 ± 4	0.25 ± 0.01	9.4 ± 1.1	$0.25 - 0.31$	20.7 ± 1.5	46 ± 2	$0.67^{+0.15}_{-0.08}$	1543
GRO J1655-40 (N. Sco 94)	F6IV	$62.920 \pm 3\,10^{-3}$	226.1 ± 0.8	2.73 ± 0.09	6.0 ± 0.4	0.42 ± 0.03	69 ± 2	86^{+3}_{-4}	0.70 ± 0.10	1655
SAX J1 819.3-2525 (V4641 Sgr)	B9III	$67.615\,2 \pm 2\,10^{-4}$	211 ± 3	2.7 ± 0.1	6.40 ± 0.60	$0.63 - 0.70$	72 ± 4	100.9 ± 0.8	—	1819

Notes. This table is an updated and completed version from Corral-Santana et al. (2016). Spin values are taken from Reynolds (2021). References: **1354**: Casares et al. (2004, 2009); **1543**: Orosz et al. (1998); Buxton & Bailyn (2004); **1655**: Bailyn et al. (1995); Hynes et al. (1998); Chaty et al. (2002); Dubus & Chaty (2006); **1819**: Chaty et al. (2003); Orosz et al. (2001).

7.1.2 Evolution of IMXB

Podsiadlowski et al. (2002) studied the evolution of IMXB, comparing it to the evolution of LMXB, using a stellar evolution code and a model for binary interactions, with initial mass of the secondary ranging from 0.6 to 7 $M_\odot$, and initial orbital period from 4 hours to 100 days. Most IMXB briefly become detached, immediately after a phase of mass transfer on a thermal timescale, in agreement with Tauris et al. (2000). Systems in which the secondary possesses a radiative envelope are stable against dynamical mass transfer for initial masses up to $4M_\odot$. For higher initial masses, they experience a delayed dynamical instability after a stable phase of mass transfer lasting up to 10^6 yr. Systems that have initial orbital period below the bifurcation period of ~ 18 hr evolve toward extremely short ~ 10 mn orbital periods, becoming ultra compact X-ray binaries (UCXB, see Section 5.2.6). Chen & Podsiadlowski (2016) confirmed these results, using the public 1-D hydrodynamical stellar evolution code *MESA*. Due to the overall evolution of binaries, and particularly to the transfer of mass, it is very likely that a significant fraction of accreting binaries presently identified as low-mass X-ray binaries descend from IMXB. These systems, although they are poorly studied because of the selection effect mentioned above, are thus an important link between these two binary populations.

We also mention that IMXB can be detected outside of the Milky Way, such as the X-ray pulsar located in the galaxy M31, of spin period $P_{\rm spin} \sim 1.2$ s, orbiting in $P_{\rm orb} \sim 1.27$ d around a companion star of mass $M \geqslant 1.5 M_\odot$, thus falling into the realm of IMXB systems, also suggested by the presence of a highly magnetized neutron star (Karino 2016). Finally, as we will see below, IMXB are also important sources in the context of extragalactic X-ray sources.

7.2 Ultra-luminous X-ray Sources (ULX)

Extragalactic compact X-ray sources are detected in a number of nearby galaxies, characterized by extreme bolometric luminosities in the range from $L \sim 10^{39}$ up to $L \sim 10^{41}$ erg s^{-1}. These unresolved sources, called *ultra-luminous X-ray sources* (ULX), significantly exceed the Eddington luminosity of an accreting neutron star, or even of a stellar-mass black hole of mass $\gtrsim 70 M_\odot$, by a factor of ~ 100 for the most luminous systems (Kaaret et al. 2017). ULX are located outside the nucleus of their host galaxy, some being likely associated to ionized nebulae of diameter up to ~ 100 pc (Kaaret & Corbel 2009). Their origin is not clear, and there are two main possibilities, depending on the nature of their X-ray emission, anisotropic or isotropic, that we now review.

7.2.1 Anisotropic Emission by an Accreting Compact Object

7.2.1.1 Anisotropic Emission by an Accreting Stellar-mass Black Hole
The first possibility to consider is that ULX are powered by conventional stellar-mass black holes ($\sim 3 \lesssim M \lesssim 20 M_\odot$), within intermediate and high-mass X-ray

binaries, radiating X-rays non-isotropically, either intrinsically or through relativistic aberration (King et al. 2001). By assuming mild X-ray beaming, the ULX phenomenon may take place during the short thermal-timescale mass transfer stage in the evolution of such binaries. Since higher black hole mass implies less extreme accretion luminosities compared to the Eddington limit, theoretical models have naturally focused on black hole rather than neutron star binaries.

7.2.1.2 Anisotropic Emission by an Accreting Neutron Star

However, the detection of X-ray pulsations in ULX unambiguously showed that at least some ULX are X-ray binary systems hosting an accreting neutron star, emitting above their Eddington luminosity. Bachetti et al. (2014) detected pulsations of period 1.37 s emanating from an ULX—M82 X-2—located in the nuclear region of the galaxy M82, identified as the spin period of the magnetized neutron star, and a 2.5 d sinusoidal modulation associated to the orbital period of the system. This ULX emits an X-ray luminosity ranging from $L_X \sim 5 \times 10^{39}$ to 2×10^{40}erg s^{-1}, thus more than 10 times brighter than any known accreting pulsar, corresponding to 100 times the Eddington limit for a $1.4 M_\odot$ neutron star. Later, an accreting pulsar, spinning at $P_{\rm spin} \sim 0.42$ s, was detected in the variable ULX NGC 7793 P13 ($L_X \sim 2 - 5 \times 10^{39}$erg s^{-1}), orbiting around a $18 - 23 M_\odot$ B9Ia companion ($P_{\rm orb} = 64$ d). This brought another unambiguous proof of a magnetized neutron star, with $B \sim 10^{13}$ G, accreting at super-Eddington rates (Israel et al. 2017). To justify such observations, Misra et al. (2020) computed with MESA the evolution of LMXB and IMXB with donor star in the mass range $0.92 < M < 8.0 M_\odot$, exploring the initial parameters leading to the formation of dynamically stable non-conservative mass-transfer ULX hosting a neutron star, with super-Eddington X-ray luminosities in the range $10^{39} \lesssim L_X \lesssim 10^{41}$erg s^{-1}, assuming geometrical beaming, on a timescale of up to ~1 Myr for the lower luminosities. Both these observations and modeling show that LMXB and IMXB hosting neutron stars might be common in the ULX population, at least in the pulsating sources (see also Marchant et al. 2017, for the link between ULX and NSBH mergers, from very massive close binaries at low metallicity).

An alternative, proposed by Karino (2021), is that at least some ULX could be members of the BeHMXB family, since some ULX occupy the same region as BeHMXB on the Corbet diagram, which suggests a connection between spin and orbital periods of these systems (see Figure 7.2). They use the framework of mass-accretion models for BeHMXB, to investigate the conditions under which neutron stars can achieve mass-accretion rates beyond the Eddington limit, making them potentially observable as ULX. They show that a BeHMXB has the possibility to become a ULX, in case certain conditions are fulfilled, such as the magnetic field of the neutron star, and the density of the Be disk. However, even if a strong magnetic field increases the brightness of a neutron star from a Be, it cannot exceed more ~50 times the Eddington limit. Finally, Karino (2021) propose a scenario in which normal BeHMXB may become an ULX, when the donor evolves into a giant.

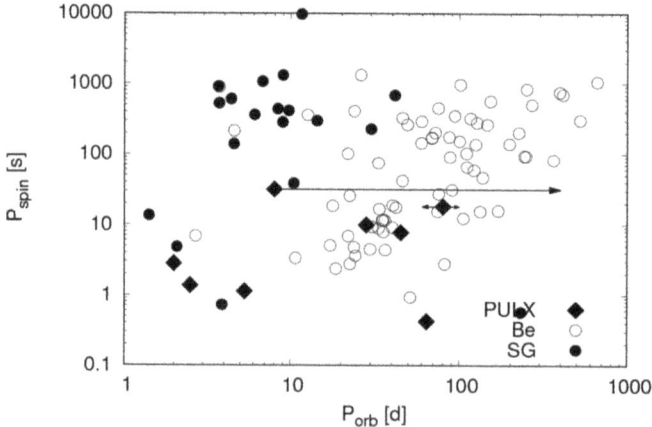

Figure 7.2. Corbet diagram, showing the relationship between P_{spin} and P_{orb} for neutron star systems. sgHMXB are indicated by black circles, and BeHMXB by open circles. Black diamonds show the position of ULX harboring neutron stars. Reprinted with permission from Karino (2021).

7.2.1.3 A Link with Galactic Sources?

Observations of ULX also suggest a link with Galactic jet sources, such as the neutron star system A0535-668 ($L_X \sim 1.2 \times 10^{39}$erg s^{-1}), and the black hole systems GRS 1915+105 (2.4×10^{39}erg s^{-1}) and V4641 Sgr (6.2×10^{39}erg s^{-1}). Since these sources reach super-Eddington luminosity at their maximum of emission, they might be viewed as ULX candidates (McClintock & Remillard 2006). In this context, the short lifetimes of HMXB could even explain the association of ULX with episodes of star formation (King et al. 2001). We can also add to this list the microquasar SS 433, which produces radio jets of high inclination, with kinetic luminosities of more than $\sim 10^{39}$erg s^{-1}, which would be super-Eddington for a $10M_\odot$ black hole (see Chapter 8). As stated by Kaaret et al. (2017), many aspects of SS 433, such as the surrounding radio nebula (W50) with an extent of about 100 pc along the radio/ optical jet axis, are similar to optical and radio counterparts exhibited by ULX, supporting interpretation of ULX as super-Eddington accretors.

7.2.2 Isotropic Emission by an Intermediate-mass Black Hole

The second possibility to consider, if their X-ray emission is isotropic, without any geometric nor relativistic beaming toward the line of sight, is that some ULX may host accreting *intermediate-mass black holes* (IMBH) of mass range $10^2 M_\odot \leqslant M \leqslant 10^5 M_\odot$, in order to remain below the Eddington limit (Miller & Colbert 2004). Several candidates for IMBH are suggested by electromagnetic observations, but still lack conclusive confirmation. Observational evidence of their existence includes direct kinematical measurement of the mass of the central black hole in massive stellar clusters and galaxies, and empirical mass scaling relations of QPO in luminous X-ray sources (Remillard & McClintock 2006). The most massive IMBH candidate among these is HLX-1, an hyper-luminous variable X-ray source, suggesting an IMBH mass of $\sim 3 - 300 \times 10^3 M_\odot$, at the edge-on spiral galaxy ESO243-49 (see Figure 7.3), with a

Figure 7.3. HLX-1 counterpart on the edge of the galaxy ESO 243-49 HLX-1. Image Credit: NASA, ESA, Z. Levay (STScI) and S. Farrell (Sydney Institute for Astronomy, University of Sydney).

conservative lower limit for the mass of the black hole of $\sim 500\,M_\odot$ (Farrell et al. 2009; Mapelli et al. 2012). An IMBH candidate was detected outside the nucleus of a large lenticular galaxy, within a dense cluster made of massive stars, through a luminous X-ray outburst due to a tidal disruption event (TDE), and subsequent accretion of stars closely approaching the black holes (Lin et al. 2018). The luminosity peaked at $L \sim 10^{43}\,\mathrm{erg\ s^{-1}}$ and then decayed on a 10 year-timescale, supporting the identification as a TDE. The ultra-soft X-ray emission was consistent with a standard thermal disk, which cooled significantly as the luminosity decreased, a signature of accreting stellar-mass black hole, with a mass $\sim 10^4\,M_\odot$. This detection suggests that IMBH are associated to very dense stellar regions. IMBH have also been invoked in the detection of gravitational waves coming from the massive binary black hole merger GW 190521 (see Section 6.6.2, and Abbott et al. 2020a, 2020b). Finally, there is a claim of an IMBH detection through a gravitationally lensed γ-ray burst (GRB 950830) located at cosmological distances, after an analysis of 2700 GRB lightcurves with millisecond-to-second time delays. The inferred lens mass is $5.5 \times 10^4\,M_\odot$, in the realm of IMBH (Paynter et al. 2021).

IMBH formation requires particular channels distinct from those leading to the formation of X-ray binaries. The main channel invokes the direct collapse of massive, and low-metallicity, Population III first-generation stars at high redshift ($z = 25 - 30$, see Section 6.5.4.3), following the hierarchical fragmentation of primordial molecular clouds into pre-galactic star formation, leading to IMBH of mass $M \sim 150\,M_\odot$, and then becoming incorporated through a series of mergers into larger and larger IMBH, naturally leading to tidal captures and disruption of ordinary stars (Madau & Rees 2001). An alternative formation channel is the runaway growth of IMBH in young dense star clusters due to multiple and hierarchical stellar collisions, in which the majority of collisions occur with the same star, eventually becoming an IMBH of mass amounting up to 0.1% of the mass of the entire star cluster (Portegies Zwart & McMillan 2002). In this scenario, star clusters older than ~5 Myr and with present-day half-mass relaxation times ≤100 Myr are expected to contain an IMBH. Mapelli (2016) studied the competition between the mass lost by stellar wind from low-metallicity stars, with dynamics and stellar evolution timescales, until it leads to the formation of a stellar remnant. At solar metallicity, the mass of the final merger product spans from few solar masses, up to ~$30M_\odot$. At low metallicity ($0.01 \leqslant Z \leqslant 0.1$), the maximum remnant mass is ~$250M_\odot$, right in the range of IMBH masses (see discussion in Section 6.6.2.5).

Understanding the formation of IMBH is fundamental, because hierarchical merger of IMBH systems in dense environments is among the putative formation channels for supermassive black holes (Volonteri 2010), providing the missing link between stellar-mass (~$3 \lesssim M \lesssim 20M_\odot$) black holes resulting from core collapse of massive stars, and supermassive (~$10^6 \lesssim M \lesssim 10^{10}M_\odot$) black holes located in the center of most, if not all, galaxies (see Figure 7.4). If a significant IMBH population exists, it could thus provide the seeds for the growth of supermassive black holes in the early universe.

Earnshaw et al. (2019) have compiled a catalog of 1314 extragalactic non-nuclear X-ray sources in nearby galaxies, by correlating the 3XMM-DR4 data release of the XMM-Newton Serendipitous Source Catalogue, with the Third Reference Catalogue of Bright Galaxies and the Catalogue of Neighbouring Galaxies. Among these X-ray sources, 384 are candidate ULX with L_X up to 5×10^{40}erg s^{-1}, the ULX population being more luminous in star-forming host galaxies than in non-star-forming galaxies.

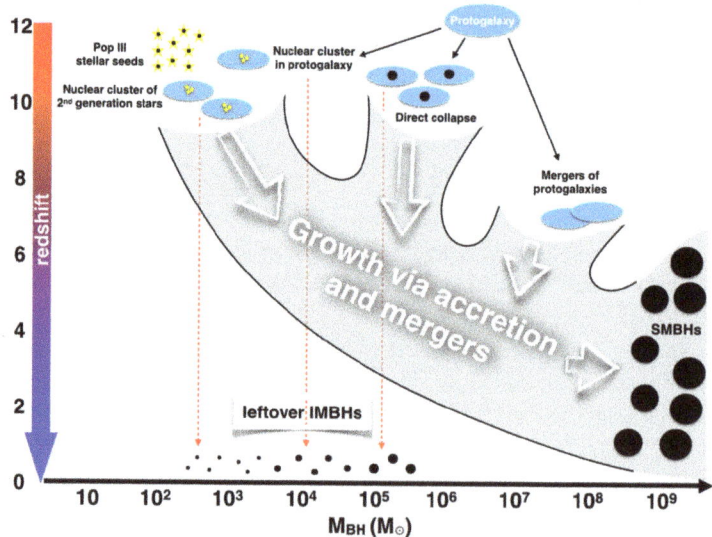

Figure 7.4. Scenario of formation of intermediate-mass black hole, the missing link between stellar-mass (~$3 \lesssim M \lesssim 20M_\odot$) black holes resulting from core collapse of massive stars, and supermassive (~$10^6 \lesssim M \lesssim 10^{10}M_\odot$) black holes at the center of galaxies. Reprinted with permission from World Scientific from Mezcua (2017).

7.3 Gamma-Ray Binaries

Some HMXB do not fit entirely in any of the classes described in Chapter 6, we already mentioned the two microquasars likely hosting neutron stars: SS 433 and Cyg X-3. Cyg X-3, with $P_{orb} = 4.8$ hr, likely represents a later evolutionary phase of a wide HMXB (Tauris & van den Heuvel 2006). It has been detected in γ-rays by the Fermi γ-ray Space Telescope (Abdo et al. 2009c). Cyg X-1, also detected by Fermi (Abdollahi et al. 2020), presents the particularity to be an sgHMXB microquasar hosting a black hole. For completion, the Fermi Large Area Telescope 4th source catalog includes a search of LMXB and HMXB emitting at high energy (Abdollahi et al. 2020).

In addition to these HMXB, mainly emitting in X-rays, there exists γ-ray binaries, which are rare systems mainly emitting in γ-rays (Dubus 2006),[1] though binary evolution models predict dozens in our Galaxy. Their high-energy emission and variability patterns are dominated by inverse Compton scattering of electrons accelerated close to the black hole, or in the interaction regions between the companion stellar winds and pulsar winds (Dubus 2013), or even microquasar jets (see Figure 8.1). There are two such systems hosting a Be companion star: PSR B1259-63/LS 2883 (Abdo et al. 2011) and LS I +61 °303 (Abdo et al. 2009a), and two systems hosting an O companion star: LS 5039 (O6.5 V spectral type secondary; Abdo et al. 2009b), and 1FGL J1018.6-5856 (O6V((f)) secondary), for which periodic γ-ray emission has been detected by Fermi with a 16.6 day period (Ackermann et al. 2012). Concerning PSR B1259-63, a radio pulsar in an eccentric orbit around a Be star, γ-ray emission of more than 100MeV was also detected by Fermi, increasing after the passage at periastron of the pulsar, on its eccentric orbit (Abdo et al. 2011; Chernyakova ct al. 2014).

Other galactic sources, such as magnetars, microquasars, novae, or even pulsar wind nebulae flares, display transient emission in X-rays, and potentially also in the high-energy domain, at MeV–GeV energies. However, none of these galactic transients have yet been detected in the very high-energy (VHE, $E \geqslant 100$ GeV) domain. Such objects will be among the target sources to be observed with the Cherenkov Telescope Array (CTA) observatory (López-Oramas et al. 2022).

Because their γ-ray emission is mainly due to inverse Compton scattering of electrons at the interface between stellar winds and the compact object, and not due to accreting process, we will not describe these sources in more detail.

7.4 Reviews, Catalogs, Database and References

7.4.1 Reviews

- Reviews including IMXB: Tauris & van den Heuvel (2006); Casares et al. (2017);
- Review on ULX: Kaaret et al. (2017).

---

[1] The title of this paper is: "Gamma-ray binaries: Pulsars in disguise?"

7.4.2 Catalogs

- IMXB sources are included in the catalog of HMXB: Liu et al. (2006);
- Catalog of ULX: Earnshaw et al. (2019).

7.4.3 Database

We give in Table 7.1 the list of dynamically-determined binary parameters and effective spin of black hole systems within IMXB.

References

Abbott, R., Abbott, T. D., Abraham, S., et al. 2020a, PhRvL, 125, 101102

Abbott, R., Abbott, T. D., Abraham, S., et al. 2020b, ApJL, 900, L13

Abdo, A. A., Ackermann, M., Ajello, M., et al. 2011, ApJL, 736, L11

Abdo, A. A., Ackermann, M., Ajello, M., et al. 2009a, ApJL, 701, L123

Abdo, A. A., Ackermann, M., Ajello, M., et al. 2009b, ApJL, 706, L56

Abdo, A. A., Ackermann, M., Ajello, M., et al. 2009c, Sci, 326, 1512

Abdollahi, S., Acero, F., Ackermann, M., et al. 2020, ApJS, 247, 33

Ackermann, M., Ajello, M., Ballet, J., et al. 2012, Sci, 335, 189

Bachetti, M., Harrison, F. A., Walton, D. J., et al. 2014, Natur, 514, 202

Bailyn, C. D., Orosz, J. A., Girard, T. M., et al. 1995, Natur, 374, 701

Buxton, M. M., & Bailyn, C. D. 2004, ApJ, 615, 880

Casares, J., Jonker, P. G., & Israelian, G. 2017, in Handbook of Supernovae (Berlin: Springer), 1499

Casares, J., Orosz, J. A., Zurita, C., et al. 2009, ApJS, 181, 238

Casares, J., Zurita, C., Shahbaz, T., Charles, P. A., & Fender, R. P. 2004, ApJL, 613, L133

Chaty, S., Charles, P. A., Martí, J., et al. 2003, MNRAS, 343, 169

Chaty, S., Mirabel, I. F., Goldoni, P., et al. 2002, MNRAS, 331, 1065

Chen, W-C., & Podsiadlowski, P. 2016, ApJ, 830, 131

Chernyakova, M., Abdo, A. A., Neronov, A., et al. 2014, MNRAS, 439, 432

Corral-Santana, J. M., Casares, J., Muñoz-Darias, T., et al. 2016, A&A, 587, A61

Dubus, G. 2006, A&A, 456, 801

Dubus, G. 2013, A&ARv, 21, 64

Dubus, G., & Chaty, S. 2006, A&A, 458, 591

Earnshaw, H. P., Roberts, T. P., Middleton, M. J., Walton, D. J., & Mateos, S. 2019, MNRAS, 483, 5554

Farrell, S. A., Webb, N. A., Barret, D., Godet, O., & Rodrigues, J. M. 2009, Natur, 460, 73

Hynes, R. I., Haswell, C. A., Shrader, C. R., et al. 1998, MNRAS, 300, 64

Israel, G. L., Papitto, A., Esposito, P., et al. 2017, MNRAS, 466, L48

Kaaret, P., & Corbel, S. 2009, ApJ, 697, 950

Kaaret, P., Feng, H., & Roberts, T. P. 2017, ARA&A, 55, 303

Karino, S. 2016, PASJ, 68, 93

Karino, S. 2021, MNRAS, 507, 1002

King, A. R., Davies, M. B., Ward, M. J., Fabbiano, G., & Elvis, M. 2001, ApJL, 552, L109

Klochkov, D. K., Shakura, N. I., Postnov, K. A., et al. 2006, AstL, 32, 804

Lin, D., Strader, J., Carrasco, E. R., et al. 2018, NatAs, 2, 656

Liu, Q. Z., van Paradijs, J., & van den Heuvel, E. P. J. 2006, A&A, 455, 1165

López-Oramas, A., Bulgarelli, A., Chaty, S., et al. 2022, ICRC (Berlin), 37, 784

Madau, P., & Rees, M. J. 2001, ApJL, 551, L27

Mapelli, M. 2016, MNRAS, 459, 3432

Mapelli, M., Annibali, F., & Zampieri, L. 2012, in Half a Century of X-ray Astronomy, ed. I. Georgantopoulos, & M. Plionis, 130

Marchant, P., Langer, N., Podsiadlowski, P., et al. 2017, A&A, 604, A55

McClintock, J. E., & Remillard, R. A. 2006, in Cambridge Astrophysics Ser. 39, Compact Stellar X-Ray Sources, ed. W. Lewin, & M. van der Klis (Cambridge: Cambridge Univ. Press), 157

Mezcua, M. 2017, IJMPD, 26, 1730021

Miller, M. C., & Colbert, E. J. M. 2004, IJMPD, 13, 1

Misra, D., Fragos, T., Tauris, T. M., Zapartas, E., & Aguilera-Dena, D. R. 2020, A&A, 642, A174

Orosz, J. A., Jain, R. K., Bailyn, C. D., McClintock, J. E., & Remillard, R. A. 1998, ApJ, 499, 375

Orosz, J. A., Kuulkers, E., van der Klis, M., et al. 2001, ApJ, 555, 489

Paynter, J., Webster, R., & Thrane, E. 2021, NatAs, 5, 560

Podsiadlowski, P., & Rappaport, S. 2000, ApJ, 529, 946

Podsiadlowski, P., Rappaport, S., & Pfahl, E. D. 2002, ApJ, 565, 1107

Portegies Zwart, S. F., & McMillan, S. L. W. 2002, ApJ, 576, 899

Remillard, R. A., & McClintock, J. E. 2006, ARA&A, 44, 49

Reynolds, C. S. 2021, ARA&A, 59

Tauris, T. M., & van den Heuvel, E. P. J. 2006, in Cambridge Astrophysics Ser. 39, Compact Stellar X-ray Sources, ed. W. Lewin, & M. van der Klis (Cambridge: Cambridge Univ. Press), 623

Tauris, T. M., van den Heuvel, E. P. J., & Savonije, G. J. 2000, ApJ, 530, L93

Trümper, J., Pietsch, W., Reppin, C., et al. 1978, ApJL, 219, L105

Volonteri, M. 2010, A&ARv, 18, 279

Chapter 8

Jets in Accreting Binaries

Everything else aside,
Microquasars are to Quasars, what Paris is to Earth

—Yvain Dibuz, *2006*

We describe in this chapter the jet properties related to all types of accreting stellar binaries. We first give a bit of history, the jet main properties, and we give details on how accretion and ejection phenomena are coupled. We then describe jets in low-mass X-ray binaries and high-mass X-ray binaries, with a few emblematic jet sources, such as SS 433, Cyg X-1, GRS 1915 + 105, and more generally with the so-called microquasars, hosting both low-mass and high-mass stars. At the end of this chapter, we report useful reviews, catalogs, database, and references.

8.1 Jets in Celestial Sources

Collimated ejections are found in distinct classes of celestial objects such as young stars, binary systems (hosting a white dwarf, neutron star or black hole), and also AGN hosting a supermassive black hole (Blandford et al. 2019). We first focus on ejection phenomena such as mass ejection and jet collimation, before describing the accretion/ejection coupling, via multi-wavelength observations and numerical simulations.

8.1.1 A Bit of History

Historically, the first jet associated to a celestial source was detected in a 5 mn exposure in the optical domain in the field of view of the galaxy M87 in 1918 by Curtis (1918).[1]

---

[1] Curtis (1918) gives the following description: *"A curious straight ray lies in a gap in the nebulosity in position angle 20°, apparently connected with the nucleus by a thin line of matter. The ray is brightest at its inner end, which is 11″ from the nucleus.".*

Jets in galactic objects were first detected—nearly simultaneously—in young stars, so-called Herbig–Haro objects (Cudworth & Herbig 1979; Herbig & Jones 1981), and in a compact binary, SS 433, exhibiting jets consisting of matter moving at relativistic speeds (Margon et al. 1979), with a title clearly showing the astonishment of the observers at this time: *The bizarre spectrum of SS 433*. Jets, in the form of radio lobes from the synchrotron plasma ejected by the central source, are thereafter discovered in radio in other galactic binaries hosting a neutron star or a black hole, such as the *microquasars* located within the Galactic bulge, 1E 1 740.7-2942 (Mirabel et al. 1992) and GRS 1758-258 (Rodríguez et al. 1992), so-called in reference to their large cousins, the quasars.

On the one hand, jets in microquasars (see Figure 8.1) differ from those in quasars by two main parameters: first, the mass of the central source (neutron stars or black holes weigh a few $M_\odot$ while supermassive black holes in active galactic nuclei weigh

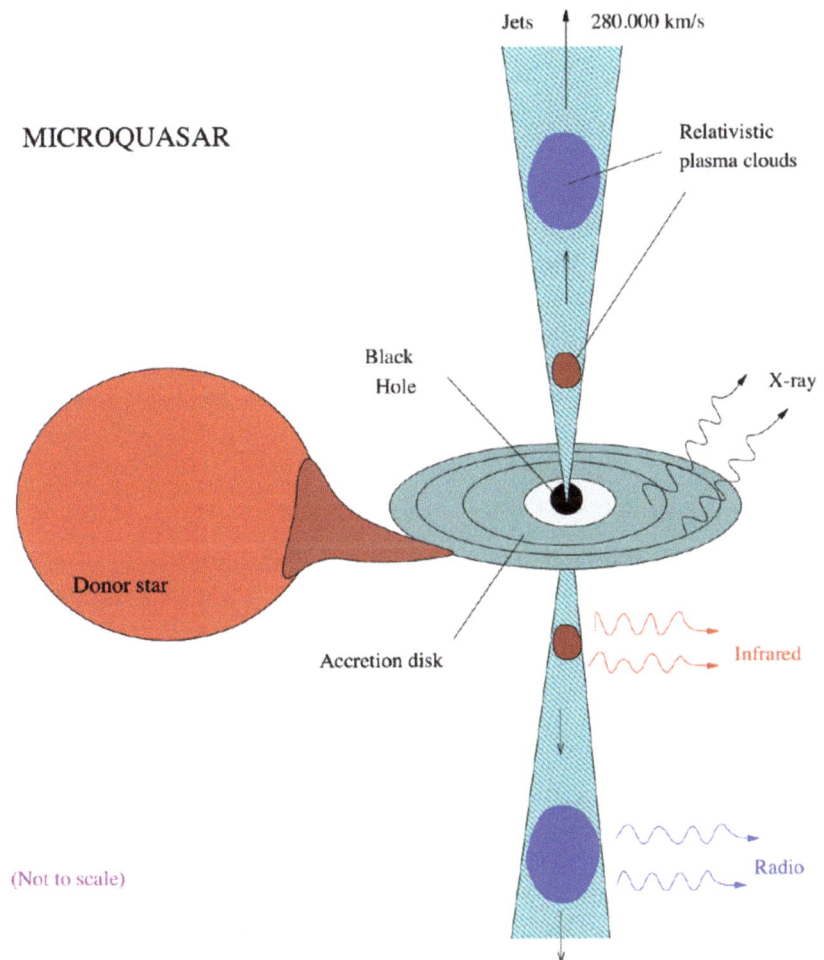

Figure 8.1. Figure showing a generic microquasar (LMXB or HMXB). Credit: Sylvain Chaty.

from 10^6 to $10^9 M_\odot$); and second, their length (10^{17}cm for microquasars compared to 10^{24}cm for quasars). On the other hand, microquasars and quasars share many properties, in particular at high energy (Healey et al. 2008; Abdo et al. 2009, 2010; Ackermann et al. 2013; Tomsick et al. 2020, 2021), not only in term of spectra, but also in morphology. One of the most intriguing case is the extended source IGR J17448-3232, discovered by the INTEGRAL satellite, identified as a blazar—a quasar exhibiting jets pointing toward us—hidden beyond the Scorpius galaxy cluster (Barrière et al. 2015). Galactic and extragalactic jet sources also share similar properties at other wavelengths, such as the same morphology in radio (Martí et al. 1998b; Di Cocco et al. 2004; Foschini et al. 2005; Combi et al. 2006; Zurita Heras et al. 2009; Curran et al. 2011, 2012). But the most important, is that microquasars and quasars share a number of fundamental properties, and among them the velocity of their jets, relativistic in both cases, reflecting the common fundamental physics underlying the acceleration of such ejections (Türler et al. 2004).

Bipolar outflow components, in the form of transient collimated winds, were also detected in tight binaries hosting accreting white dwarfs (mainly luminous supersoft X-ray sources), via observations of satellite lines blue- and redshifted with respect to central lines (Hα, Hβ, HeII, etc.), exhibiting velocities ranging from ~800 to ~6000 km s^{-1} (Southwell et al. 1996; Motch 1998; Ogley et al. 2002).

Finally, we can add to this list the so-called *tidal disruption event* (TDE), in which a star is gravitationally captured by a quiescent supermassive black hole of mass in the range $10^6 - 10^8 M_\odot$, lurking in the nucleus of an active galaxy, until eventually torn apart by intense tidal forces. When this happens, a short-lived accretion disk forms around the supermassive black hole, consisting of the matter of the bound star, powering jets similar to the ones produced in microquasars and quasars. These events, predicted by Rees (1988), are now observed nearly routinely, as shown in the review by Komossa (2015), with a list of candidate TDE identified from X-ray, UV, and optical observations.

8.1.2 Jet Main Properties

It seems natural that galactic and extra-galactic jets exhibit different properties, their velocity and emission processes depending on the nature of accreting object. However, the fact that ejection seems ubiquitous around accretion disks surrounding a central gravitational potential (we already cited young stars, accreting binaries hosting white dwarfs, neutron stars and black holes, and AGN), suggests a common fundamental mechanism linked to accretion of matter (Livio 1997). Young stars exhibit slow jets on a parsec scale, with characteristic velocities of a few hundreds (100–350) km s^{-1} for objects such as Herbig–Haro, radiating thermal emission and heated by shocks. Microquasars exhibit relativistic jets between 0.3 and 0.9 c, on a parsec scale, radiating synchrotron emission via electrons accelerated along magnetic field lines. On the other opposite, AGN exhibit jets on a kpc scale.

There is an interesting property of jets to point out. Both stellar-mass and supermassive black holes share the same engine—a black hole—characterized by the Schwarzschild radius property ($R_S = \frac{2GM}{c^2}$, see Section 2.1.5). Their dynamical variation timescale, related to the time required for ejected matter to travel a distance corresponding to the event horizon radius, is proportional to the mass of the black hole:

$$\frac{t_{\text{microquasar}}}{t_{\text{AGN}}} = \frac{M_{\text{microquasar}}}{M_{\text{AGN}}} \sim 10^{-5} \text{ to } 10^{-8}. \tag{8.1}$$

It follows that a timescale of a few years for a $\sim 10^6 M_\odot$ AGN reduces to a few minutes for a $10 M_\odot$ microquasar, and even drops to a few seconds compared to a $\sim 10^9 M_\odot$ AGN. It is thus possible to study accretion–ejection phenomena in Galactic sources on a second-to-minute timescale, which would range from month-to-year-timescale in AGN, too long to be accessible to human timescale.

As described in Livio (1997), a fundamental role of jets is to evacuate the angular momentum: $\dot{j} = \dot{M}_W \Omega r_a^2$, with $\dot{M}_W$ the wind mass loss rate, r_a the Alfvén radius, and Ω the angular velocity. To allow for accretion to occur at the rate $\dot{M}_{\text{acc}}$ at the distance r from the center of the disk, the angular momentum to evacuate is: $\dot{j}_{\text{acc}} = \frac{1}{2} \dot{M}_{\text{acc}} \Omega r^2$, implying:

$$\frac{\dot{M}_W}{\dot{M}_{\text{acc}}} = \frac{1}{2} \left(\frac{r}{r_a} \right)^2. \tag{8.2}$$

In the inner part of the disk (corresponding to $r = \frac{r_a}{10}$), a small wind mass loss $\dot{M}_W$ corresponding to only 1/100 of the accretion mass loss $\dot{M}_{\text{acc}}$ is enough to get rid of the angular momentum. It follows from this simple calculation that the jet process allows to evacuate a substantial part of angular momentum.

8.1.3 Accretion and Ejection Coupling

For all sources exhibiting jets, the velocity of the jet appears to be of the order of the release velocity of the central object, suggesting that jets take their origin in inner parts of the accretion disk. For instance, for binaries hosting white dwarfs, v_{release} corresponds to the high range of velocity observed:

$$v_{\text{release}} = \left(\frac{2GM}{R} \right)^{1/2} = \left(\frac{M}{M_\odot} \right)^{1/2} \left(\frac{10^7 m}{R} \right)^{1/2} 5100 \text{ km s}^{-1}. \tag{8.3}$$

Jets are likely accelerated and collimated by magneto-centrifugal forces. In the model proposed by Blandford & Payne (1982), a bipolar magnetic field, either of interstellar or dynamo origin, is anchored in the accretion disk, turning with it, with the plasma following the magnetic field lines. When these lines form an angle with the disk smaller than 60°, plasma movements along these field lines become unstable. The centrifugal force then begins to dominate the gravitational force, propulsing the plasma with strong equatorial component, at least up to the Alfvén sphere, when kinetic energy equals magnetic energy. Jet collimation continues at larger distance, for instance through the so-called *hoop stress* process, when the magnetic field spirals and leads to a centripetal Lorentz force, thus allowing collimation (Sakurai 1985).

An alternative leading mechanism to power jets from rotating Kerr black holes is the effect proposed by Blandford & Znajek (1977), exploiting the existence of an ergosphere, to extract rotation energy directly from the black hole, via magnetic field lines, brought by accretion of matter.

8.2 Jets in Low-mass and High-mass X-ray Binaries

The ejection phenomenon seems ubiquitous in low-mass X-ray binaries (LMXB) hosting accretion disks, at least on the basis of radio observations: most accreting binaries emit in the radio domain, with a flat spectrum of synchrotron origin. While collimated jets are spatially resolved only in a small number of cases (see radio properties of various types of sources in Table 8.1), it is likely that radio emission emanates from *compact jets*, collimated outflows or even winds.

Concerning neutron star binaries, while Atoll sources rarely emit in radio, Z-type sources exhibit intense radio luminosity, likely via synchrotron emission, suggesting the presence of a jet. Accreting X-ray pulsars do not exhibit radio emission, probably due to the absence of accretion disk. Accreting black holes exhibit strong radio emission in their low-hard state, likely via synchrotron emission, thus also suggesting the presence of a jet (Cadolle Bel et al. 2007, 2009). On the other hand, their radio emission is weak or nonexistent in their high-soft state (Corbel et al. 2000). Transient accretion phenomena in black holes also produce discrete ejections during state transitions, at flaring onset, from absence of accretion to high and

Table 8.1. Galactic Sources of Relativistic Jets.

Source	Object	V_{app} (c)	V_{int} (c)	$i(°)$	D (kpc)	Ref
CI Cam	NS?	~0.15	~0.15	>70	~5	h98a, b19
Circinus X-1	NS?	⩾0.1	⩾0.1	>70	9.4	f98
SS 433	NS?	0.26	0.26	79	5.5	d98, b04
Cygnus X-3	NS?	~0.3	~0.3	>70	~7	m01
MAXI J1348-630	BH	~1.4	?	25 — 32	$2.2^{+0.5}_{-0.6}$	c21
GRO J1655-40	BH	1.1	0.92	69 ± 2	3.5	h95
1E 1740.7-2942	BH				~8.5	m00
XTE J1748-288	BH	0.9–1.5	>0.9		~8.5	h98b
GRS 1758-258	BH				~8.5	m98
GRS 1915+105	BH	1.2–1.7	0.92–0.98	60 ± 5	~11	m94, f99

Notes. V_{app} is the apparent velocity of the fastest component of ejecta, V_{int} is the intrinsic velocity of ejecta, and i is the inclination of the jet (angle between jet and line of sight). This table is an updated and completed version from Mirabel & Rodríguez (1999). References: **h98a**: Hjellming et al. (1998a); **b19**: Bartlett et al. (2019); **f98**: Fender et al. (1998); **d98**: Dubner et al. (1998); **b04**: Blundell & Bowler (2004); **m01**: Martí et al. (2001); **c21**: Carotenuto et al. (2021); **h95**: Hjellming & Rupen (1995); **m00**: Martí et al. (2000); **h98b**: Hjellming et al. (1998b); **m98**: Martí et al. (1998a); **m94**: Mirabel & Rodríguez (1994); **f99**: Fender et al. (1999).

luminous state (Kuulkers et al. 1999). We now review some of the most peculiar cases, putting together low-mass and high-mass binaries, and we will close this section by giving a list of sources exhibiting relativistic jets.

8.2.1 The Drilling Case of SS 433

The first jets to be discovered in a Galactic binary were emanating from the peculiar source SS 433 (Margon et al. 1979; Margon 1984), a high-mass microquasar in an advanced evolutionary stage (Cherepashchuk et al. 2020), located at a distance of 5kpc, right at the center of a supernova remnant, W50 (see Figures 8.2 and 8.3). The nature of its compact object remains unknown, though a neutron star is favored. The optical spectrum of SS 433, characterized by a strong Balmer Hα emission line complex, exhibits large variable Doppler blueshifts and redshifts $\frac{\Delta\lambda}{\lambda}$, due to matter ejected at the relativistic velocity $v \sim 0.26$ c, along oppositely directed jets strongly collimated with a $\sim$4° angle, inclined by 78.8°, and launch axis precessing with a cone angle of $\sim$19.8°, with a periodicity of $\sim$162 days. Its jet is entirely revealed in the radio domain, with a zigzag/corkscrew structure due to its changing orientation with respect to the line of sight (reported in Stirling et al. 2004; Blundell & Bowler 2004; Blundell et al. 2018), suggesting a ballistic post-launch motion, instead of jet deceleration. Ionized atoms (such as O, S, Si, Ar, Ca, Fe, Ni) are detected in the jet spectra, suggesting an efficient acceleration process, the central source ejecting baryonic matter in a ballistic way. A model describing thermal X-ray emission emanating from the baryonic jet has been developed especially for SS 433 (see Figure 8.4; Khabibullin et al. 2016).

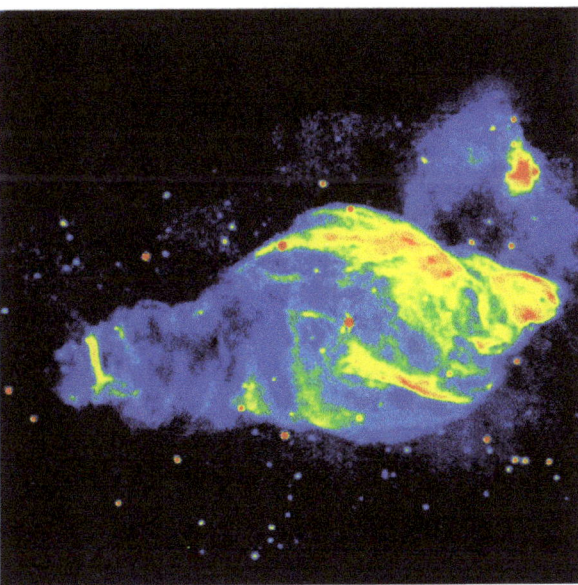

Figure 8.2. Large field radio image (VLA and Bonn-100m) of the SNR nebula W50 surrounding the HMXB SS 433 at the center. Reprinted with permission of the AAS from Dubner et al. (1998).

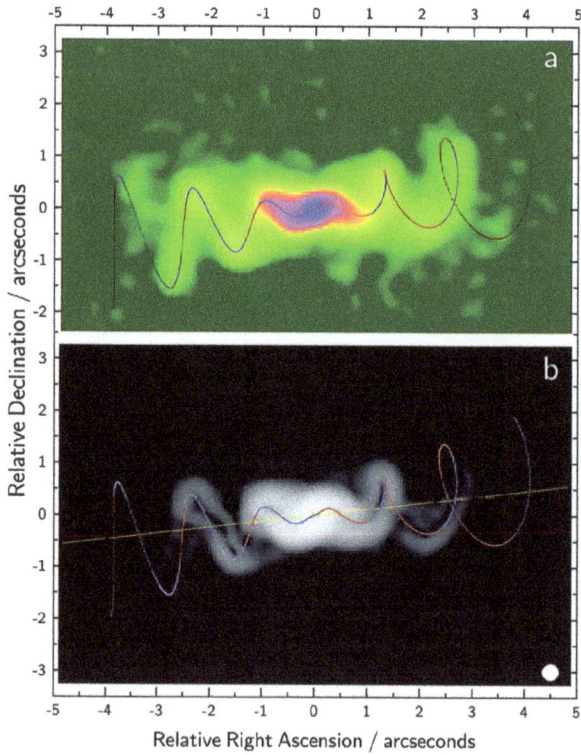

Figure 8.3. VLA image of compact jets close to the central engine of SS 433, showing the effect of the precession of the jets (see Blundell et al. 2018 for ALMA observations of the same region). Reprinted with permission of the AAS from Blundell & Bowler (2004).

The radiative lifetime of relativistic electrons is greater than 1000 years, thus remaining relativistic until they collide with the surrounding matter constituting W50. At greater distance ($\sim$100 pc from the central source), X-ray emission reappears, due to ejected matter colliding with the SNR W50, which seems drilled at all the impacted locations (see Figure 8.3; Dubner et al. 1998). SS 433 exhibits a low X-ray luminosity of $L_X \sim 5 \times 10^{35}$ erg s^{-1}. Since the kinetic energy at the base of the jet is of the same order of magnitude than the energy stored in the surrounding jet-powered optical nebula ($\sim 10^{40}$ erg s^{-1}), the accretion energy seems wholly injected in jet acceleration, with a mass loss rate of $\dot{M} \sim 10^{-5}$ to $10^{-6} M_\odot$/year.

8.2.2 The Impacting Case of Cyg X-1

Although Cyg X-1 is an high-mass X-ray binary (O9.7I supergiant star and $21 M_\odot$ black hole; Miller-Jones et al. 2021), it exhibits intense ejection phenomenon, with jets launched by the same accretion–ejection process described in this section, requiring the presence of an accretion disk to eject matter. Cyg X-1, by undergoing accretion, releases the gravitational potential energy of the matter falling toward its central black hole. This source not only produces highly collimated and energetic

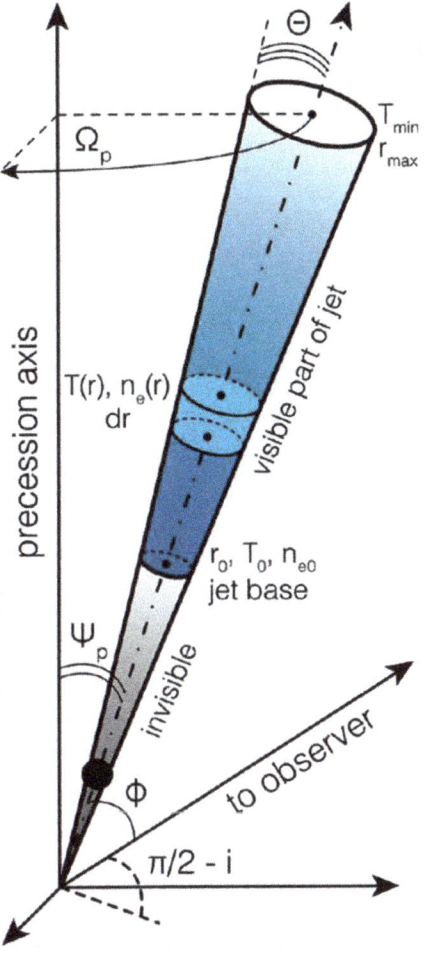

Figure 8.4. Schematic view of the jet model developed for the microquasar SS 433. The jet geometry is treated as an axi-symmetrical ballistic flow, directed away from the compact object. Reprinted with permission from Khabibullin et al. (2016).

jets of particles resolved in radio, flowing out of the system at relativistic velocities (Stirling et al. 2001), but also is surrounded by a large-scale (diameter ∼5 pc) ring-like structure visible in radio, infrared, and optical Hα, in alignment with the inner compact radio jets, and thus apparently inflated and powered by this inner jet (Gallo et al. 2005; Russell et al. 2007). The radio emission was interpreted as originating from bremsstrahlung radiation, from the shock-compressed gas just behind the radiative shock front, as described by models of jet and ISM interactions, with $9 \times 10^{35} \leqslant P_{jet} \leqslant 10^{37}$erg s^{-1} (Russell et al. 2007). Gallo et al. (2005) estimate that the source transfers to its surrounding interstellar medium (ISM) an amount of kinetic energy that can be as high as its bolometric X-ray luminosity (see Figure 8.5).

Cyg X-1 constitutes one of the best proof, with SS 433, that jets—both compact and transient—dissipate the bulk of the released accretion power in the form of

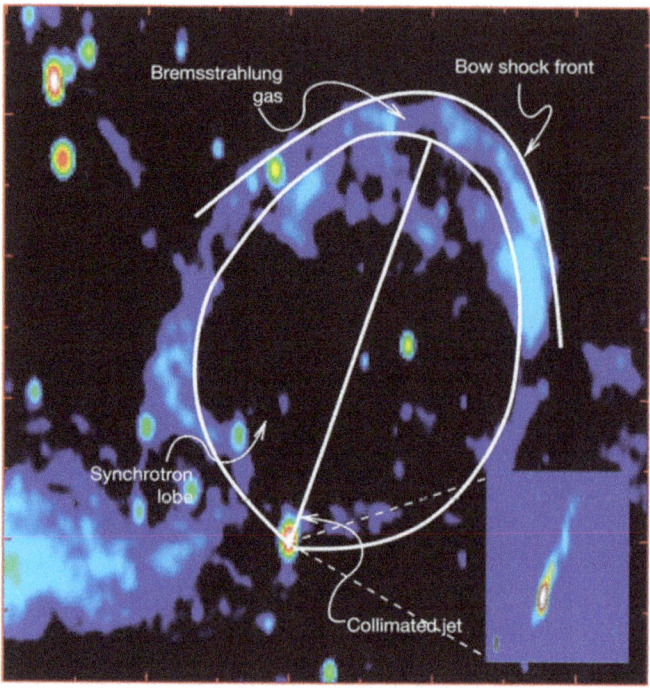

Figure 8.5. Collimated jets of Cyg X-1 impacting and transferring energy to its surrounding interstellar medium, creating a jet-powered nebulae. Reprinted with permission from Gallo et al. (2005).

radiatively inefficient and relativistic outflows, eventually energizing their surrounding ISM, and producing jet-powered nebulae, in a similar way to large scale lobes of FR II radio galaxies, powered by AGN. Since energy from such jets is transferred to the ISM, their power can be constrained by quantifying their interaction with the surrounding ISM, for instance by searching for optical line emission produced by jet-ISM interactions within thermal shells. Other candidate jet-powered nebulae, similar to Cyg X-1, include the binaries LMC X-1, GRS 1009-45 and the microquasar GRO J1655-40 (Russell et al. 2006).

8.2.3 The Superluminal Case of GRS 1915+105

The second source in which relativistic jets are discovered is the microquasar GRS 1915+105, a highly variable and luminous low-mass X-ray binary hosting a $\sim$12.4 $M_\odot$ black hole ($L_X \sim 3 \times 10^{38}$ erg s^{-1}) accreting from a low-mass star, located at the distance of $\sim$8.6 kpc (Reid et al. 2014), in which two jets are detected in radio, at apparent velocities of $v = 1.25$ c and $v = 0.65$ c. These velocities, one of them measured as superluminal, correspond to ejection velocity of $v = 0.92$ c in the frame of the central source (see Figure 8.6; Mirabel & Rodríguez 1994). Other galactic sources presenting relativistic jets are reported in Table 8.1.

Multi-wavelength observations of Galactic jet sources are primordial to understand the mechanisms of jet formation, and to disentangle the connection between

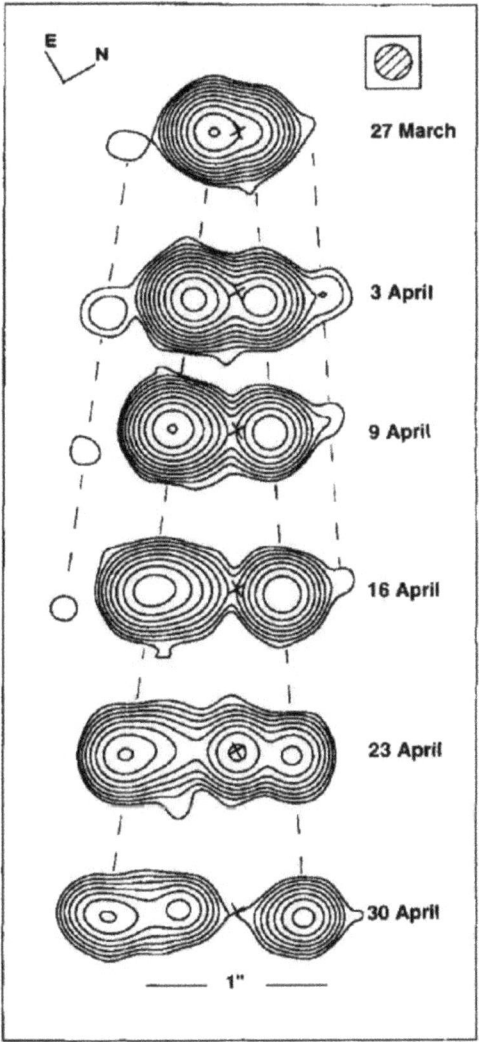

Figure 8.6. Superluminal ejections emanating from the microquasar GRS 1915+105. Reprinted with permission from Mirabel & Rodríguez (1994).

accretion and ejection of matter. On the one hand, X-ray emission traces accretion of matter, with a complex temporal and spectral variability on timescales of a few seconds to a few minutes. Such rapid variations are interpreted as subsequent increase and decrease of inner radius of the accretion disk (see Section 5.3.2), on a minute-timescale. On the other hand, radio and infrared domains trace ejection of matter, through synchrotron emission of relativistic jets. A breakthrough occurred with the first successful campaign of simultaneous multi-wavelength observations of

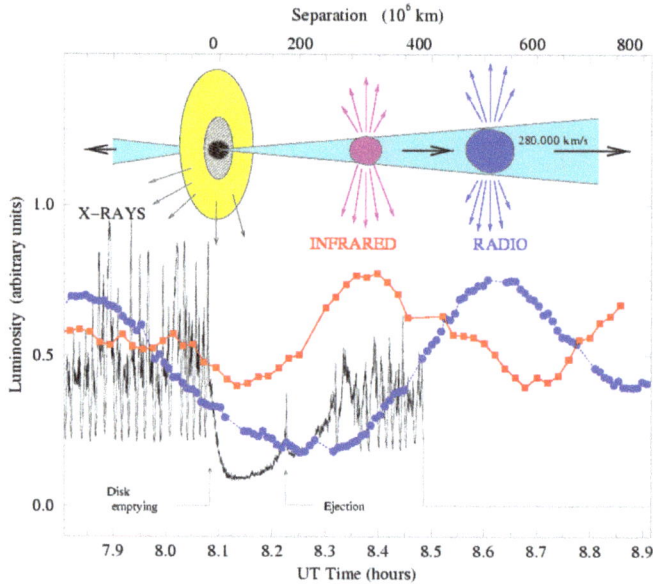

Figure 8.7. The microquasar GRS 1915+105 seen during a simultaneous X-ray-infrared-radio campaign, during which is clearly seen an accretion–ejection cycle (Mirabel et al. 1998). Credit: Sylvain Chaty, adapted from Chaty (1998).

GRS 1915+105 (see Figure 8.7 from Chaty 1998; Mirabel et al. 1998), later followed by additional campaigns (Fuchs et al. 2003; Rahoui et al. 2010; Ueda et al. 2002, 2010). The accretion disk first empties, and a few minutes later, a small peak of X-ray emission seems to indicate the ejection of a synchrotron bubble. Then X-ray emission increases, as soon as the inner part of the disk replenishes. The ejected bubble first emits in infrared at 2.2μm, and then in radio. Times of maxima and relative fluxes in infrared and radio confirm the ejection at a velocity $v \sim 0.92$ c, every 10 mn, of a plasma bubble of size $\sim 10^{11}$ m and mass $\sim 10^{16}$ kg, adiabatically expanding with a power-law coefficient of the electron energy distribution ~ 0. Assuming equipartition, the magnetic field is ~ 10 G. Fifteen minutes after the ejection, the synchrotron luminosity reaches $L_{\mathrm{synch}} \sim 10^{36}$ erg s^{-1}, the total energy emitted in the plasma $\sim 10^{40}$ erg, and the total mechanical jet energy $10^{3} L_{\odot}$, smaller than the jet energy of SS 433, while its X-ray luminosity is much higher. Inspired by these multi-wavelength observations of accretion and ejection coupling in GRS 1915 +105 (Chaty 1998; Mirabel et al. 1998), a scenario was developed, applicable to other jet sources since it is basically related to black hole spectral states (see Figure 8.8, Section 5.3.3.2, and Fender & Belloni 2004).

With such as wealth of nearly-continuous ejections at various levels, an impact on the surrounding interstellar medium around GRS 1915+105 could be expected. After an extensive multi-wavelength search (Chaty et al. 2001), while a hint of possible interaction has been discovered, no conclusive evidence of clear interaction can be drawn from these observations.

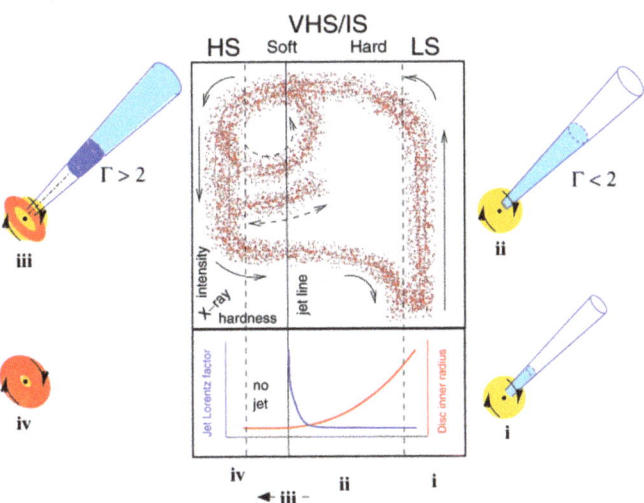

Figure 8.8. A model of coupling between accretion and ejection, initially developed for GRS 1915+105 (Fender & Belloni 2004), and subsequently applied to other jet sources. Reprinted with permission from Fender et al. (2004).

8.3 Reviews, Catalogs, Database and References

8.3.1 Reviews

- Review on microquasars and other jet sources: Mirabel & Rodríguez (1999);
- Review on the microquasar GRS 1915+105 and the associated disk-jet coupling phenomenon: Fender & Belloni (2004);
- Review on Tidal Disruption Event (TDE): Komossa (2015).

8.3.2 Catalogs

- Jet sources, included in the catalog of LMXB: (Liu et al. 2007);
- Jet sources, included in the catalogs of HMXB: (Liu et al. 2006) (according to the nature of the source).

8.3.3 Database

We give in Table 8.1 the list of relativistic jet sources, along with their main parameters: velocity, inclination, and distance.

References

Abdo, A. A., Ackermann, M., Agudo, I., et al. 2010, ApJ, 716, 30

Abdo, A. A., Ackermann, M., Ajello, M., et al. 2009, ApJ, 699, 817

Ackermann, M., Ajello, M., Allafort, A., et al. 2013, ApJ, 765, 54

Barrière, N. M., Tomsick, J. A., Wik, D. R., Chaty, S., & Rodriguez, J. 2015, ApJ, 799, 24

Bartlett, E. S., Clark, J. S., & Negueruela, I. 2019, A&A, 622, A93

Blandford, R., Meier, D., & Readhead, A. 2019, ARA&A, 57, 467

Blandford, R. D., & Payne, D. G. 1982, MNRAS, 199, 883

Blandford, R. D., & Znajek, R. L. 1977, MNRAS, 179, 433

Blundell, K. M., & Bowler, M. G. 2004, ApJL, 616, L159

Blundell, K. M., Laing, R., Lee, S., & Richards, A. 2018, ApJL, 867, L25

Cadolle Bel, M., Prat, L., Rodriguez, J., et al. 2009, A&A, 501, 1

Cadolle Bel, M., Ribó, M., Rodriguez, J., et al. 2007, ApJ, 659, 549

Carotenuto, F., Corbel, S., Tremou, E., et al. 2021, MNRAS, 504, 444

Chaty, S. 1998, PhD thesis, University Paris XI and Service d'Astrophysique, CEA Saclay, France

Chaty, S., Rodríguez, L. F., Mirabel, I. F., Geballe, T., & Fuchs, Y. 2001, A&A, 366, 1041

Cherepashchuk, A., Postnov, K., Molkov, S., Antokhina, E., & Belinski, A. 2020, NewAR, 89, 101542

Combi, J. A., Ribó, M., Martí, J., & Chaty, S. 2006, A&A, 458, 761

Corbel, S., Fender, R. P., Tzioumis, A. K., et al. 2000, A&A, 359, 251

Cudworth, K. M., & Herbig, G. 1979, AJ, 84, 548

Curran, P. A., Chaty, S., Zurita Heras, J. A., Tomsick, J. A., & Maccarone, T. J. 2011, MNRAS, 417, L26

Curran, P. A., Chaty, S., Zurita Heras, J. A., Tomsick, J. A., & Maccarone, T. J. 2012, MNRAS, 426, L106

Curtis, H. D. 1918, PLicO, 13, 9

Di Cocco, G., Foschini, L., Grandi, P., et al. 2004, A&A, 425, 89

Dubner, G. M., Holdaway, M., Goss, W. M., & Mirabel, I. F. 1998, AJ, 116, 1842

Fender, R., & Belloni, T. 2004, ARA&A, 42, 317

Fender, R. P., Belloni, T. M., & Gallo, E. 2004, MNRAS, 355, 1105

Fender, R. P., Garrington, S. T., McKay, D. J., et al. 1999, MNRAS, 304, 865

Fender, R. P., Spencer, R., Tzioumis, T., et al. 1998, ApJ, 506, L121

Foschini, L., Chiaberge, M., Grandi, P., et al. 2005, A&A, 433, 515

Fuchs, Y., Rodriguez, J., Mirabel, I. F., et al. 2003, A&A, 409, L35

Gallo, E., Fender, R., Kaiser, C., et al. 2005, Natur, 436, 819

Healey, S. E., Romani, R. W., Cotter, G., et al. 2008, ApJS, 175, 97

Herbig, G. H., & Jones, B. F. 1981, AJ, 86, 1232

Hjellming, R., & Rupen, M. 1995, Natur, 375, 464

Hjellming, R. M., Mioduszewski, A. J., Ghigo, F., Rupen, M. P., & Waltman, E. B. 1998a, AAS Meeting 192, 78.05

Hjellming, R. M., Rupen, M. P., Mioduszewski, A. J., et al. 1998b, AAS Meeting 193, 103.08

Khabibullin, I., Medvedev, P., & Sazonov, S. 2016, MNRAS, 455, 1414

Komossa, S. 2015, JHEAp, 7, 148

Kuulkers, E., Fender, R. P., Spencer, R. E., Davis, R. J., & Morison, I. 1999, MNRAS, 306, 919

Liu, Q. Z., van Paradijs, J., & van den Heuvel, E. P. J. 2006, A&A, 455, 1165

Liu, Q. Z., van Paradijs, J., & van den Heuvel, E. P. J. 2007, A&A, 469, 807

Livio, M. 1997, in ASP Conf. Ser. 121, IAU Colloq. 163: Accretion Phenomena and Related Outflows 121 ed. D. T. Wickramasinghe, G. V. Bicknell, & L. Ferrario (San Francisco, CA: ASP), 845

Margon, B. 1984, ARA&A, 22, 507

Margon, B., Ford, H. C., Katz, J. I., et al. 1979, ApJL, 230,

Martí, J., Mereghetti, S., Chaty, S., et al. 1998a, A&A, 338, L95

Martí, J., Mirabel, I. F., Chaty, S., & Rodríguez, L. F. 1998b, A&A, 330, 72

Martí, J., Mirabel, I. F., Chaty, S., & Rodríguez, L. F. 2000, A&A, 363, 184

Martí, J., Paredes, J. M., & Peracaula, M. 2001, A&A, 375, 476

Miller-Jones, J. C. A., Bahramian, A., Orosz, J. A., et al. 2021, Sci, 371, 1046

Mirabel, I., & Rodríguez, L. 1999, ARA&A, 37, 409

Mirabel, I. F., Dhawan, V., Chaty, S., et al. 1998, A&A, 330, L9

Mirabel, I. F., & Rodríguez, L. F. 1994, Natur, 371, 46

Mirabel, I. F., Rodríguez, L. F., Cordier, B., Paul, J., & Lebrun, F. 1992, Natur, 358, 215

Motch, C. 1998, A&A, 338, L13

Ogley, R. N., Chaty, S., Crocker, M., et al. 2002, MNRAS, 330, 772

Rahoui, F., Chaty, S., Rodriguez, J., et al. 2010, ApJ, 715, 1191

Rees, M. J. 1988, Natur, 333, 523

Reid, M. J., McClintock, J. E., Steiner, J. F., et al. 2014, ApJ, 796, 2

Rodríguez, L. F., Mirabel, I. F., & Martí, J. 1992, ApJ, 401, L15

Russell, D., Fender, R., Gallo, E., Miller-Jones, J. C. A., & Kaiser, C. R. 2006, in VI Microquasar Workshop: Microquasars and Beyond, 59

Russell, D. M., Fender, R. P., Gallo, E., & Kaiser, C. R. 2007, MNRAS, 376, 1341

Sakurai, T. 1985, A&A, 152, 121

Southwell, K. A., Livio, M., Charles, P. A., O'Donoghue, D., & Sutherland, W. J. 1996, ApJ, 470, 1065

Stirling, A. M., Spencer, R. E., Cawthorne, T. V., & Paragi, Z. 2004, MNRAS, 354, 1239

Stirling, A. M., Spencer, R. E., de la Force, C. J., et al. 2001, MNRAS, 327, 1273

Tomsick, J. A., Bodaghee, A., Chaty, S., et al. 2020, ApJ, 889, 53

Tomsick, J. A., Coughenour, B. M., Hare, J., et al. 2021, ApJ, 914, 48

Türler, M., Courvoisier, T. J-L., Chaty, S., & Fuchs, Y. 2004, A&A, 415, L35

Ueda, Y., Honda, K., Takahashi, H., et al. 2010, ApJ, 713, 257

Ueda, Y., Yamaoka, K., Sánchez-Fernández, C., et al. 2002, ApJ, 571, 918

Zurita Heras, J. A., Chaty, S., & Tomsick, J. A. 2009, A&A, 502, 787

Chapter 9

How to Study the Stellar Couples

À cheval sur la lune
j'ai vu le soleil
la nuit
Pendant que dans ton lit
tu rêvais des étoiles

Riding the Moon
I saw the Sun
the night
While in your bed
you dreamed of stars

—Guy Chaty, À Cheval sur la Lune, 2017

We compile in this chapter the various observing facilities of accreting stellar binaries, at various wavelengths (multi-wavelength astrophysics, from radio to γ-rays), and even at various messenger (multimessenger astrophysics, from photon to gravitational wave, via neutrino and cosmic ray) domains. We also compile virtual observatory facilities and astronomical tools. We finally mention numerical facilities focused on the study of accreting stellar binaries, with useful links to all facilities and references.

9.1 Multimessenger Astrophysics

Declined in multi-wavelength (photon), gravitational wave, neutrino and cosmic ray facilities, we give here an overview of main facilities (see also Figures 9.1 and 9.2).

Figure 9.1. NASA and ESA astronomy missions, including collaborative missions with partner agencies (NASA/ESA Hubble Space Telescope and NASA/ESA/CSA James Webb Space Telescope), upcoming missions (JAXA/NASA Xrism and ESA Euclid). Credit: NASA (https://www.nasa.gov/ames/science/science-projects).

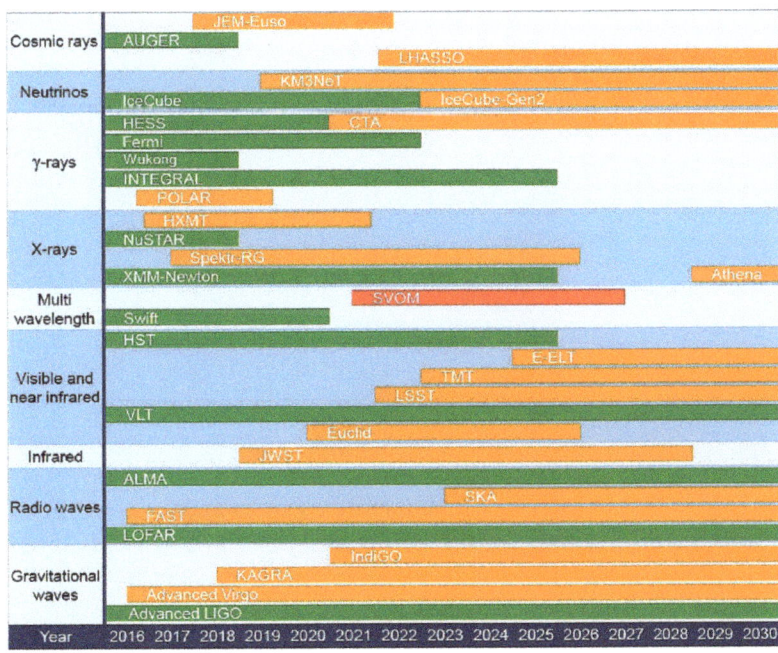

Figure 9.2. Multimessenger astronomic facilities, including radio, infrared, visible, and gamma-ray telescopes, advanced gravitational-wave interferometers and neutrino detectors of the km^3 class. Figure reproduced from Wei et al. (2016). Credit: SVOM collaboration.

9.1.1 Multi-wavelength Observing Facilities

All ground-based and satellite observatories can be used to study stellar accreting binaries, here we give a—non-exhaustive—(and in chronological order) list of the facilities mentioned throughout this book, either currently available, for which archival data are publicly available, or forthcoming (see Figure 9.1).

Radio
- ALMA (Atacama Large Millimeter Array): https://www.eso.org/public/unitedkingdom/teles-instr/alma
- ATNF (Australia Telescope National Facility): https://www.atnf.csiro.au
- SKA (Square Kilometer Array): https://www.skatelescope.org
- VLA (Very Large Array): https://public.nrao.edu/telescopes/vla
- VLBA (Very Long Baseline Array): https://public.nrao.edu/telescopes/vlba

Optical and Infrared
- ESO (European Southern Observatory, Chile): https://www.eso.org Archive: http://archive.eso.org/cms.html
- HST (Hubble Space Telescope, NASA/ESA, 1990, see Figure 9.1): https://www.stsci.edu/hst Archive: https://archive.stsci.edu/missions-and-data/hst
- Herschel (ESA, 2009–2013): https://sci.esa.int/web/herschel Archive: http://archives.esac.esa.int/hsa/whsa
- SST Spitzer Space Telescope (NASA, 2003–2020, see Figure 9.1): https://www.spitzer.caltech.edu Archive: https://irsa.ipac.caltech.edu/data/SPITZER/docs/spitzerdataarchives
- JWST (James Webb Space Telescope, NASA/ESA, see Figure 9.1): https://www.stsci.edu/jwst/
- *Gaia* (astrometry satellite, ESA, 2013–): https://www.cosmos.esa.int/web/gaia Archive: https://gea.esac.esa.int/archive

Ultraviolet
- EUVE (Extreme Ultraviolet Explorer, NASA, 1992–2001): https://heasarc.gsfc.nasa.gov/docs/euve/euve.html Archive: https://archive.stsci.edu/euve

X-Rays
- ROSAT (Germany, UK, US, 1990–1999): https://heasarc.gsfc.nasa.gov/docs/rosat/rosat.html Archive: https://heasarc.gsfc.nasa.gov/docs/rosat/rhp_archive.html ROSAT all-sky bright source catalog (1RXS, Voges et al. 1999).
- ASCA (Japan, US, 1993–2001): https://heasarc.gsfc.nasa.gov/docs/asca/asca.html Archive: https://heasarc.gsfc.nasa.gov/docs/asca/ahp_archive.html
- RXTE (US, 1995–2012): https://heasarc.gsfc.nasa.gov/docs/xte/rxte.html Archive: https://heasarc.gsfc.nasa.gov/docs/xte/archive.html
- Beppo-SAX (ASI, 1996–2002): https://heasarc.gsfc.nasa.gov/docs/sax/sax.html Archive: https://heasarc.gsfc.nasa.gov/docs/sax/shp_archive.html

- Chandra X-ray Observatory (NASA, 1999–, see Figure 9.1): https://chandra.harvard.edu Archive: https://cxc.harvard.edu/cda Chandra source catalog 2.0 (Evans et al. 2010)
- XMM–Newton X-ray Multi-Mirror Mission (ESA, 1999–, see Figure 9.1): https://www.cosmos.esa.int/web/xmm-newton Archive: https://www.cosmos.esa.int/web/xmm-newton/xsa XMM–Newton slew survey (XMMSL1, Saxton et al. 2008) XMM–Newton serendipitous survey (4XMM, Webb et al. 2020)
- Swift Neil Gehrels Swift Observatory (NASA, 2004–, see Figure 9.1): https://swift.gsfc.nasa.gov Archive: https://swift.gsfc.nasa.gov/archive Swift/XRT point-source catalog (2SXPS, Evans et al. 2020)
- Suzaku (JAXA, 2005–2015): https://heasarc.gsfc.nasa.gov/docs/astroe/astroe2.html Archive: https://heasarc.gsfc.nasa.gov/docs/suzaku/aehp_archive.html
- NICER Neutron star Interior Composition ExploreR (NASA, 2017–, see Figure 9.1): https://heasarc.gsfc.nasa.gov/docs/nicer Archive: https://heasarc.gsfc.nasa.gov/docs/nicer/nicer_archive.html
- NuStar Nuclear Spectroscopic telescope array (CalTech, 2012–, see Figure 9.1): https://www.nustar.caltech.edu Archive: https://heasarc.gsfc.nasa.gov/docs/nustar/nustar_archive.html
- eROSITA (Germany, 2019-) https://www.mpe.mpg.de/eROSITA
- SVOM https://www.svom.eu Multi-wavelength observatory, primarily designed to study γ-ray bursts, but which will also detect and observe accreting binaries (Yu et al. 2020)

Gamma-Rays
- INTEGRAL INTErnational Gamma-Ray Astrophysics Laboratory (ESA, 2002–, observing in the range 3 keV–10 MeV): https://sci.esa.int/web/integral (Kuulkers et al. 2021) Archive: https://www.isdc.unige.ch/integral/archive INTEGRAL/IBIS 15-year hard X-ray all-sky survey (Krivonos et al. 2021) Multimessenger astronomy with INTEGRAL (Ferrigno et al. 2021)
- FGST Fermi Gamma-ray Space Telescope (NASA, 2008–, observing in 10 keV–300 GeV, see Figure 9.1): https://fermi.gsfc.nasa.gov Archive: https://fermi.gsfc.nasa.gov/ssc/data/access (Ackermann et al. 2013; Abdollahi et al. 2020)
- CTA (Cerenkov Telescope Array, will observe in the 20 GeV–300 TeV band) https://www.cta-observatory.org (Abdalla et al. 2021; Acharyya et al. 2019, 2021)

Transient Database
- ASAS SN transients https://www.astronomy.ohio-state.edu/asassn/transients.html: V-band/g-band magnitudes of the source, representing the excess flux compared to the reference image
- Astro-COLIBRI: a new platform for real-time multimessenger astrophysics (Schüssler et al. 2022): https://astro-colibri.com
- Gaia Photometric Science Alerts http://gsaweb.ast.cam.ac.uk/alerts: The table provides links to the source alert pages, including light-curves and BP/RP spectra

- Master Global Robotic Net http://observ.pereplet.ru/MASTER_OT.html: The List of Optical Transients discovered by the Russian MASTER Global Robotic Net auto-detection
- TNS https://www.wis-tns.org: Transient Name Server
- ZTF https://www.ztf.caltech.edu: Zwicky Transient Facility, Systematic exploration of the dynamic sky

9.1.2 Gravitational Waves

LIGO–Virgo–Kagra

We list now the current ground-based interferometers for gravitational-wave (GW) astronomy (see Figure 9.3):

- LIGO (Laser Interferometer Gravitational-wave Observatory) (Hanford and Livingston, and LIGO-India in construction), arms of length $L = 4$ km https://www.ligo.caltech.edu
- Virgo (EGO, European Gravitational Observatory) (Pisa, Italy), arms of length $L = 3$ km https://www.virgo-gw.eu
- GEO600 (Germany) https://www.geo600.org
- KAGRA (Kamioka Gravitational-Wave Detector, Japan) https://gwcenter. icrr.u-tokyo.ac.jp/en
- Pulsar timing array (PTA): array of radio telescopes for timing shifts in spatially-separated pulsars, science target: $\sim 10^7 - 10^{10} M_\odot$ BBH at 1 Gpc (lower-frequency GW $f < 10^{-6}$ Hz) http://ipta4gw.org

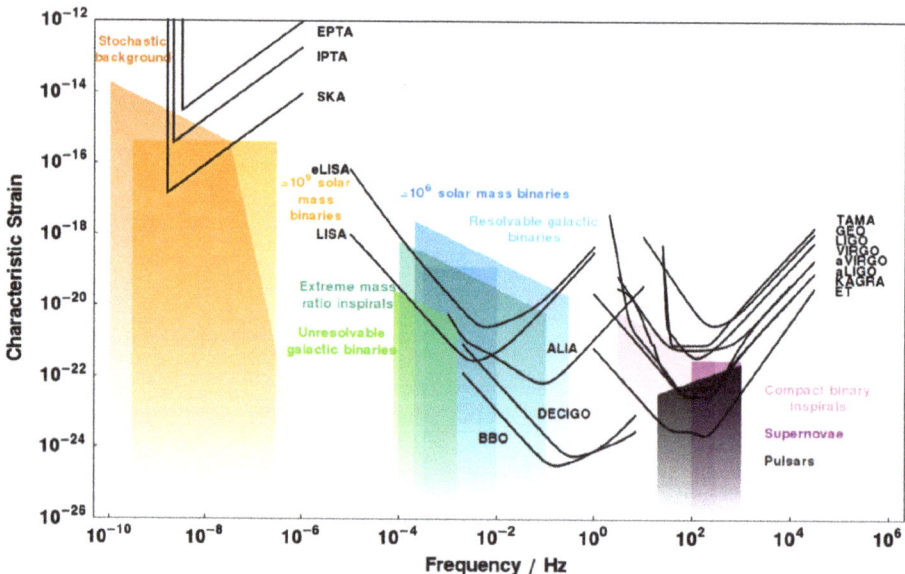

Figure 9.3. Characteristic strain versus frequency for various GW detectors and potential sources. Reprinted with permission from Moore et al. (2015), Creative Commons Attribution 3.0 License.

Future Ground-based and Satellite Projects
- LISA (ESA, see Figure 9.1): space laser interferometers with 2.5×10^6 km arms (launch in mid-2030s): science target $10^3 - 10^7 M_\odot$ BBH at 100 Gpc ($f \sim 10^{-3 to -2}$ Hz) for stellar mass compact binaries, and massive BH mergers, inspiral and merger https://sci.esa.int/web/lisa
- Einstein telescope: 10 km arms http://www.et-gw.eu
- Cosmic explorer: 20–40 km arms https://cosmicexplorer.org

9.1.3 Neutrinos
- Icecube https://icecube.wisc.edu
- KM3Net https://www.km3net.org

9.1.4 Cosmic Rays
- Pierre Auger Observatory https://www.auger.org

9.2 Virtual Observatory

9.2.1 Virtual Observatory Facilities

All Virtual observatory (VO) applications and services are listed in the website https://www.ivoa.net, we give here a long but non-exhaustive list:
- 3D View http://3dview.irap.omp.eu: 3D visualization of spacecraft position and attitude, planetary ephemerides, as well as scientific data (observations, simulations and models) representation
- Aladin https://aladin.u-strasbg.fr: Search and visualization of images, spectra and catalogs, cross-correlation, footprint Service, fixing WCS
- APERICubes http://voparis-apericubes.obspm.fr/apericubes/js9/demo.php: a tool for exploring spectral cubes
- AppLauncher http://www.jmmc.fr/applauncher: VO Dock for Astronomers
- CASSIS http://cassis.irap.omp.eu: Search and visualization of spectra
- CDS Xmatch Service http://cdsxmatch.u-strasbg.fr/xmatch: Cross-correlation
- Data Discovery Tool https://mast.stsci.edu/portal/Mashup/Clients/Mast/portal.html: Search for Images, Spectra and Catalogs
- Filter Profile Service http://svo2.cab.inta-csic.es/theory/fps: Filter curves
- Iris http://cxc.cfa.harvard.edu/iris/latest/index.html: SED analysis tool
- Montage http://montage.ipac.caltech.edu: astronomical image mosaic engine
- Octet http://www.cadc-ccda.hia-iha.nrc-cnrc.gc.ca/cvo
- pyVO https://pyvo.readthedocs.io/en/latest: Python VO
- Seleste http://cda.cfa.harvard.edu/seleste/seleste/guiguide.jsp: Query Databases
- Simbad http://simbad.u-strasbg.fr/simbad/: SIMBAD Astronomical Database—CDS (Strasbourg)
- SkyView https://skyview.gsfc.nasa.gov: Search and visualization of images
- Specview https://www.specview.com: Search and visualization of spectra
- SPLAT https://splat.physics.ucsd.edu/splat/quickstart.html: Search and visualization of spectra

- SVO Discovery Tool http://sdc.cab.inta-csic.es/SVODiscoveryTool/jsp/search-form.jsp: Search for Images, spectra, and catalogs
- TAPHandle http://saada.u-strasbg.fr/taphandle: Query Databases
- TAPsh http://soft.g-vo.org//tapsh: Query Databases
- TESELA http://sdc.cab.inta-csic.es/tesela/index.jsp: Search for a catalog of blank regions
- Time Series Search Tool https://irsa.ipac.caltech.edu/irsaviewer/timeseries: Search for Time Series Data
- TOPCAT http://www.star.bris.ac.uk/%7Embt/topcat and STILTS http://www.star.bris.ac.uk/%7Embt/stilts: Search and visualization of catalogs, Cross-correlation, Scatter, 3D plots and histograms, table format conversion, Query Databases
- VESPA http://vespa.obspm.fr
- VisIVO http://palantir7.oats.inaf.it/visivoweb
- VizieR http://vizier.cfa.harvard.edu/index.gml: library of published astronomical catalogs—tables and associated data—with verified and enriched data, accessible via multiple interfaces.
- VOConvert http://voi.iucaa.in/voi/VOConvert.htm: table format conversion
- VODesktop http://www.astrogrid.org/vodesktop.html: Search for images and catalogs
- VOPlot http://voi.iucaa.in/voi/voplot.htm: Catalogs visualization, Scatter, 3D plots and histograms
- VOSA http://svo2.cab.inta-csic.es/theory/vosa: SED building
- VOServices http://www.voservices.net/services: Search and visualization of spectra, Footprint Service, Filter curves
- VOSpec http://www.cosmos.esa.int/web/esdc/vospec: Search and visualization of spectra, SED building

9.2.2 Astronomical Tools

- astrocatalogs/astrocats https://github.com/astrocatalogs/astrocats: astrocatalogs github
- Astropy https://www.astropy.org: The Astropy project
- Astroquery https://astroquery.readthedocs.io/en/latest: set of tools for querying astronomical web forms and databases
- Beautiful soup https://www.crummy.com/software/BeautifulSoup/bs4/doc/: Python library for pulling data out of HTML and XML files

9.3 Numerical Simulations

Different codes are available to study in detail the evolution of stellar couples, involving transfer of mass and angular momentum, and more generally to perform population-synthesis studies. We list here the codes mentioned throughout this book.

- Binstar https://adsabs.harvard.edu/full/2011ASPC..447..339S: a binary stellar evolution code based on the single stellar evolution STAREVOL

- BSE https://astronomy.swin.edu.au/~jhurley/bsedload.html: Binary Stellar Evolution population-synthesis code, based on the single star stellar evolution SSE
- COMPAS https://compas.science: Compact Object Mergers: Population Astrophysics and Statistics code combines tools for statistical analysis and model selection with rapid population synthesis, to compute stellar and binary evolution. The COMPAS code is publicly available at https://github.com/TeamCOMPAS/COMPAS
- MESA http://mesa.sourceforge.net: a public 1-D hydrodynamical stellar evolution code, including a binary package
- StarTrack: a private stellar population-synthesis code (Belczynski et al. 2002)

References

Abdalla, H., Abe, H., Acero, F., et al. 2021, JCAP, 02(2021)048

Abdollahi, S., Acero, F., Ackermann, M., et al. 2020, ApJS, 247, 33

Acharyya, A., Adam, R., Adams, C., et al. 2021, JCAP, 01(2021)057

Acharyya, A., Agudo, I., Angüner, E.O., et al. 2019, APh, 111, 35

Ackermann, M., Ajello, M., Allafort, A., et al. 2013, ApJ, 765, 54

Belczynski, K., Bulik, T., & Kluźniak, W. 2002, ApJL, 567, L63

Evans, I.N., Primini, F.A., Glotfelty, K.J., et al. 2010, ApJS, 189, 37

Evans, P.A., Page, K.L., Osborne, J.P., et al. 2020, ApJS, 247, 54

Ferrigno, C., Savchenko, V., Coleiro, A., et al. 2021, NewAR, 92, 101595

Krivonos, R.A., Bird, A.J., Churazov, E.M., et al. 2021, NewAR, 92, 101612

Kuulkers, E., Ferrigno, C., Kretschmar, P., et al. 2021, NewAR, 93, 101629

Moore, C.J., Cole, R.H., & Berry, C. P. L. 2015, CQGra, 32, 015014

Saxton, R.D., Read, A.M., Esquej, P., et al. 2008, A&A, 480, 611

Schüssler, F., Alkan, A.K., Lefranc, V., & Reichherzer, P. 2022, ICRC (Berlin), 37, 935

Voges, W., Aschenbach, B., Boller, T., et al. 1999, A&A, 349, 389

Webb, N.A., Coriat, M., Traulsen, I., et al. 2020, A&A, 641, A136

Wei, J., Cordier, B., Antier, S., et al. 2016, arXiv:1610.06892

Yu, S.-J., Gonzalez, F., Wei, J.-Y., Zhang, S.-N., & Cordier, B. 2020, ChA&A, 44, 269

Accreting Binaries
Nature, formation, and evolution
Sylvain Chaty

Chapter 10

Conclusion

It is now time to conclude this book, in which we have covered all astrophysical basics concerning the evolution of isolated stars, from their formation to their final endpoint (Chapter 2), and all astrophysical basics concerning accreting binary systems (Chapter 3). After this introduction, we have covered the main aspects of the nature, formation, and evolution of cataclysmic variables (Chapter 4), low-mass X-ray binaries (Chapter 5), high-mass X-ray binaries (Chapter 6), peculiar binaries (Chapter 7), and jet sources (Chapter 8). We finally summarized the various observing facilities, tools, and codes, useful to study all types of accreting stellar binaries (Chapter 9). Of course, this book is not exhaustive, but we hope that the reader now has a glimpse of the wealth of knowledge offered by accreting binaries, and will, from this reading entry, know where to find useful information.

We will now step back for a moment, to realize how far we have gone from the detection, by Giacconi et al. (1962), of Scorpius X-1, the brightest X-ray source beyond the solar system, and the first accreting stellar binary. Six decades later, we have gone from opening a bit the window of the study of these energetic sources, to let the X-rays enter, to widely opening it by observing through the whole electromagnetic domain (Chaty 2013), and even adding to the photonic view the detection of cosmic rays, neutrinos, and finally of gravitational waves (Abbott et al. 2016). We have now just entered this multimessenger era, first with the simultaneous detection with gravitational waves and electromagnetic spectrum of the *kilonova* and merger GW 170817 (Abbott et al. 2017), and then with the detection by IceCube of a high-energy neutrino emission (IceCube-170922A). This latter detection was simultaneous to cosmic-ray acceleration of the flaring blazar TXS 0506+056, in the frame of a multi-wavelength campaign with Swift, NuSTAR, Fermi satellites and VLT/X-shooter instruments, combined with numerical models to constrain the blazar particle acceleration processes (Keivani et al. 2018).

Very bold is the one who dares say now what will be the situation in six more decades from now... But what is certain is that multimessenger astronomy will further develop, and the photon has become just one way among others, to reveal the full nature of celestial sources constituting our universe. We now know, at least for the next decade, what is the future of photon observations from the ground, in optical and infrared with the ESO 38m-ELT telescope, and in space, with the James Webb Space Telescope, and in radio with the Square Kilometer Array (SKA), split in two distinct arrays, in Australia and South Africa. The Cherenkov Telescope Array (CTA), also divided in two arrays, in La Palma Island (Spain) and Paranal (Chile), is the major next-generation observatory for ground-based very-high-energy γ-ray astronomy, improving the sensitivity of current instruments by a factor up to twenty, and at the same time their angular and energy resolutions over four decades in energy, from 20 GeV to 300 TeV (Acharyya et al. 2019, 2021; Abdalla et al. 2021). Neutrinos will be observed with IceCube and KM3Net, and cosmic rays are observed with the Pierre Auger Observatory.[1] The future of gravitational waves is also bright, with funded projects like the space-born LISA interferometer constituted of three satellites following the Earth on its orbit around the Sun, and the ground-based Einstein Telescope and Cosmic explorer.

As we saw all along this book, accreting binaries are among the best celestial objects to study at the same time stellar evolution and high-energy astrophysics. Studies on accreting binaries will mainly focus on the nature of accretion disks around white dwarfs, neutron stars and black holes. Astronomers will also continue their quest, not only trying to reveal the equation of state of neutron stars, but also to constrain the condition of matter approaching the central engine, by getting extensive observations of all type of short-time variability, including quasi-periodic oscillations, following up on the *timing* revolution brought by the RXTE satellite.[2] They will also be dedicated to the evolution of accreting binaries, leading to the formation of double neutron stars, as described in Tauris et al. (2017). In parallel, the study of black holes will expand, focused both on their short-term (QPO) and long-term (outbursts) variability, without forgetting the study of energetic phenomena occurring in their surroundings, involving accretion–ejection coupling. Studies will dive even closer to the gravitational well, revealing properties intimately linked to fundamental physics (such as their formation, presence of event horizon, spin, etc.), all of which eventually revealed only by combining multi-wavelength with multimessenger observations, gravitational waves playing here a key role, as described in Barack et al. (2019).

Most instruments are now conceived from the beginning as both multi-wavelength and promptly reacting, in order to efficiently reveal and study transient sources, with facilities such as SVOM satellite,[3] linked to a battery of ground-based telescopes, to allow for prompt follow-up observations in optical and infrared of transient sources.

---

[1] We also mention here the Jem-EUSO mission, designed to detect ultra-high-energy cosmic rays: https://www.jemeuso.org.

[2] For instance with instruments similar to the proposed Large Observatory for X-ray Timing (LOFT) instrument (Feroci et al. 2012): https://sci.esa.int/web/loft.

[3] We also mention here Theseus, a satellite that had been shortlisted by ESA: https://www.isdc.unige.ch/theseus.

The future of transient sources is ahead of us, with instruments covering multi-wavelength domains—from SKA observatory in radio, to the future Athena satellite in X-rays, and CTA array in γ-rays—and linked to multimessenger facilities—from LIGO–Virgo–KAGRA in gravitational waves, to KM3Net and IceCube in neutrinos.

For millenia, the sky was observed with the naked eye only. Then, for 350 years, from Galileo to roughly the Second World War, the telescope allowed to go further and deeper, but still restricted to the optical domain. For 50 years, from the development of radio astronomy after the second world war, until the beginning of the 21st century, the sky was fully revealed, by opening windows after windows, until covering the whole electromagnetic spectrum. Now, from 2015, we have for the first time in the history, full access to all messengers sent by the universe: photon, gravitational wave, cosmic ray, and neutrino. By combining multimessenger instruments and heavy numerical simulations, the future of accreting binaries is bright, allowing us to discover the hidden and unknown part of the iceberg, now fully revealed to our wide-open eyes, capable of multimessenger vision!

References

Abbott, B. P., Abbott, R., Abbott, T. D., et al. 2016, PhRvL, 116, 061102

Abbott, B. P., Abbott, R., Abbott, T. D., et al. 2017, PhRvL, 119, 161101

Abdalla, H., Abe, H., Acero, F., et al. 2021, JCAP, 02(2021)048

Acharyya, A., Adam, R., Adams, C., et al. 2021, JCAP, 01(2021)057

Acharyya, A., Agudo, I., Angüner, E. O., et al. 2019, APh, 111, 35

Barack, L., Cardoso, V., Nissanke, S., et al. 2019, CQGra, 36, 143001

Chaty, S. 2013, AdSpR, 52, 2132

Feroci, M., Stella, L., van der Klis, M., et al. 2012, ExA, 34, 415

Giacconi, R., Gursky, H., Paolini, F. R., & Rossi, B. B. 1962, PhRvL, 9, 439

Keivani, A., Murase, K., Petropoulou, M., et al. 2018, ApJ, 864, 84

Tauris, T. M., Kramer, M., Freire, P. C. C., et al. 2017, ApJ, 846, 170

Printed in the USA
CPSIA information can be obtained
at www.ICGtesting.com
LVHW062308201223
767001LV00010B/30